Costs of Ammonia Abatement and the Climate Co-Benefits

Stefan Reis • Clare Howard • Mark A. Sutton
Editors

Costs of Ammonia Abatement and the Climate Co-Benefits

 Springer

Editors
Stefan Reis
Centre for Ecology & Hydrology (CEH)
Penicuik, Midlothian, United Kingdom

Clare Howard
NERC Centre for Ecology and Hydrology
Edinburgh Research Station
Penicuik, United Kingdom

Mark A. Sutton
Edinburgh Research Station
Centre for Ecology and Hydrology
Penicuik, Midlothian
United Kingdom

ISBN 978-94-017-9721-4   ISBN 978-94-017-9722-1   (eBook)
DOI 10.1007/978-94-017-9722-1

Library of Congress Control Number: 2015939271

Springer Dordrecht Heidelberg New York London

Printed on acid-free paper

Springer Science+Business Media B.V. Dordrecht is part of Springer Science+Business Media (www.springer.com)

Foreword

Ammonia emissions represent a key emerging challenge for European environmental and agricultural policies. Ammonia contributes to several environmental problems, including threats to human health through the formation of fine particulate matter (PM) in the atmosphere and threats to biodiversity through nitrogen deposition to sensitive ecosystems. It causes both nitrogen saturation and soil acidification, with losses of key plant species. At the same time, ammonia contributes to greenhouse gas emissions through indirect contributions to nitrous oxide and to water pollution, where deposited nitrogen causes eutrophication of both freshwater and coastal ecosystems.

If these problems were not enough, ammonia emissions also represent a huge loss of nitrogen from farming systems. Ammonia losses can account for as much as 50 % of the added nitrogen when spreading animal manure or urea fertilizer. The result is substantial economic loss for farmers, while also wasting the energy used to produce fertilizers in the first place. As 1–2 % of total world energy goes to the manufacture of ammonia-based fertilizers, this is far from trivial.

In this context, the UNECE Convention on Long-Range Transboundary Air Pollution (CLRTAP) has put substantial effort to reaching agreements that reduce ammonia emissions. This includes new emission ceilings in the Gothenburg Protocol, which was recently revised in 2012, and complements actions in the European Union to revise its National Emission Ceilings Directive.

In order to make progress in these agreements, it has been essential to demonstrate that there is a substantial economic and environmental benefit to be gained from reducing ammonia emissions. From a wide 'societal view' of the Green Economy, it needs to be shown that the environmental, health and agronomic benefits outweigh the costs. Similarly, from a 'farmers view' of the Green Economy, it needs to be shown that measures are not prohibitively expensive, and in many cases can pay for themselves. The costs data derived can then be included in the integrated assessment that supports decision making by the CLRTAP and the European Union.

This book provides a key resource to support this process, which has been prepared as part of the work of the CLRTAP Task Force on Reactive Nitrogen

(TFRN). Starting with an expert workshop in Paris (25–26 October 2010), the contributors have since worked to bring together the key evidence to prepare the present synthesis. The work has benefited from financial support to TFRN from the UK Department for Environment, Food and Rural Affairs (Defra) and from dissemination activities within the ÉCLAIRE project, funded by the European Commission.

The outcome delivers a very clear message. Expressed per kg of nitrogen, abatement of ammonia emissions is rather cheap compared with further abatement of nitrogen oxides (NO_x). Substantial progress has already been made for NO_x emission reduction, but the remaining measures start to become increasingly expensive. By comparison, with a very little ammonia abatement accomplished todate, the 'low-hanging fruit' of low-cost measures is still available.

Since they can deliver nitrogen savings for farmers at the same time, such ammonia measures should become increasingly attractive for policy makers as they consider the next generation of international air pollution agreements.

Task Force on Reactive Nitrogen   Mark A. Sutton
UNECE Convention on Long-range   Tommy Dalgaard
Transboundary Air Pollution
Palais des Nations, 8-14, avenue de la Paix
CH-1211 Geneva 10, Switzerland

Contents

1 Overview, Aims and Scope 1
Stefan Reis, Mark A. Sutton, and Clare Howard

2 Economic Costs of Nitrogen Management in Agriculture 7
Oene Oenema, Steen Gyldenkaerne, and Jouke Oenema

3 Economics of Low Nitrogen Feeding Strategies 35
Ad M. van Vuuren, Carlos Pineiro, Klaas W. van der Hoek,
and Oene Oenema

4 Ammonia Abatement by Animal Housing Techniques 53
Gema Montalvo, Carlos Pineiro, Mariano Herrero, Manuel Bigeriego,
and Wil Prins

**5 Ammonia Abatement with Manure Storage
and Processing Techniques** 75
Andrew VanderZaag, Barbara Amon,
Shabtai Bittman, and Tadeusz Kuczyński

**6 Cost of Ammonia Emission Abatement from Manure
Spreading and Fertilizer Application** 113
J. Webb, John Morgan, and Brian Pain

**7 Co-benefits and Trade-Offs of Between Greenhouse Gas
and Air Pollutant Emissions for Measures Reducing
Ammonia Emissions and Implications for Costing** 137
Vera Eory, Cairistiona F.E. Topp, Bronno de Haan,
and Dominic Moran

8   Country Case Studies   169
     Mark A. Sutton, Clare Howard, Stefan Reis, Diego Abalos,
     Annelies Bracher, Aleksandr Bryukhanov, Rocio Danica Condor-Golec,
     Natalia Kozlova, Stanley T.J. Lalor, Harald Menzi, Dmitry Maximov,
     Tom Misselbrook, Martin Raaflaub, Alberto Sanz-Cobena,
     Edith von Atzigen-Sollberger, Peter Spring, Antonio Vallejo,
     and Brian Wade

**9   Estimating Costs and Potential for Reduction of Ammonia
     Emissions from Agriculture in the GAINS Model**   233
     Zbigniew Klimont and Wilfried Winiwarter

**10   Costs of Ammonia Abatement: Summary, Conclusions
     and Policy Context**   263
     Clare Howard, Mark A. Sutton,
     Oene Oenema, and Shabtai Bittman

Index   283

Contributors

Diego Abalos ETSI Agrónomos, Technical University of Madrid, Madrid, Spain

Eduardo Aguilera Universidad Pablo de Olavide, Ctra. de Utrera, Sevilla, Spain

Barbara Amon Leibniz Institute for Agricultural Engineering Potsdam-Bornim e.V. (ATB), Potsdam, Germany

Manuel Bigeriego Spanish Ministry of Agriculture and Environment, Madrid, Spain

Shabtai Bittman Agriculture and Agri-Food Canada, Agassiz, Canada

Annelies Bracher School of Agricultural, Forest and Food Sciences HAFL, Bern University of Applied Sciences (BFH), Zollikofen, Switzerland

Institute for Livestock Sciences, Agroscope, Posieux, Switzerland

Aleksandr Bryukhanov North-West Research Institute of Agricultural Engineering and Electrification (SZNIIMESH), St-Petersburg-Pavlovsk, Russia

Rocio Danica Condor-Golec Istituto Superiore per la Protezione e la Ricerca Ambientale, ISPRA, Rome, Italy

Food and Agriculture Organization of the United Nations (FAO), Rome, Italy

Bronno de Haan PBL Netherlands Environmental Assessment Agency, Bilthoven, The Netherlands

Vera Eory Land Economy, Environment & Society, Scotland's Rural College (SRUC), Edinburgh, UK

Steen Gyldenkaerne Institute of Environmental Science, Aarhus University, Roskilde, Denmark

Mariano Herrero Feaspor, La Lastrilla, Segovia, Spain

Clare Howard NERC Centre for Ecology & Hydrology (CEH), Penicuik, UK

School of GeoSciences, University of Edinburgh, Grant Institute, Edinburgh, UK

Zbigniew Klimont International Institute for Applied Systems Analysis (IIASA), Laxenburg, Austria

Natalia Kozlova North-West Research Institute of Agricultural Engineering and Electrification (SZNIIMESH), St-Petersburg-Pavlovsk, Russia

Tadeusz Kuczyński University of Zielona Góra, Zielona Góra, Poland

Stanley T.J. Lalor Grassland AGRO, Limerick, Ireland

Formerly Teagasc, Johnstown Research Centre, Wexford, Ireland

Luis Lassaletta Ecotoxicology of Air Pollution, CIEMAT, Madrid, Spain

Dmitry Maximov North-West Research Institute of Agricultural Engineering and Electrification (SZNIIMESH), St-Petersburg-Pavlovsk, Russia

Harald Menzi School for Agricultural, Forestry and Food Sciences (HAFL), Bern University of Applied Sciences (BFH), Zollikofen, Switzerland

Agroscope, Posieux, Switzerland

Tom Misselbrook Rothamsted Research, Okehampton, Devon, UK

Gema Montalvo PigCHAMP Pro Europa S.L., Segovia, Spain

Dominic Moran Land Economy, Environment & Society, Scotland's Rural College (SRUC), Edinburgh, UK

John Morgan Creedy Associates, Great Gutton Farm, Shobrooke, Crediton, Devon, UK

Jouke Oenema Plant Research International, Agrosystems Research, Wageningen University, Wageningen, The Netherlands

Oene Oenema Environmental Sciences Group, Alterra, Wageningen University, Wageningen, The Netherlands

Brian Pain Creedy Associates, Great Gutton Farm, Shobrooke, Crediton, Devon, UK

Carlos Pineiro PigCHAMP Pro Europa S.L., Segovia, Spain

Wil Prins Ministry of Infrastructure and the Environment, Directorate for Climate, Air and Noise, The Hague, The Netherlands

Martin Raaflaub School of Agricultural, Forestry and Food Sciences HAFL, Bern University of Applied Sciences (BFH), Zollikofen, Switzerland

Stefan Reis NERC Centre for Ecology & Hydrology (CEH), Penicuik, UK

University of Exeter Medical School, Knowledge Spa, Truro, UK

Alberto Sanz-Cobena ETSI Agrónomos, Technical University of Madrid, Madrid, Spain

Peter Spring School of Agricultural, Forestry and Food Sciences HAFL, Bern University of Applied Sciences (BFH), Zollikofen, Switzerland

Mark A. Sutton NERC Centre for Ecology & Hydrology (CEH), Penicuik, UK

Cairistiona F.E. Topp Crop & Soil Systems, Scotland's Rural College (SRUC), Edinburgh, UK

Antonio Vallejo College of Agriculture, Technical University of Madrid, Madrid, Spain

Ad M. van Vuuren Wageningen UR, Livestock Research, Wageningen, The Netherlands

Klaas W. van der Hoek RIVM, Bilthoven, The Netherlands

Andrew VanderZaag Agriculture and Agri-Food Canada, Ottawa, Canada

Edith von Atzigen-Sollberger School of Agricultural, Forestry and Food Sciences HAFL, Bern University of Applied Sciences (BFH), Zollikofen, Switzerland

Brian Wade Agrotain International, AGROTAIN International LLC, St. Louis, USA

J. Webb Ricardo-AEA, Harwell, Didcot, UK

Wilfried Winiwarter International Institute for Applied Systems Analysis (IIASA), Laxenburg, Austria

Acronyms & Abbreviations

BNF	Biological nitrogen fixation
CAFÉ	Clean Air for Europe
CAP	Common Agricultural Policy of the European Union
CBA	Cost-benefit analysis
CEA	Cost-effectiveness analysis
CBD	UN Convention on Biological Diversity
CLE	Critical level
CL	Critical load
DIN	Dissolved inorganic nitrogen
DON	Dissolved organic nitrogen
EMEP	European Monitoring and Evaluation Programme of the LRTAP Convention
GAINS	Greenhouse Gas and Air Pollution Interactions and Synergies Model developed by IIASA
GAW	Global Atmospheric Watch
GHG	Greenhouse Gas – includes carbon dioxide (CO_2), nitrous oxide (N_2O), methane (CH_4), ozone (O_3), water vapour and various other gases
GWP	Global warming potential
GPNM	Global Partnership on Nutrient Management – established under the lead of UNEP
IAM	Integrated assessment model(ling)
IIASA	International Institute for Applied Systems Analysis
ICP	International Cooperative Programme of the LRTAP Convention
IPCC	Intergovernmental Panel on Climate Change
LRTAP	UNECE Convention on Long-Range Transboundary Air Pollution
N_2	Di-nitrogen – unreactive nitrogen gas making up 78 % of the atmosphere
N_2O	Nitrous oxide – a greenhouse gas
NEC(D)	National Emission Ceilings (Directive) of the European Union

NH_3	Ammonia – a reactive gas air pollutant
NH_4^+	Ammonium – ion present in aerosols and precipitation
NH_x	Collective term for NH_3 and NH_4^+, inorganic reduced nitrogen
NO	Nitric oxide – a reactive gas air pollutant
NO_2	Nitrogen dioxide – a reactive gas air pollutant
NO_2^-	Nitrite – ion present in water samples
NO_3^-	Nitrate – ion present in aerosols, precipitation and water samples
NO_x	Nitrogen oxides (the sum of NO and NO_2)
NO_y	Collective term for inorganic oxidized nitrogen, including NO_x, NO_3^-, HONO, HNO_3, etc.
N_r	Reactive nitrogen – collective term for all nitrogen forms except for unreactive di-nitrogen (N_2), including NH_x
NUE	Nitrogen use efficiency
O_3	Ozone – tropospheric ozone (ozone in the lowest 10–20 km of the atmosphere) unless specified in text
$PM_{2.5}/PM_{10}$	Particulate Matter – aerosol mass contained in particles with an aerodynamic diameter below 2.5 (or 10 for PM_{10}) micrometre, measured with a reference technique
SIA	Secondary inorganic aerosol
SOA	Secondary organic aerosol
TFEIP	Task Force on Emission Inventories and Projection of the LRTAP Convention
TFIAM	Task Force on Integrated Assessment Modelling of the LRTAP Convention
TFRN	Task Force on Reactive Nitrogen of the LRTAP Convention
UN	United Nations
UNECE	United Nations Economic Commission for Europe
UNEP	United Nations Environment Programme
VOCs	Volatile organic compounds
WGE	Working Group on Effects of the LRTAP Convention
WGSR	Working Group on Strategies and Review of the LRTAP Convention
WMO	World Meteorological Organization

Chapter 1
Overview, Aims and Scope

Stefan Reis, Mark A. Sutton, and Clare Howard

Abstract This chapter presents an overview of the volume, introducing the background and setting out the aims and scope of the workshop and this book. Ammonia emissions primarily originate from agricultural sources and present a substantial contribution to a wide range of environmental problems (see as well Sutton et al., Atmospheric ammonia – detecting emission changes and environmental impacts – results of an expert workshop under the convention on long-range transboundary air pollution. Springer, Heidelberg, 2009; Managing the European nitrogen problem: a proposed strategy for integration of European research on the multiple effects of reactive nitrogen. Centre for Ecology & Hydrology, Edinburgh, 2009; The European nitrogen assessment. Cambridge University Press, Cambridge, 2011; Our nutrient world: the challenge to produce more food and energy with less pollution. Global overview of nutrient management. Centre for Ecology & Hydrology, Edinburgh, 2013; Philos Trans R Soc London, Ser B 368(1621):20130166, 2013), ranging from the deposition of acidifying substances and excess nutrients on soils, the formation of secondary inorganic aerosols, climate change and nutrient loads for freshwater and coastal ecosystems (Galloway et al., Bioscience 53:341–356, 2003). Yet, ammonia emissions have to date not been subject to stringent emission control policies, in contrast to sulphur dioxide or nitrogen oxides. As a consequence, ammonia emissions and the agricultural activities they originate from are discussed in detail, with the aim to identify the most promising emission sources and policy options to reduce their harmful environmental effects.

S. Reis (✉)
NERC Centre for Ecology & Hydrology (CEH), Bush Estate, Penicuik EH26 0QB, UK

University of Exeter Medical School, Knowledge Spa, Truro TR1 3HD, UK
e-mail: srei@ceh.ac.uk

M.A. Sutton
NERC Centre for Ecology & Hydrology (CEH), Bush Estate, Penicuik EH26 0QB, UK
e-mail: ms@ceh.ac.uk

C. Howard
NERC Centre for Ecology & Hydrology (CEH), Bush Estate, Penicuik EH26 0QB, UK

School of GeoSciences, University of Edinburgh, Grant Institute, Kings Buildings Campus, West Mains Road, Edinburgh EH9 3JW, UK
e-mail: cbritt@ceh.ac.uk

© Springer Science+Business Media Dordrecht 2015
S. Reis et al. (eds.), *Costs of Ammonia Abatement and the Climate Co-Benefits*,
DOI 10.1007/978-94-017-9722-1_1

Keywords Ammonia emissions • Transboundary air pollution • Agriculture • Ecosystem effects • Health effects

1.1   Overview

Emissions of ammonia (NH_3) into the atmosphere contribute substantially to local, regional and transboundary air pollution effects. Ambient concentrations and the deposition of reactive Nitrogen (N_r) contribute to a range of adverse effects on human health and ecosystems. Ammonia emissions, stemming mainly from agricultural sources (Fig. 1.1), have remained relatively stable (Fig. 1.2) in contrast to for instance sulphur dioxide or nitrogen oxide emissions in the last decades, the relative contribution of ammonia to future impacts of nitrogen and acidity on terrestrial ecosystems in Europe can be expected to increase. At the same time, ammonia is contributing an increasing share to the formation of secondary inorganic aerosols (SIA), a major constituent of particulate matter, with associated human health risks.

Recent episodes of high levels of ambient levels of fine particulate matter ($PM_{2.5}$) in the UK (Vieno et al. 2014) and in France have been to a large extent due to long-range transport of ammonium nitrates originating from spring manure spreading and fertiliser application in the agricultural regions of Europe. To date, there is no robust scientific evidence identifying specific components of $PM_{2.5}$ as less or not harmful to human health, policy measures aim at a reduction of human exposure to all components of $PM_{2.5}$, including secondary inorganic aerosols (SIA), which comprise ammonium sulphates and nitrates.

By 2020, it is estimated that NH_3 will be the largest single contributor to the deposition of acidifying substances and nutrients and thus the challenges posed by acidification, eutrophication and secondary particulate matter formation in Europe. This increasing share reflects the success of European policies in reducing SO_2 and NO_x emissions and thus the contributions of anthropogenic emission sectors such as power generation and road transport. As a consequence, NH_3, which is mainly emitted from agricultural sources (Fig. 1.1) which have so far not been subject to equally stringent regulations, is increasingly dominating nitrogen and acidifying inputs. In this context, reducing ammonia emissions and the associated environmental impacts remain major challenges for the future (Fowler et al. 2013).

This book is the result of an Expert Workshop held under the auspices of the UNECE Convention on Long-range Transboundary Air Pollution (CLRTAP) and organised by the Task Force on Reactive Nitrogen (TFRN). It summarises the current state-of-the-art regarding abatement measures, their associated costs and implications from the co-benefits for greenhouse gas emissions arising from reducing ammonia emissions from agricultural sources.

The Expert Workshop was organised in Paris from 25th–26th of October 2010 and reported to the 5th meeting of the Task Force on Reactive Nitrogen on the following day. The findings of this workshop have informed the development of documents

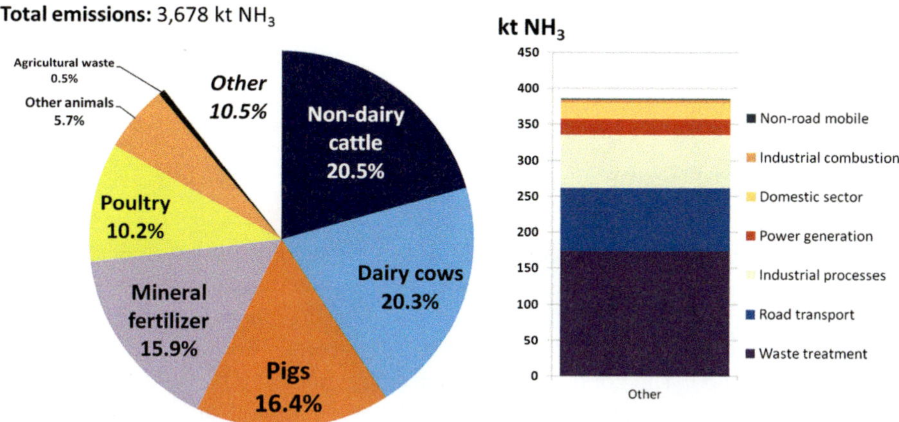

Fig. 1.1 Share of anthropogenic source sectors in total ammonia emissions in the year 2010 for the EU28 (Source: IIASA)

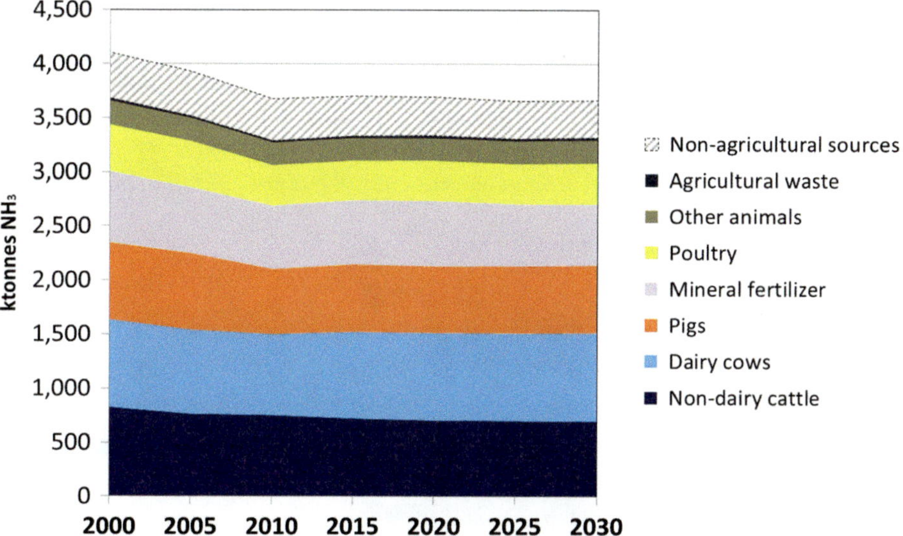

Fig. 1.2 Projected development of EU28 ammonia emissions (in kT NH$_3$) from 2000 to 2030 (Source: IIASA)

supporting the revision of the Gothenburg Protocol (UNECE 2013) under the Convention on Long-Range Transboundary Air Pollution. In addition to that, updated and revised cost information emerging from the workshop have been included in the GAINS (Greenhouse Gas and Air Pollution Interactions and Synergies) model (Klimont and Winiwarter 2015), which has been developed by the International Institute of Applied Systems Analysis (IIASA) and widely applied to conduct integrated assessment model analysis in support of the Gothenburg Protocol revision (Reis et al. 2012). The workshop has thus significantly improved the understanding and provided vital new data and information into the CLRTAP.

1.2   Aims and Scope

The aims of this book are to summarise the current state-of-the-art in determining best available techniques to reduce ammonia emissions from agricultural practises at every stage, starting from animal feed and housing, including the storage of liquid and solid manure and the application of mineral fertiliser and manure to the fields (Fig. 1.3). The complexity of controlling ammonia from these sources is that nitrogen conserved at each stage is available for volatilisation of NH_3 in the next stage and measures need to consider the knock-on effects on downstream emissions.

In each of the Chaps. 2, 3, 4, 5 and 6, the book addresses one of the agricultural production stages, the measures available to control emissions, issues of their implementation and related costs. In Chap. 7, the relationship between ammonia control and greenhouse gas emissions is explored and in Chap. 9, the implications of the revised abatement cost figures for integrated assessment modelling and resulting cost-effective control strategies, including environmental effects of these strategies, are discussed. Chapter 8 provides examples and case studies for

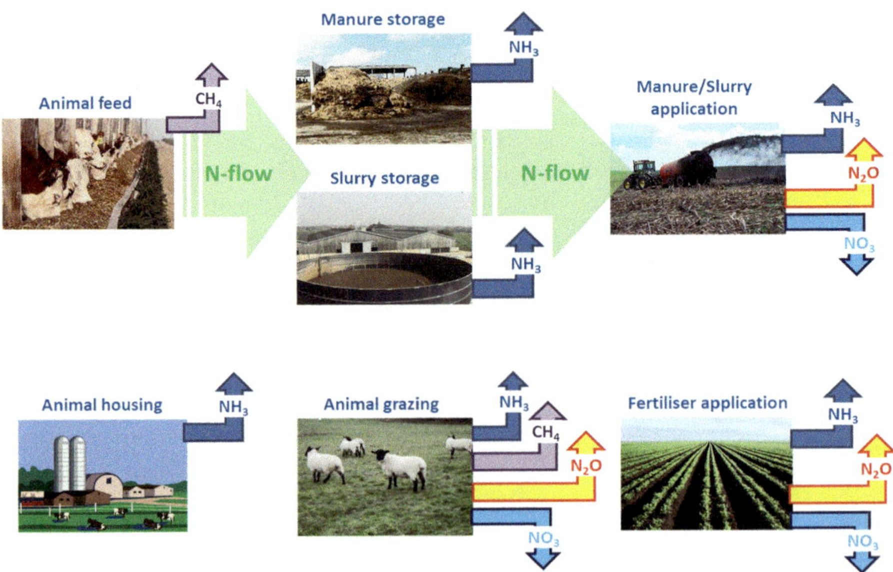

Fig. 1.3 Schema of different stages of nitrogen management in the agricultural production process with illustrations of control points for different forms of N_r emissions

selected countries, where detailed studies of ammonia abatement costs and the effectiveness of implementing control measures have been assessed under different conditions. Finally, in Chap. 10, the outcomes of the development of a guidance document under the United Nations Economic Commission for Europe's Convention on Long-range Transboundary Air Pollution, in the context of the revision of the Gothenburg Protocol, are summarised.

Acknowledgements We gratefully acknowledge financial support from the UK Department for Environment, Food and Rural Affairs (Defra), the European Commission through the ÉCLAIRE project and the UK Natural Environment Research Council.

References

Fowler D, Coyle M, Skiba UM, Sutton MA, Cape JN, Reis S, Sheppard LJ, Jenkins A, Grizzetti B, Galloway JN, Vitousek P, Leach A, Bouwman AF, Butterbach-Bahl K, Dentener F, Stevenson D, Amann M, Voss M (2013) The global nitrogen cycle in the 21st century. Philos Trans R Soc London Ser B 368(1621):20130164

Galloway JN, Aber JD, Erisman JW, Seitzinger SP, Howarth RW, Cowling EB, Cosby BJ (2003) The nitrogen cascade. Bioscience 53:341–356

Klimont Z, Winiwarter W (2015) Estimating costs and potential for reduction of ammonia emissions from agriculture in the GAINS model. In: Reis S, Howard CM, Sutton MA (eds) Costs of ammonia abatement and the climate co-benefits. Springer Publishers, Amsterdam

Reis S, Grennfelt P, Klimont Z, Amann M, ApSimon H, Hettelingh J-P, Holland M, LeGall A-C, Maas R, Posch M, Spranger T, Sutton MA, Williams M (2012) From acid rain to climate change. Science 338(6111):1153–1154

Sutton MA, Reis S, Baker SMH (eds) (2009a) Atmospheric ammonia – detecting emission changes and environmental impacts – results of an expert workshop under the convention on long-range transboundary air pollution. Springer, Heidelberg

Sutton MA, Oenema O, Erisman JW, Grennfelt P, Beier C, Billen G, Bleeker A, Britton C, Butterbach Bahl K, Cellier P, van Grinsven H, Grizzetti B, Nemitz E, Reis S, Skiba U, Voss M, de Vries W, Zechmeister-Boltenstern S (2009b) Managing the European nitrogen problem: a proposed strategy for integration of European research on the multiple effects of reactive nitrogen. Centre for Ecology & Hydrology, Edinburgh

Sutton MA, Howard C, Erisman JW, Billen G, Bleeker A, Grenfelt P, van Grinsven H, Grizzetti B (eds) (2011) The European nitrogen assessment. Cambridge University Press, Cambridge

Sutton MA, Bleeker A, Howard CM, Bekunda M, Grizzetti B, de Vries W, van Grinsven HJM, Abrol YP, Adhya TK, Billen G, Davidson EA, Datta A, Diaz R, Erisman JW, Liu XJ, Oenema O, Palm C, Raghuram N, Reis S, Scholz RW, Sims T, Westhoek H, Zhang FS (2013a) Our nutrient world: the challenge to produce more food and energy with less pollution. Global overview of nutrient management. Centre for Ecology & Hydrology, Edinburgh

Sutton MA, Reis S, Riddick SN, Dragosits U, Nemitz N, Theobald MR, Tang YS, Braban CF, Vieno M, Dore AJ, Mitchell RF, Wanless S, Daunt F, Fowler D, Blackall TD, Milford C, Flechard CR, Loubet B, Massad R, Cellier P, Coheur PF, Clarisse L, Van Damme M, Ngadi Y, Clerbaux C, Ambelas Skjøth C, Geels C, Hertel O, Wickink Kruit R, Pinder RW, Bash JO, Walker JD, Simpson D, Horvath L, Misselbrook TH, Bleeker A, Dentener F, de Vries W (2013b) Toward a climate-dependent paradigm of ammonia emission & deposition. Philos Trans R Soc London Ser B 368(1621):20130166

UNECE (2013) 1999 Protocol to abate acidification, eutrophication and ground-level ozone to the convention on long-range transboundary air pollution, as amended on 4 May 2012 ECE/EB. AIR/114

Vieno M, Heal MR, Hallsworth S, Famulari D, Doherty RM, Dore AJ, Tang YS, Braban CF, Leaver D, Sutton MA, Reis S (2014) The role of long-range transport and domestic emissions in determining atmospheric secondary inorganic particle concentrations across the UK. Atmos Chem Phys 14:8435–8447

Chapter 2
Economic Costs of Nitrogen Management in Agriculture

Oene Oenema, Steen Gyldenkaerne, and Jouke Oenema

Abstract Nitrogen (N) management is one of the measures of Annex IX of the revised Gothenburg Protocol and described in detail in the Guidance Document (Bittman et al., Options for ammonia mitigation: guidance from the UNECE task force on reactive nitrogen. Centre for Ecology & Hydrology, Edinburgh, 2014). The measures of Annex IX aim at the abatement of ammonia (NH_3) emissions from agricultural sources. This chapter reviews literature dealing with the economic costs of N management, aimed at decreasing the N surplus and increasing N use efficiency (NUE) at farm level.

Nitrogen balances are important tools for N management; they are prerequisites for monitoring, reporting and verification. They have been implemented in practice in Denmark and The Netherlands, and are used in many other countries as research tool. The economic costs of making N balances at farm level range between 200 and 500 € per farm per year. Possible additional costs relate to comparing and discussing these balances with other farmers. Also governments make costs for verification and control, estimated at 50–500 € per farm per year.

Management activities related to decreasing the N surplus and increasing NUE at farm level are diverse and the economic costs of these activities vary greatly, depending on farming type and site-specific conditions. Conveniently, a distinction should be made between crop, mixed and landless animal production systems, also because the N management activities will differ between these systems. Relatively cheap measures (providing net benefits) include proper timing of activities, selecting high-yielding varieties and breeds, increasing N fertilizer replacement

O. Oenema (✉)
Environmental Sciences Group, Alterra, Wageningen University, P.O. Box 47, NL-6700 AA Wageningen, The Netherlands
e-mail: oene.oenema@wur.nl

S. Gyldenkaerne
Institute of Environmental Science, Aarhus University, Frederiksborgvej 399, 4000 Roskilde, Denmark
e-mail: sgy@envs.au.dk

J. Oenema
Plant Research International, Agrosystems Research, Wageningen University, P.O. Box 16, NL-6700 AA Wageningen, The Netherlands
e-mail: jouke.oenema@wur.nl

© Springer Science+Business Media Dordrecht 2015

S. Reis et al. (eds.), *Costs of Ammonia Abatement and the Climate Co-Benefits*, DOI 10.1007/978-94-017-9722-1_2

value of manure with a concomitant lowering of fertilizer use, proper matching N demand by the crop and N supply by manures and fertilizers, precision fertilization and precision feeding, and optimization of crop husbandry and animal husbandry. Relatively expensive measures (leading to net cost) include (see also next chapters) fertilizer application far below the economic optimum, low-emissions animal housing, leak-tight and covered manure storages, and long-distance manure transport and manure treatment on landless animal farms.

For dairy farms it was observed that decreasing N surplus by 1 kg decreases NH_3 emissions on average by 0.25 kg. All of these farms had implemented low-emission manure storage and manure application techniques, indicating that decreasing N surplus has additional NH_3 emissions abatement effect.

Estimates of the economic costs of N management measures tend to decrease over time, because of learning effects and dynamic effects that improve the overall agronomic and environmental performance of farms. A plea is made for more comparative and longitudinal studies at farm level in different countries, because there is a relative paucity of accurate assessments over long periods.

Keywords Ammonia emissions • Economic costs • Nitrogen management • Agriculture • Nitrogen use efficiency

2.1   Introduction

Management is commonly defined as 'a coherent set of activities undertaken to achieve objectives' (e.g., Drucker 1954). This definition applies to all businesses and sectors of the economy, including agriculture. The application of the management concept to nitrogen (N) in agriculture dates from the early 1990s. Nitrogen management is defined as 'a coherent set of activities related to the allocation and handling of N in agriculture to achieve agronomic and environmental/ecological objectives' (e.g., Oenema and Pietrzak 2002). Common agronomic objectives relate to crop yield, crop quality and animal performance, while environmental/ecological objectives commonly relate to minimizing N losses and to increasing N use efficiency (NUE). Condor-Golec (2015) provides an overview of research, demonstration and dissemination activities related to N management in Italy. The objectives of N management are often region-, watershed-, site-, farm-, and/or field-specific (e.g., Hatch et al. 2004; Mosier et al. 2004; Hatfield and Follett 2008; Schepers and Raun 2008). Evidently, N management is evaluated as being successful when the specified objectives are being achieved.

Nitrogen management is one of the measures of Annex IX of the revised Gothenburg Protocol of the UN-ECE Convention on Long-Range Transboundary Air Pollution (CLRTAP) (http://live.unece.org/env/lrtap/multi_h1.html). This Annex lists the measures for the control of ammonia (NH_3) emissions from agricultural sources. The objective of the measure *'Nitrogen management, taking into account the full nitrogen cycle'* is that 'all available on-farm N sources and external N inputs are used effectively'. For that purpose, N input-output balances have to be established on farms, and the success of N management is evaluated on the basis of the decrease of the N surplus and the increase of the N use efficiency over 5-year periods. It is meant to prevent pollution swapping, i.e. decreasing NH_3 emissions but increasing other unwanted N emissions, such as nitrate leaching and nitrous oxide emissions (Bittman et al. 2014).

This chapter reviews literature dealing with the economic costs of N management measures aimed at decreasing the N surplus and increasing NUE at farm level. Unfortunately, there is not much empirical information about the economic cost of N management in agriculture, because of its recent nature, its limited implementation and experience in practice (mainly in Western Europe and US), and because of the regulatory character of the policy measures dealing with N management (Oenema et al. 2011a). This review draws heavily on experiences and literature from countries with some 20 years of experience with N management (Netherlands and Denmark).

Various measures and farm activities may contribute to decreasing N surplus and increasing NUE, including all the measures of the Annex IX. However, the specific cost of these technical measures (low-emission animal feeding, animal housing, manure storage, manure spreading and fertilizer spreading) are discussed in the subsequent chapters of this book. Here, N management is perceived in terms of decreasing N surplus and increasing NUE in sensu stricto, and the economic cost of N management in agriculture is perceived as (i) the economic costs of making an N input-output balance sheet of a farm, and (ii) the economic costs of decreasing N surplus and increasing NUE through optimization of N use. Where possible, we account for economic benefits in agriculture associated with improvements in N management. However, we do not address the benefits of improvements in air and water quality for human health, biodiversity and climate change (see Brink et al. 2011).

2.2 Economic Cost of Establishing N Input-Output Balances at Farm Level

Estimating the cost of making an N input-output balance sheet of a farm requires the recording of all N inputs into a farm and all N outputs out of a farm. There are various procedures for making N input-output balances, namely gross nitrogen balance, soil-surface balance, farm balance and farm-gate balance, which may slightly differ in outcome, accuracy and in efforts needed to establish the input-output balance (e.g., Oenema et al. 2003; OECD 2007; Leip et al. 2011). Hence, it is important to use standardized formats for making N input-output balances.

In situations with proper farm accountancy data, it is relatively easy to establish farm N balances, because the N input and output data can be assessed on the basis of these accountancy data. Further, it is easier to establish and interpret N balances of specialized farms than mixed farms.

Information from countries that have implemented farm N balances in practice (e.g. Denmark, The Netherlands) indicates that farmers learn easily to interpret such N balances. They may also easily learn to compile these N balances on the basis of records of the farm economic administration combined with tabulations of N mass fractions in the inputs and outputs of a farm. However, in many cases N balances are compiled by accountancy offices. If done on a routine basis, i.e. for a number of farms each year, it takes on average half a day for compiling an N balance and a phosphorus (P) balance. Accountancy offices in Denmark and The Netherlands charge farmers on average about 250–500 euro per farm per year for farm N & P balances (e.g., Jacobsen et al. 2005a, b). The net costs to the farmers are less than 250–500 euro, because of the deduction of these costs from the taxable farm income. In Czech Republic, the estimated costs of farm N & P gate balances are also in the range of 500 euro per farm per year (personal communication Dr Pavel Cermak, October 2010). In Czech Republic, it takes more time to establish a balance than in Denmark and Netherlands, because of less experience, but this is compensated by lower rates per hour.

2.3 Economic Cost of the Verification and Control of N Input-Output Balances

In addition to the direct costs to farmers, national governments have costs too, for supporting and establishing the knowledge infrastructure and for supporting the control and verification of N balances. Unfortunately, there is not much empirical information about the economic costs by governments (Parties) for establishing a knowledge base and infrastructure for enabling the verification and control of N balances of farms.

A mandatory N and P accounting system MINAS at farm level had been implemented in The Netherlands, between 1998 and 2003 (Schroder and Neeteson 2008). All farms ($>$80,000) had to submit each year N and P balances (farm-gate balances) to a governmental office. This office checked all balances and verified and reported to all individual farms whether balances were made correctly and whether or not the N and P surpluses surpassed levy-free surpluses. The MINAS regulatory accounting system was effective in lowering N and P surpluses, and N and P leaching losses; it contributed also to decreases in NH_3 and N_2O emissions. MINAS was most effective in dairy farming. The economic cost of the registration and control by the governmental office ranged from 7 to 36 million euro per year during the period 1998–2003 (MNP 2004; Table 7.2), which translates roughly to a mean of 80–500 euro per farm per year. The wide range is caused in part by 'initial learning' problems.

The economic costs related to the verification and control of fertiliser plans and N and P balances is less in Denmark than in The Netherlands, because of 'self-control' and 'continuous dialogue' between government, researchers and farmers union in Denmark (Jacobsen et al. 2005a, b; Mikkelsen et al. 2010).

2.4 Economic Cost of Decreasing N Surplus and Increasing NUE

2.4.1 Some General Considerations

Estimating the economic costs of N management activities can be done at field level, farm compartment level, farm level, sector level and national/society levels. Estimates at farm level provide an integral account; these integral estimates tend to be (much) lower than the summed costs at field and/or farm compartment levels, due to compensation effects. Estimates at sector level include the indirect economic effects for suppliers and processing industries, which can be significant when N management activities at the farm significantly change farm inputs and/or outputs. Finally, cost-benefit analyses at national or society level basically integrate all effects, the cost of the N management activities as well as the benefits to society of lower N surpluses and higher NUE (e.g., Brink et al. 2011; Jensen et al. 2011). In this chapter, the focus is mainly on field and farm levels.

In principle, estimating the economic costs of N management activities can be done through longitudinal comparisons of one or few similar farms over time or through comparisons of different farms with and without improved N management. Effects of N management activities are monitored over time in the first case, while differences between farms in N management activities are analyzed statistically in the second case. Both types of studies are useful.

Farms of similar types may differ a lot in management. Farms with poor nutrient management often have high Nsurplus and low NUE; lowering the Nsurplus and increasing NUE is often economic beneficial because of decreased resource use and/or increased yields (Ondersteijn et al. 2003). On the other hand, efficiently managed farms generally have good economic and environmental performances, and may not easily decrease N surplus and increase NUE further. Evidently, the law of diminishing returns applies also to N management in agriculture.

It has to be understood that the relationship between N surplus and NUE is not linear. The Nsurplus and NUE are defined as

$$\text{Nsurplus} = (\text{N input}) - (\text{N output}) \tag{2.1}$$

$$\text{NUE} = (\text{N output})/(\text{N input}) \tag{2.2}$$

$$\text{Nsurplus} = (1 - \text{NUE}) \times (\text{N input}) \tag{2.3}$$

The Nsurplus, N input and N output are expressed in kg per ha per year; NUE is expressed either as a dimensionless fraction or as percentage. Evidently, Nsurplus can be decreased by increasing N output and/or decreasing N input and NUE can be increased by increasing N output and/or decreasing N input. Roughly, increasing N output can be achieved by increasing the yield (produce) and/or the N content of the produce (including animal wastes) exported from the farm. Decreasing N input can be achieved by lowering the import of N via fertilizers, animal feed, manure and other possible sources into the farm. In general, N surplus will decrease if NUE increases; however, a change to more productive and N-responsive crop varieties and/or animal varieties may lead to both increases in NUE, N surplus, N output and N input (see Appendix 1).

Possible N management activities greatly depend on farm type and a distinction should be made between (a) specialized arable and vegetable farms, (b) mixed farms, with livestock and cropped land for producing animal feed, and (c) specialized animal farms, with little or no land. The principle difference between these categories is the difference in the type of N inputs and outputs, the on-farm transformation processes, and the ease with which these inputs, outputs and processes can be modified. The economic cost of N management for each of these farming types is discussed in the next three sections.

2.4.2 Economic Costs of N Management Activities on Arable and Vegetable Farms

Management activities aimed at decreasing N surplus and increasing NUE on specialized arable farms relate to maximizing the N output (i.e., yield) and maximizing the utilization of available N sources, using the right method, time and amount of application. Maximizing yield involves using the proper genetic crop materials and optimal crop husbandry, including irrigation, pest and disease management. Maximizing the utilization of available N sources is also known as the 4R Nutrient Stewardship concept (IPNI 2012); i.e. the right source, right method, right amount and right time of application. To put this in other words, lowering the N inputs while increasing the effectiveness of the available N sources through choosing the right method and time of application. The economic cost of selecting the appropriate timing, method and rate are relatively small, but the implementation of these best management practices is still modest. In 2006, 65 % of surveyed cropland in US was in need of improved N management (Ribaudo et al. 2011). Sheriff (2005) examined why farmer perceptions of agronomic advice, input substitutability, hidden opportunity costs, uncertainty, and risk aversion can make it economically rational to "waste" fertilizer by applying it above agronomically recommended rates.

On specialized arable and vegetables farms, fertilizer N is often the main N source. The costs of N fertilizer in proportion to the total production cost may range from 20 to 30 % on large cereals farms to 1–5 % on farms specialized in growing seed potatoes, vegetables and flowers (Pederson et al. 2005; Van Dijk et al. 2007; Jensen et al. 2011). The relatively low cost of N use for high-value crops is one of the reasons for its liberal use in these crops, and for the relatively high N surpluses and low NUE (Jensen et al. 2011).

Estimating the cost of decreasing Nsurplus and increasing NUE can be done on the basis of analyzing (i) yield curves and (ii) farm accountancy data of whole farms. Yield curves or 'doses-response relationships' provide insight in the possible management actions that are needed to decrease Nsurplus and increase NUE (Jensen et al. 2011; Appendix 1). For crop land, we distinguish the following management activities to decrease Nsurplus and increase NUE:

- Decrease over-fertilization, i.e. lower the N input, taking into account also the amounts of N delivered by soil, atmosphere, crop residues, and leguminous crops;
- Increase the effectiveness of N applied via fertilizers, manures, composts, i.e., use the right method and time of application;
- Increase the yield of the crop through selection of high-yielding crop varieties and optimal crop husbandry, including optimal pest and disease management, irrigation and drainage management, soil cultivation and weeding management, as well as a proper supply of all 14 essential mineral nutrients for plant growth and development.

The gross economic cost of these activities may relate to the use of soil (mineral N) and crop analyses, better fertilizer and manure spreaders, crop monitoring, advice, and training. The gross benefits relate to decreased fertilizer costs and possibly increased yields. The net costs are highly depending on crop type, but are usually in the range of -0.5 to $+2$ euro per kg N saved, which translates to -5 to 25 euro per ha (e.g., Van Dijk et al. 2007; Mikkelsen et al. 2010).

Van Dijk et al (2008) estimated the economic cost of applying economic sub-optimal N applications to a range of crops, using the results of numerous field experiments and regression analyses (second degree polynomial and exponential models). Financial losses due to sub-optimal N fertilization were estimated relative to the crop-specific recommended N fertilization rates. Results for some crops are summarized in Table 2.1. As an example, Fig. 2.1 shows the relationships between crop yield and N application for four experiments with spinach. Evidently, the response varied greatly from one experiment to another.

Financial losses of sub-optimal N fertilization varied also greatly between crops, experimental years (differences between minimum and maximum losses), and soil type. Financial losses progressively increased with a lowering of the N application rate. Losses were relatively small for crops like silage maize and starch potatoes, which generally show a relatively small response to N application. Losses are relatively large for high-value crops like vegetables (e.g., spinach) and flowers (e.g. lily, Liliaceae), especially when responsive to N application. Results indicate that farmers may benefit from a 5–10 % decrease of the N application rate to silage maize and starch potatoes, considering the fact that fertilizer N savings (5 to 20 € per ha) have not been included in the estimates of Table 2.1. However, lowering the

Table 2.1 Financial loss due to sub-optimal N fertilization for selected crops and soil types, in euro per ha

Crop	Soil type		Financial loss of suboptimal N fertilization, euro per ha				
			N application, % of the recommended amount				
			50	60	70	80	90
Potato	Sand	Mean	415	305	205	125	55
		Min	105	70	45	20	5
		Max	910	690	490	305	145
Potato	Loess	Mean	695	500	335	200	90
		Min	310	195	115	55	15
		Max	1,090	845	615	395	195
Starch potato	Sand	Mean	120	80	45	20	5
		Min	−5	−5	−10	−10	−5
		Max	315	230	155	90	40
Silage maize	Sand	Mean	105	75	50	25	10
		Min	5	−5	−10	−15	−10
		Max	485	385	290	190	95
Spinach	Clay	Mean	1,295	830	475	210	70
		Min	300	−20	−210	−265	−175
		Max	2,295	1,680	1,155	685	315
Lily	Sand	Mean	2,070	1,425	910	505	205
		Min	−1,510	−1,265	−990	−690	−360
		Max	6,365	4,350	2,755	1,520	615

Mean, minimum and maximum losses are derived from the statistical analyses of field experiments, the number of which varied per crop. Note that fertilizer savings are not included in the assessments (After Van Dijk et al. 2008)

N application rate for other crops to below the recommended level likely costs money, depending on crop type. The assessments presented in Table 2.1 do not include the effects of precision fertilization techniques. Model calculations made by Tavella et al. (2011) show that savings in input and variable costs, and increases in yields depend on the technology applied. However, gross margins and total returns were for all precision technologies positive, and higher compared to conventional farming. Introduction of Controlled Traffic Farming and Auto Guidance led to the most profitable results (Tavella et al. 2011).

2.4.3 Economic Costs of N Management Activities on Dairy Farms

Mixed farming systems produce crop and animal products; the crops produced are often used as feed for animal production, though some farms sell both crop and animal products. Mixed farming systems may also derive income from non-agricultural activities, but these types of mixed systems are not discussed here

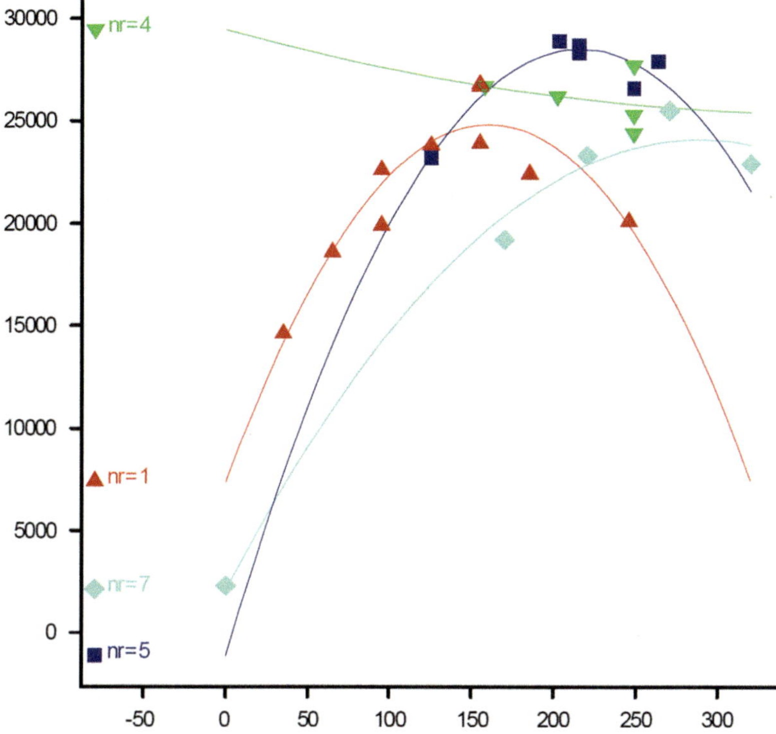

Fig. 2.1 Response curves of spinach to N fertilizer application (X-axis, in kg N per ha). Marketed fresh yield (Y-axis) is expressed in kg per ha (Van Dijk et al. 2008)

further. Common mixed farming systems include dairy and beef production systems, pig production systems and poultry production systems, which grow a significant fraction of the feed on the land of the farming system. As livestock density increases, the need for importing additional feed increases, and thereby the amount of N imported increases. The most dominant mixed farming system in Europe is dairy farming, which covers roughly 20 % of the surface area of utilized agricultural land. This section therefore focusses on dairy farms, because of their dominance.

Quantitative information about economic cost of measures to decrease N surplus and increase NUE is available for dairy farms in NL, where policy measures have been implemented to lower the N surplus at farm level from 1998. This policy has been successful especially on dairy farms, in part because the N surpluses were high initially, but also because of the many possible management activities that can be utilized to lower the N surplus. Total NH_3 losses decreased by about 50 % between 1990 and 2006, although a significant fraction of the decrease in NH_3 losses has been attributed to other measures of Annex IX than just N management (MNP 2004).

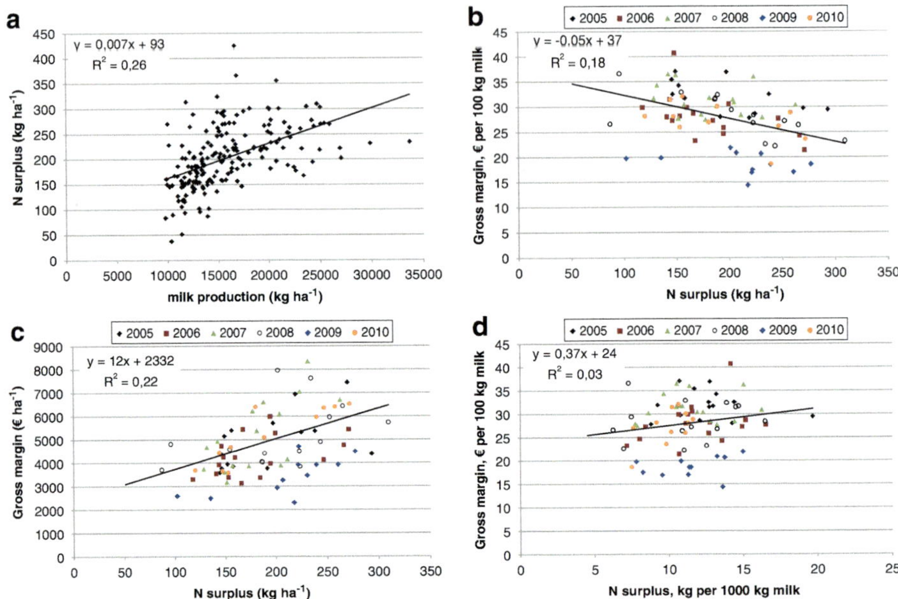

Fig. 2.2 Relationships between (**a**) milk production intensity (kg milk per ha per year) and N surplus (kg per ha), (**b**) N surplus (kg per ha) and gross margin (euro per 100 kg milk), (**c**) N surplus (kg per ha) and gross margin (euro per ha), and (**d**) N surplus (kg per 1,000 kg of milk) and gross margin (euro per 100 kg milk) for dairy farms of the project Cows & Opportunities in the Netherlands during the period 2005–2010 (Oenema et al. 2012)

Figure 2.2a shows the large scatter in practice in the relationship between mean milk yield per ha and N surplus at farm level. It shows that a high milk yield per ha can be achieved at a low and high N surplus. During the period 1997 and 2007, N surpluses at farm level decreased by on average 200 kg per ha, without loss of milk yield per ha, but the scatter remained high (see also Van den Ham et al. 2010). The relationship between gross financial result, in euro per 100 kg milk produced and N surplus tend to be negative although the scatter is large again (Fig. 2.2b). The trend in the relationship between gross financial result and N surplus is reversed when the gross margin is expressed in euro per ha (Fig. 2.2c) or N surplus is expressed in terms of kg per 1,000 kg of milk. However, the scatter is large; it suggests that a high gross margin can be obtained at both low and high N surplus.

Empirical information on the relationship between farm management, N surplus and financial consequences on dairy farms have been collected also by Rougoor et al. (1997), Ondersteijn et al. (2003), Doornewaard et al. (2007) and Daatselaar et al. (2010). A major conclusion of these studies is that improved management leads to improved efficiency and to improved financial results, though within certain boundaries. Similar conclusions have been reached by Powell et al. (2009, 2010) and Rotz (2004) for dairy farms in the USA. Improving the utilization of nutrients from manure while decreasing the use of synthetic fertilizers is cost-effective measure to decrease N surplus and increase NUE.

Developments in technical and economical performances over time of two groups of dairy farms between 1999 and 2004 are compared in Table 2.2. The 16 dairy farms of Cows & Opportunities were guided to lower N surpluses and to be ahead of the reference group (Doornewaard et al. 2007). Both the Cows-&-Opportunities farms and the reference groups decreased N surplus; as expected the N surplus was lower for the Cows-&-Opportunities than for the reference groups. Also the NUE increased significantly on the Cows-&-Opportunities farms (Oenema et al. 2011b). Both groups increased farm area considerably between 1999 and 2004; investments in land and buildings were larger for the Cows-&-Opportunities. Fertilizer cost are only a small percentage of the total allocated costs (<10 %) and non-allocated costs (<3 %). A significant fraction of contractor costs is related to low-emission slurry spreading, which varied between 2.5 and 3.5 € per m^3 in 2010, depending on contractor and transport distance.

Figure 2.3 shows large scatter in the relationship between N surplus and NH_3 emissions on dairy farms. The NH_3 emissions include the emissions from animal housing, manure storage, manure applied to land, and manure from grazing animals. The scatter is related to differences between farms in livestock density, animal feeding practices, grazing strategy, manure management and N fertilizer use. The linear relationship suggests that lowering N surplus by 1 kg N per ha decreases NH_3 emissions by on average 0.25 kg per ha. Evidently, this relationship will hold only for dairy farms that used low-emission slurry application techniques, covered slurry storage systems and to some extent also low-protein animal feeding strategies. On average 25 % of the N surplus on these dairy farms is lost via NH_3 volatilization, while the remaining 75 % will be lost through N leaching and denitrification. Probably more than 25 % of the N surplus is lost via NH_3 volatilization on dairy farms that have not implemented NH_3 emission abatement techniques.

2.4.4 Costs of N Management Activities on Specialized Pig and Poultry Farms

Specialized pig and poultry farms basically have a yard with animal housings, animal feed storages and manure storages, but no land. These farms import all animal feed and export animal products and manure. Activities related to improving N management on these farms include (i) low protein, phase-feeding, (ii) general animal herd management (genetic selection, reproduction, disease management, management of young stock, etc.), (iii) low-emission housing system, and (iv) low-emission manure storage, treatment and export. Most of these activities are addressed in other chapters of this book (e.g. low-protein feeding, low-emission housing, low-emission manure storage) and are not repeated here.

Table 2.2 Comparison of two groups of dairy farms in technical and economic performances between 1999 and 2004

	Cows & Opportunities		Reference groups	
	1999	2004	1999	2004
Area, ha	41	52	42	51
Dairy cows	76	97	75	92
Milk yield, Mg/ha	15.6	15.2	15.4	15.1
Milk yield, Mg/cow	8.1	7.9	8.1	8.0
Milk fat & protein, g/kg	78.4	79.6	78.6	79.4
Young stock, number per cow	0.8	0.64	0.83	0.71
Concentrates, kg/cow	2,098	2,256	2,079	2,004
N surplus, kg/ha	275	165	333	212
P surplus	15	3	10	5
Economic results, €/100 kg milk				
Revenues milk	33.5	33.4	33.0	33.2
Revenues cattle	4.2	5.8	4.1	4.7
Total allocated costs	9.0	10.1	8.9	10.0
Concentrates	*4.5*	*5.4*	*4.6*	*5.4*
Veterinary assistance	*1.1*	*1.0*	*0.8*	*0.9*
Fertilizers	*0.5*	*0.7*	*0.7*	*0.9*
Total non-allocated costs	40.4	41.1	40.7	39.7
Labor	*12.8*	*11.9*	*13.5*	*13.4*
Contractors	*2.3*	*3.3*	*1.9*	*2.4*
Machines	*4.7*	*5.5*	*4.8*	*5.5*
Land & buildings	*9.5*	*10.3*	*9.3*	*9.3*
Milk quota	*7.9*	*7.0*	*8.1*	*6.6*
Energy and water	*1.0*	*1.2*	*0.8*	*1,1*
General costs	*2.2*	*1.9*	*2.3*	*1.5*
Net operating result	−11.7	−12.0	−12.5	−11.8

The 16 dairy farms of Cows & Opportunities were guided to lower N surpluses, farms of the reference group (about 500 farms) not (Doornewaard et al. 2007)

Here we focus on manure transport (export) and manure treatment. Basically, all manure produced has to be exported from the farm, as there is no land for manure disposal. In case neighboring farms are able and willing to accept the manure, the transport and handling costs are likely small. However, landless pig and poultry production systems tend to agglomerate in certain areas, because of location-specific cost advantages, and in this case neighboring farms will not have sufficient land for manure disposal (Menzi et al. 2010). Long-distance transport and/or manure processing are needed then, and these activities are costly. Moreover, the value of the manure becomes negative when the supply is (much) larger than the demand for manure nutrients; it becomes waste instead of manure.

Figure 2.4 shows that manure disposal cost off farm in NL varied significantly between years and tended to increase over time. For pig slurry, with an average dry matter content of about 10 %, the long-distance (50–150 km) transport costs were

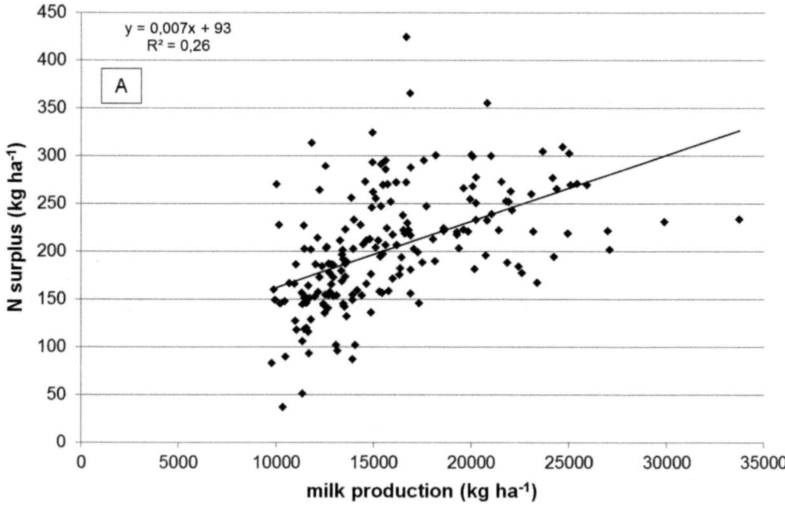

Fig. 2.3 Relationships between N surplus and total NH_3 emissions at farm level. Results are based on 16 Cows and Opportunities dairy farms during the period 2003–2009, when all farms complied to the ammonia abatement measures listed in Annex IX of the Gothenburg Protocol (Oenema et al. 2012)

on average 6 € per m^3, but depending on distance and fuel prices. In addition, there were transaction costs related to administration, manure analyses and goodwill fees for the receiver of the slurry. These transaction costs varied from 0 to 15 € per m^3 during 1995–2009, and mainly depended on the height of the goodwill fee (personal communication, Harry Luesink, LEI, 2011). Before the 1990s, pig farmers did not pay a goodwill fee, but instead got paid for the slurry, as the fertilizer value of the nutrients was equivalent to 5–10 € per m^3. Note that land-based pig and poultry farming systems only have cost for manure spreading, which will be in the range of 2–4 € per m^3, which give them a competitive advantage relative to landless systems. Disposal cost were higher for poultry manure than for pig slurry, but dry matter content is also much higher for poultry manure (about 60 %) than for pig slurry (about 10 %). The drop in manure disposal costs after 2008, especially for poultry manure, coincide with the increases in fertilizer prices. Also, a poultry manure incineration plant became operational by 2009.

Manure treatment is gaining popularity again, fueled by increasing phosphorus fertilizer prices and subsidies on biofuel production, but also by the high manure disposal costs for landless pig and poultry farms (Fig. 2.4). Pig slurry treatment may involve a series of steps, including anaerobic digestion for biogas generation, separation of solids from liquids, drying of solids followed by incineration or pyrolysis for bioenergy production and recovery of phosphorus, and finally ultrafiltration and reverse osmosis of the liquid fraction (e.g., Menzi et al. 2010; Schoumans et al. 2010). The costs involved in pig slurry treatment depend on the number of steps, and varied from 2 to 4 € per m^3 for separation of solids from liquids, to more than 20 € per m^3 for the full treatment scheme (Schoumans et al. 2010).

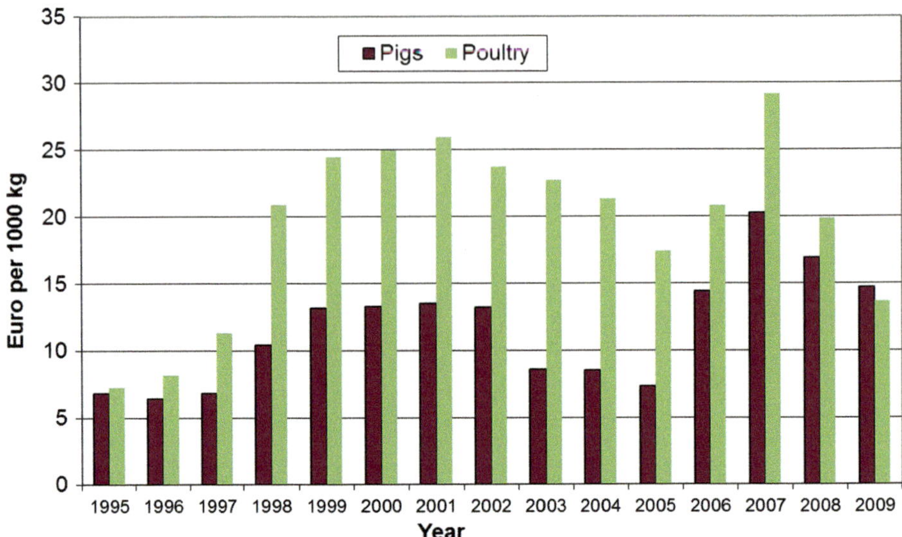

Fig. 2.4 Changes in the cost of pig and poultry manure disposal (off farm cost) in The Netherlands during the period 1995 and 2009, in € per 1,000 kg (for pig slurry, 1 € per 1,000 kg is roughly equivalent to 1 € per m^3 slurry) (Source: Agricultural Economics Institute LEI, Farm Accountancy Data Network, 2011)

Hence, treatment costs are similar to, or higher than, the pig slurry disposal costs depicted in Fig. 2.5. Treatment costs for poultry manure are modest, because of its relatively high initial dry matter content; the treatment usually involves only drying and pelleting at 5–10€ per ton.

The manure disposal cost in pig production (0.08€ per kg slaughter weight) were about 5 % of the total production costs in 2007, which translates to about 90.000 € per farm per year (Hoste 2011). The cost of low-emission housing (air scrubbers) were about 0.035€ per kg slaughter weight. The manure disposal costs and the costs for low-emission housing are the environmental costs depicted in Figs. 2.5 for 2.6 EU countries. Clearly, environmental costs for pig production were highest in NL and lowest in Poland. Further, environmental costs tend to increase due to the implementation of new policy measures.

The composition of the total production costs in pig production is shown in Fig. 2.6. DK and NL had the lowest production costs in EU-27 in 2007, but these were much higher than the costs in Brazil and US. The relatively low costs in US and Brazil were related to monetary exchange rates, but also to lower environmental costs. Mean production costs vary between years; for example, production costs in France were estimated at 1.35 € per kg in 2009, i.e., 0.13 € per kg lower than in 2007 (Hoste 2011). Feed costs are roughly 50–60 % of the total costs. Feed costs were lowest for NL, FR, DE and DK, in part because of relatively low feed conversion rates, which ranged from 2.75 kg per kg in NL, to 2.9 in DK, 3.1 in

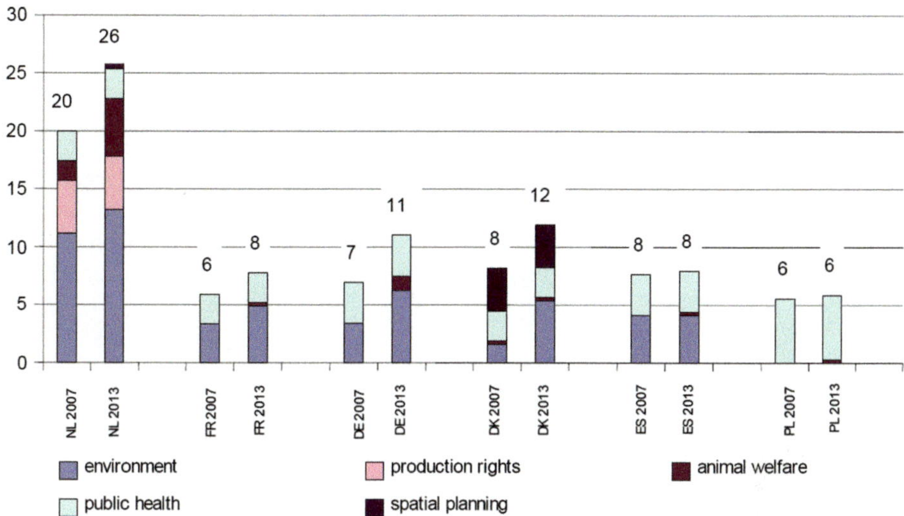

Fig. 2.5 Additional costs in pig production due to policy measures in 2007 and 2013 per country, in euro cents per kg slaughter weight (Hoste and Puister 2009). *NL* Netherlands, *FR* France, *De* Germany, *DK* Denmark, *ES* Spain, *PL* Poland

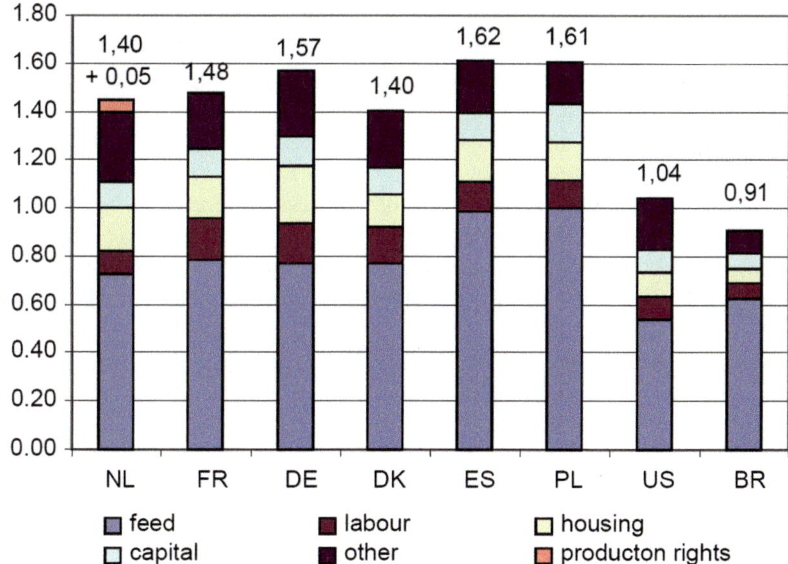

Fig. 2.6 Composition of the production costs for pigs per country in 2007, in euro per kg slaughter weight (Hoste and Puister 2009)

ES and 3.2 in PL and US. Manure disposal costs and low-emission housing air scrubbers are included in the item 'other costs' in Fig. 2.6.

Economic cost related to the implementation of N management measures increase the cost of production. However, variations in feed, labour, housing and capital costs vary greatly between countries and thereby also influence the economic competitiveness of the sector. These variations in cost tend to be larger than the cost of the N management measures. There is also much variation in production costs between farms within a country, which are mainly caused by the differences in labour costs, production results, feed prices and farm size (Hoste and Puister 2009).

2.4.5 Economic Costs of N Management Activities at Sector and National Levels

Implementing strict N management measures may have significant 'downstream' and 'upstream' effects because of changes in the production volume. The fertilizer industry will be affected directly when decreases in N surplus and increases in NUE have been achieved through lower fertilizer N use. The whole production – processing – retail – consumption chain may be affected when the N management measures lead to changes in the economic competiveness of a sector. This may happen for example when country A takes over the market share of country B. Clearly, this is the main argument for having a level playing field for environmental measures in all countries.

Some sectors may be hit more severely by the N management measures than others. For example, the stepwise lowering of the N and P surpluses in agriculture in NL has led to a transfer of money from the landless pig and poultry producers to specialized crop production systems, in exchange of animal manure (e.g., De Hoop and Stolwijk 1999; MNP 2004). Also, arable farms made less cost for the purchase of synthetic fertilizers. The gain of arable farmers was ~0.5 € per kg N per ha per year.

Governments incur costs for the monitoring, reporting and verification of the implementation of N management measures in practice. These costs may vary from less than 50 to more than 500 € per farm per year, depending on the intensity and organization of the monitoring, reporting and verification. Such costs could be transferred to the producers or to the processing industry. There is an increasing trend that the processing industry becomes involved in standard setting and in monitoring, reporting and verification, also to guaranty the safety of the products and to increase the overall transparency of the production methods.

2.5 Discussion and Conclusions

2.5.1 Economic Costs and Implementation of N Management Measures

This chapter reviewed literature dealing with the economic costs of N management measures aimed at decreasing the N surplus and increasing NUE. Economic costs and associated risks of management measures are often seen as an obstacle and/or delay for implementing such measures in practice (Sheriff 2005; Oenema et al. 2011a). Indeed, management measures require additional activities and possible changes in practices, which cost money. Thereby the competitiveness of the farms may decrease, especially when other farms or countries do not implement such measures. In an open, globalized market, it is necessary to establish a level playing field; otherwise producers will remain reluctant to fully implement measures that put them at a comparative disadvantage.

On the other hand, there are often direct and indirect economic benefits associated with improvements in N management, especially in the long term, which should not be overlooked, as they may even nullify the direct and indirect cost of the measures. There is evidence that increasing N management increases the overall agronomic and economic performance of farms, especially in mixed farming systems and specialized crop production systems (see Sects. 2.4.3 and 2.4.4). Assessing such direct and indirect benefits requires in-depth and long-term comparative and longitudinal studies involving many farms, such as those of the Farm Accountancy Data Network in EU-27. These benefits require more attention of research and policy makers.

Further, it is increasingly known that people, including farmers, are not only guided by cost-benefit analyses, but also by a series of other principles, such as tradition, reciprocity, commitment and consistency, social proof, authority, liking, and scarcity (Cialdini 2007). These principles are not discussed here further, but should not be neglected when analyzing the (lack of) progress of improved N management in practice.

2.5.2 Economic Costs of N Balances

Nitrogen balances and budgets are important tools for N management; they are prerequisites for monitoring, reporting and verification. The economic costs of making N balances at farm level likely range between 200 and 500 € per farm per year. These N balances at farm level provide information about the overall performance of the farm; which allows also to make comparisons between farms. Evidently, farmers will make additional costs when discussing the balances with other farmers and extension services. They have to meet and visit each other's farms. Moreover, additional analyses and calculations at compartment levels (feed, animal, manure, land) may be needed, to identify the best options for improving the overall performance of the farms (Oenema et al. 2001; Rotz et al. 2005). This is key to N management; the activities must be seen within the context of improving the overall agronomic and environmental performances of the farms (as emphasized also in the definition of N management in the Introduction section).

2.5.3 Assessing the Economic Costs of N Management Measures

There is a relative paucity of empirical studies about the economic costs of N management measures, despite the fact that these costs are often seen as an obstacle for its implementation. One reasons for this relative paucity is the difficulty of assessing these costs accurately, because N management activities are not well-defined and/or standardized. They require the assessment of a whole range of costs and benefits, such as learning costs, labour costs, investment costs, contractor costs, decreased or increased yields, and direct and indirect benefits. And they require a proper reference. A second reason is the large variability between farms and years, which makes it difficult to generalize and extrapolate the estimated cost of one farm to another, and from 1 year to another. A third reason of the paucity of data is that assessments of economic costs are often kept private or are published in reports of companies and farmers journals, and not in peer-reviewed publications. Evidently, there is a need for more in depth studies of the effects of N management on the economic and environmental performances of farms, and the results of these studies should be made available to others through peer-reviewed publications.

Preferably, costs of N management activities should be assessed at farm level, to be able to assess the integrated effects of direct and indirect costs and benefits (De Haan 2001; Rotz 2004; Rotz et al. 2005). These assessments can be done on the basis of simulation models (e.g. De Hoop and Stolwijk 1999; De Haan 2001) or through the analysis of farm accountancy data (e.g., Doornewaard et al. 2007; Hoste and Puister 2009; Daatselaar et al. 2010), or through a combination (e.g., Ondersteijn 2003; Ondersteijn et al. 2002, 2003; Rotz et al. 2005). Thereby, combinations of comparative and longitudinal analyses are made, i.e., empirical information is collected from different farms over time, and the performances of these different farms are compared. A special case are pilot and demonstration farms (Oenema et al. 2001). Evidently, the data are not coming from 'controlled-condition experiments with untreated control treatments'.

Most of the economic analyses relate to studies carried out in NL, DK and US, and these countries serve to some extent as pilot countries. However, farm structure and the composition of costs of production may differ largely between countries, as follows for example from the data presented in Figs. 2.5 and 2.6. Hence, the results cannot be transferred directly from one country to another country, just as results cannot be transferred directly from one farm to the other; country-specific and farm-specific analyses and assessments are needed. The information, knowledge and hardware developed in one country may be used as building blocks in other countries, but the implementation has to be made farm- and country specific.

2.5.4 Net Economic Costs of N Management Tend to Decrease Over Time

Ex-ante estimates of the economic costs of the implementation of measures to decrease N losses from agriculture tend to be larger than ex post estimates. The reason is that increased learning brings new solutions and optimization overtime (Ondersteijn et al. 2003). Farmers and companies find new ways to decrease the cost related to decreasing N surplus, through optimization of activities, cheaper hardware, increasing yields, lowering fertilizer costs, improving animal performance (e.g. decreasing feed needs per kg of milk, meat and eggs produced, less young stock per cow; lowering the N and P contents of purchased animal feed), and through up-scaling.

Increasing the utilization of N from animal manures is a key measure for improving N use efficiency at farm level. This holds for all farming systems and for all manure types. Table 2.3 presents indicative N fertilizer replacement values (NFRV, kg N per 100 kg N) for urine, slurry and solid manure, as function of application time and method. The NFRV of manures may differ up to a factor of 9, simply because of differences in timing and method of application. Applying the manure at the right time and in appropriate portions (match between N demand by the crop and N supply via animal manure), and using low-emission manure application techniques, saves fertilizer and increases NUE. Evidently, increasing the NFRV by a factor of 2 to 9 is a big leap and may not be achieved in 1 year.

Apart from the optimization of activities at operational and tactical levels, incentives to improve N management may also induce changes at strategic level of farming operations. The results presented in Sects. 2.4.3 and 2.4.4 about dairy farms and pig and poultry farms indicate that entrepreneurial forerunners make additional investments to increase productivity and to lower the costs of N management measures. Thereby, the overall economic and environmental performance of the farm increases. As a result, the actual costs of N management measures tend to decrease over time, and can be assessed accurately only after a number of years.

2.5.5 Concluding Remarks

Nitrogen management aimed at decreasing N surplus and increasing NUE at farm level may encompass a range of activities, depending on farm type and site-specific conditions. The need and feasibility of such activities can be assessed on the basis

Table 2.3 Indicative NFRV (kg N per kg N applied) of urine, slurry and solid manure in the first year after application and in the long term, as affected by the time and method of application, climatic conditions/soil conditions

Manure type	First year		Repeated use	
	Autumn application[a]	Spring application[b]	Autumn application[a]	Spring application[b]
Urine	0.10–0.20	0.50–0.90	0.10–0.20	0.50–0.90
Slurry	0.10–0.30	0.30–0.70	0.30–0.40	0.40–0.80
Solid manure	0.10–0.30	0.20–0.40	0.40–0.60	0.60–0.70

After Schröder (2005a, b)
[a]Lower values referring to wet winter and/or lightly textured soils
[b]Lower values referring to situations with considerable volatilisation losses

of N balances. When N balances have relatively low N surplus and show high NUE, there is little need for additional activities. The reverse is true for farms with high N surplus and low NUE, but again depending on farm type and site-specific conditions.

The relationship between N surplus and NH_3 emissions is shown in Fig. 2.3. for dairy farms. It indicates that decreasing N surplus by 1 kg will decrease NH_3 emissions on average by 0.25 kg. Such positive relationships are also expected to exist for other farm types, but there is little or no information for other farm types. Likely, the slope of the relationship and the scatter will differ. For arable farms, the slope of the relationship between N surplus and NH_3 emissions will be much less than 0.25, because the amount of animal manure on these farms is much smaller than on dairy farms. Focusing N management on NH_3 emission abatement measures listed in Annex IX of the Gothenburg Protocol, as described further in other chapters of this book, will increase the slope of the relationship between N surplus and NH_3 emissions.

The economic costs of N management greatly vary between measures and also between farm types. Relatively cheap measures (providing net benefits) include:

- Proper timing of activities;
- Selecting high-yielding varieties and breeds;
- Increasing N fertilizer replacement value of manure and lowering fertilizer use;
- Proper matching N demand by the crop and N supply by manures and fertilizers;
- Precision fertilization and precision feeding;
- Optimization of crop husbandry and animal husbandry.

Relatively expensive measures (leading to net cost) include (see also next chapters):

- Fertilizer application far below the economic optimum;
- Low-emissions animal housing;
- Leak-tight and covered manure storages
- Long-distance manure transport and manure treatment.

Appendix 1: Estimating Economic Cost of N Management Activities Aimed at Decreasing Nsurplus and Increasing NUE on the Basis of Yield Curves

Yield curves or 'doses-response relationships' show the relationship between crop yield (or N yield) and total N input (from soil, atmospheric deposition, biological N_2 fixation, crop residues, manures, composts, fertilizers). Figure 2.A1 shows three hypothetical yield curves, to illustrate the effects of (i) a decrease of over-fertilization, i.e. lower the amounts of N applied, (ii) an increase of the effectiveness of the N applied i.e., use the right method and time of fertilizers, manures and composts applications; and (iii) an increase of the N yield of the crop, i.e. through optimal crop varieties and crop husbandry.

Over-fertilization is defined as applying more than the economical optimum application rate; the latter is defined as the application rate where the marginal economic returns equals zero ($dy/dx = 0$). The cost of fertilizer N (including the costs of spreading) is about 1 euro per kg N, but may vary from 0.5 to 2.0 euro per kg depending on fertilizer type and the variable fertilizer prices. Because of these

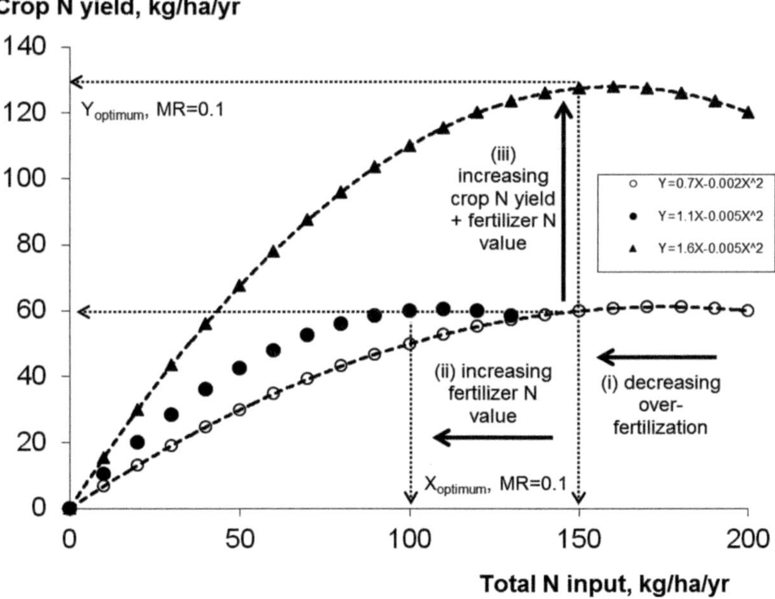

Fig. 2.A1 Three hypothetical crop N yield curves (second order polynomial), illustrating the effects of (i) decreasing 'over-fertilization', (ii) increasing the N fertilizer value at similar crop yields, and (iii) increasing crop N yield and N fertilizer value simultaneously (*solid arrows*). The broken arrows indicate the economic optimal N application rate at a Marginal Ratio of N fertilizer prize to crop N value equivalent to 0.1 (MR = 0.1). The parameter values and related N surplus and N use efficiency are shown in Table 2.A1

Table 2.A1 Parameter values of three hypothetical yield curves $(Y = A + BX + CX^2)$

$Y = A + BX + CX^2$	Yield curves of N input and N output		
	$Y = 0.7X - 0.002X^2$	$Y = 1.1X - 0.005X^2$	$Y = 1.6X - 0.005X^2$
A	0	0	0
B	0.7	1.1	1.6
C	0.002	0.005	0.005
$X_{optimum}$ (MR = 1.0)	175	110	160
$X_{optimum}$ (MR = 0.1)	150	100	150
$Y_{optimum}$ (MR = 1.0)	61	61	128
$Y_{optimum}$ (MR = 0.1)	60	60	128
Nsurplus (MR = 1.0)	114	50	32
Nsurplus (MR = 0.1)	90	40	23
NUE (MR = 1.0), %	35	55	80
NUE (MR = 0.1), %	40	60	85

The economic optimal N application rates ($X_{optimum}$) were estimated at a "Marginal Ratio" of N fertilizer prize to crop N value equivalent to 1.0 and 0.1 (MR = 1.0 and MR = 0.1, respectively). The N surplus and N use efficiency (NUE) at $X_{optimum}$ are estimated at both MR = 1.0 and MR = 0.1. See also Fig. 2.A1

variations, two assumptions have been used (see also Table 2.A1): (i) the cost of N in fertilizer equals the value of N in the harvested produce, and (ii) the cost of N in fertilizer is 10 % of the value of N in the harvested produce, denoted as MR = 1.0 and MR = 0.1, respectively. Note that the variation in MR has a small effect on the N yield ($Y_{optimum}$) but a large effect on the economic optimal N application rate (X_{optium}), N surplus and NUE (Table 2.A1). The cost of decreasing over-fertilization is roughly equivalent to the fertilizer savings, i.e., a benefit of about 1 euro per kg N.

The effect of increasing the crop N yield and the N fertilizer value (at similar N input) depends on the value of the crop; the beneficial effect of the case examined in Fig. 2.A1 is large, due to the assumed doubling in crop N yield; a net benefit > 1 euro per kg N. Decreasing the N input below the economically optimum N application progressively increases the cost of the N savings, through progressively declining crop N yields. This holds especially for high-value crops, such as vege-tables, nursery trees, flowers.

Decreasing the N input below the economically optimum N application progres-sively increases the cost of the N savings, through progressively declining crop N yields. This holds especially for high-value crops when responsive to N applica-tion (for example curve $Y = 1.6X - 0.005X^2$ in last column of Table 2.A1 and in Fig. 2.A1). A significant decrease in N input below the economically optimum N application is also not an effective strategy for decreasing Nsurplus and increasing NUE, because of the relative strong effect on N output.

Figure 2.A2 and Table 2.A2 illustrate two cases where management actions increase the N fertilizer value and crop N yield, and thereby NUE, but these improvements in NUE are not associated with decreases in Nsurplus, because the significant increases in $Y_{optimum}$ are accompanied by significant increases in $X_{optimum}$. Hence, increases in NUE are not always accompanied by decreases in

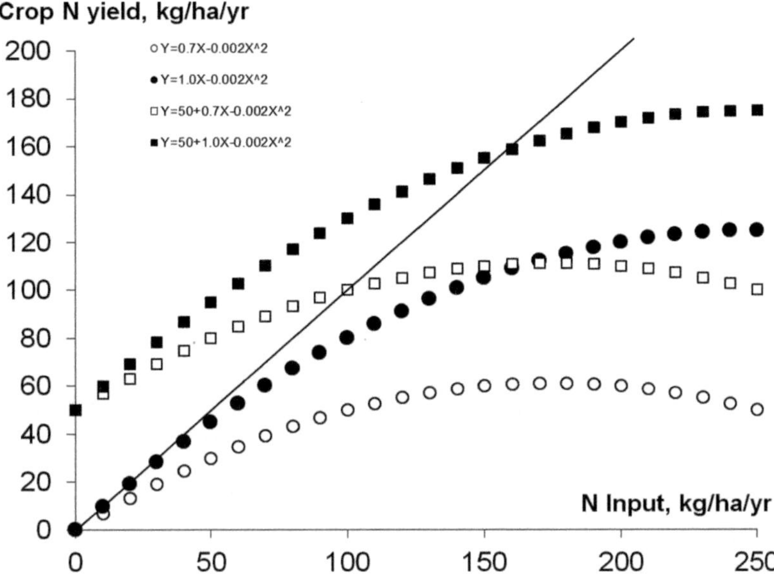

Fig. 2.A2 Four hypothetical crop N yield curves (second order polynomial), illustrating the effects of (i) increasing the N fertilizer value, (ii) increasing crop N yield and N fertilizer value simultaneously, and (iii) increasing the utilization of soil available N. The parameter values and related N surplus and N use efficiency are shown in Table 2.A2

Table 2.A2 Parameter values of four hypothetical yield curves ($Y = A + BX + CX^2$)

$Y = A + Bx + Cx^2$	Yield curves of N input and N output			
	$Y = 0.7x - 0.002x^2$	$Y = 1.0x - 0.002x^2$	$Y = 50 + 0.7x - 0.002x^2$	$Y = 50 + 1.0x - 0.002x^2$
A	0	0	50	50
B	0.7	1	0.7	1
C	0.002	0.002	0.002	0.002
$X_{optimum}$ (MR = 1.0)	175	250	175	250
$X_{optimum}$ (MR = 0.1)	150	225	150	225
$Y_{optimum}$ (MR = 1.0)	61	125	111	175
$Y_{optimum}$ (MR = 0.1)	60	124	110	174
Nsurplus (MR = 1.0)	114	125	64	75
Nsurplus (MR = 0.1)	90	101	40	51
NUE (MR = 1.0), %	35	50	64	70
NUE (MR = 0.1), %	40	55	73	77

The economic optimal N application rates ($X_{optimum}$) were estimated at a "Marginal Ratio" of N fertilizer prize to crop N value equivalent to 1.0 and 0.1 (MR = 1.0 and MR = 0.1, respectively). The N surplus and N use efficiency (NUE) at $X_{optimum}$ are estimated at both MR = 1.0 and MR = 0.1. See also Fig. 2.A2

Nsurplus. Further, Fig. 2.A2 and Table 2.A2 illustrate that increasing the utilization of soil available N (increasing the intercept of the polynomial) increases NUE and decreases N surplus. Increasing the utilization of soil available N may be achieved through proper cultivation and drainage of the soil, selection on crop varieties with long growing season, and proper timing of the planting and seeding. The simplified graphs shown in Fig. 2.A2 are meant to illustrate the possible effects; a possible decrease in N fertilizer value due to the increased utilization of soil available N is not considered here.

The direct costs of increasing NUE and decreasing N surplus in arable farming and vegetable growing are small, as long as the N application rate remains $\geq$ the economically optimum N application rate $X_{optimum}$. In practice, the farmer generally does not know which of the yield curve is applicable to a particular year. As the consequences of a suboptimal N application can be large, the farmer will opt for 'a responsive curve', and will fertilize accordingly. The consequences of such strategy are explore in Table 2.A3 and Fig. 2.A3; it results in much lower NUE (40 versus 90 %) and a higher N surplus (40 versus 10 kg per ha per yr) when the farmer opt for the middle curve but finds that the first one is applicable. Conversely, when the actual yield curve is more responsive to N application than expected, crop N yield will be lower (-13 kg N per ha per yr), N surplus much lower (6 versus -32 kg per ha per yr), and NUE much higher (96 versus 124 %). The possible yield decline is a main reason why N input in practice is often higher than $X_{optimum}$.

Table 2.A3 Parameter values of three contrasting yield curves

	Yield curves of N input and N output		
$Y = A + BX + CX^2$	$Y = 50$ $+ 0.7X - 0.002X^2$	$Y = 70$ $+ 1.4X - 0.002X^2$	$Y = 90 +$ $0.8X - 0.002X^2$
A	50	50	90
B	0.7	1.4	0.8
C	0.003	0.005	0.002
$X_{optimum}$ (MR $= 1.0$)	117	140	200
$X_{optimum}$ (MR $= 0.1$)	100	130	175
$Y_{optimum}$ (MR $= 1.0$)	91	148	170
$Y_{optimum}$ (MR $= 0.1$)	90	148	169
Nsurplus (MR $= 1.0$)	26	-8	30
Nsurplus (MR $= 0.1$)	10	-18	6
NUE (MR $= 1.0$), %	78	106	85
NUE (MR $= 0.1$), %	90	113	96

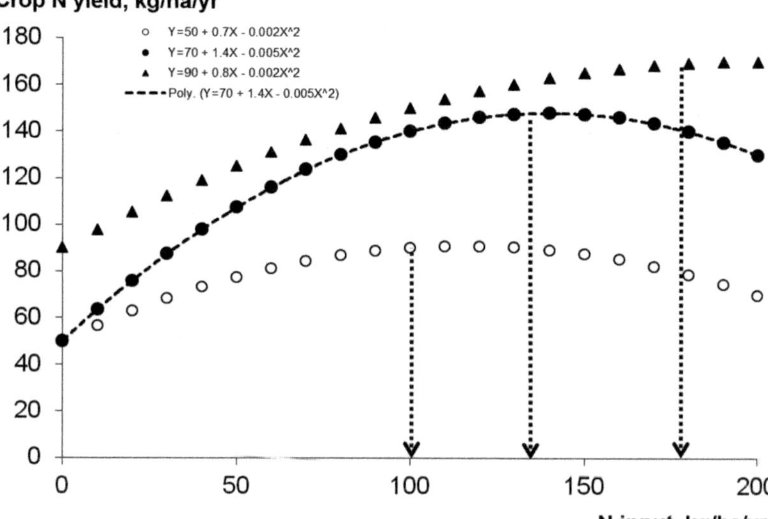

Fig. 2.A3 Three possible crop N yield curves that may occur in one site to illustrate the effect of fertilizing for one curve but getting another one. The parameter values and related N surplus and N use efficiency are shown in Table 2.A3

References

Bittman S, Dedina M, Howard CM, Oenema O, Sutton MA (eds) (2014) Options for ammonia mitigation: guidance from the UNECE task force on reactive nitrogen. Centre for Ecology & Hydrology, Edinburgh

Brink C, van Grinsven H, Jacobsen BH et al (2011) Costs and benefits of nitrogen in the environment. In: Sutton MA, Howard C, Erisman JW, Billen G, Bleeker A, Grenfelt P, van Grinsven H, Grizzetti B (eds) The European nitrogen assessment. Cambridge University Press, Cambridge

Cialdini R (2007) Influence; the psychology of persuasion. HarperCollins Publishers Inc, New York

Condor-Golec RD (2015) Research, demonstration farms and dissemination activities related to N management in Italy. In: Reis S, Howard CM, Sutton MA (eds) Costs of ammonia mitigation. Springer, Heidelberg (Chapter 8.1 of this volume)

Daatselaar CHG, Doornewaard GJ, Gardebroek C, de Hoop DW, Reijs JW (2010) Farm management, economy and environmental quality; their relationships on Dairy farms. LEI-report 2010-053, Wageningen University, The Netherlands

De Haan MHA (2001) Economics of environmental measures on experimental farm 'De Marke'. Neth J Agric Sci 49:179–194

De Hoop DW, Stolwijk HJJ (1999) Economic effects of environmental policy options for agriculture in 2002 and 2003. LEI/CPB-report 2.99.12, The Hague, The Netherlands

Doornewaard GJ, Beldman ACG, Daatselaar CHG (2007) Comparison of farm results of cows & opportunities with other dairy farms. Report 42; LEI-Report 2.07.14. Animal Sciences Group, Wageningen University, The Netherlands

Drucker PF (1954) The practice of management. HarperCollins Publishers Inc, New York

Hatch DJ, Chadwick DR, Jarvis SC, Roker JA (eds) (2004) Controlling nitrogen flows and losses. Wageningen Academic Publishers, Wageningen, p 624

Hatfield JL, Follett RF (eds) (2008) Nitrogen in the environment: sources, problems, and management, 2nd edn. Academic Press/Elsevier, Amsterdam, p 720

Hoste R (2011) Economic cost of pig production; results of InterPIG over 2009. LEI-Report 2011-012; The Hague, The Netherlands

Hoste R, Puister L (2009) Economic cost of pig production; an international comparison. LEI-report 2008-082, The Hague, The Netherlands

IPNI (2012) 4R plant nutrition: a manual for improving the management of plant nutrition. International Plant Nutrition Institute, Norcross. ISBN 978-0-9834988-1-0

Jacobsen BH, Abildtrup J, Jensen JD, Hasler B (2005a) Costs of reducing nutrient losses in Denmark – analyses of different regulation systems and cost effective measures. International Congress of the European Association of Agricultural Economists, Copenhagen, Denmark

Jacobsen BH, Abildtrup J, Orum JE (2005b) Reducing nutrient losses in Europe and implications for farming – in the light of the water framework directive. 15th Congress of the International Farm Management Association, Campinas SP, Brazil

Jensen LS, Schjoerring JK, van der Hoek K, Poulsen HD, Zevenbergen JF, Paliére C (2011) Benefits of nitrogen for food fibre and industrial production. In: Sutton MA, Howard C, Erisman JW, Billen G, Bleeker A, Grenfelt P, van Grinsven H, Grizzetti B (eds) The European nitrogen assessment. Cambridge University Press, Cambridge

Leip A, Britz W, Weiss F, de Vries W (2011) Farm, land, and soil nitrogen budgets for agriculture in Europe. Environ Pollut 159:3243–3253

Menzi H, Oenema O, Burton C, Shipin O, Gerber P, Robinson T, Franceschini G (2010) Impacts of intensive livestock production and manure management on the environment. In: Steinfeld H, Mooney HA, Schneider F, Neville LE (eds) Livestock in a changing landscape: drivers, consequences and responses, vol 1. Island Press, Washington DC, pp 139–163

Mikkelsen SA, Iversen TM, Jacobsen BH, Kjaer SS (2010) Denmark–European union; Reducing nutrient losses from intensive livestock operations. In: Gerber P, Mooney H, Dijkman J, Tarawali S, de Haan C (eds) Livestock in a changing landscape: regional perspectives, vol 2. Island Press, Washington DC, pp 140–153

MNP (2004) Mineralen beter geregeld. Evaluatie van de werking van de Meststoffenwet 1998–2003. RIVM-report 500031001, Bilthoven, p 170

Mosier AR, Syers JK, Freney JR (eds) (2004) Agriculture and the nitrogen cycle, SCOPE 65. Island Press, Washington, DC, p 296

OECD (2007) OECD and EUROSTAT gross nitrogen balances handbook. OECD, Paris, France. www.oecd.org/tad/env/indicators. Accessed Oct 2011

Oenema O, Pietrzak S (2002) Nutrient management in food production: achieving agronomic and environmental targets. Ambio 31:159–168

Oenema J, Koskamp GJ, Galama PJ (2001) Guiding commercial pilot farms to bridge the gap between experimental and commercial dairy farms; the project 'Cows & Opportunities'. Neth J Agric Sci 49:277–296

Oenema O, Kros H, de Vries W (2003) Approaches and uncertainties in nutrient budgets: implications for nutrient management and environmental policies. Eur J Agron 20:3–16

Oenema J, Van Ittersum M, Van Keulen H (2012) Improving nitrogen management on grassland on commercial pilot dairy farms in the Netherlands. Agric Ecosyst Environ 162:116–126

Oenema J, van Keulen H, Schils RLM, Aarts HFM (2011a) Participatory farm management adaptations to reduce environmental impact on commercial pilot dairy farms in the Netherlands. Njas-Wagen J Life Sci 58:39–48

Oenema O, Bleeker A, Braathen NA, Budňáková M, Bull K, Čermák P, Geupel M, Hicks K, Hoft R, Kozlova N, Leip A, Spranger T, Valli L, Velthof GL, Winiwarter W (2011b) Nitrogen in current European policies. In: Sutton MA, Howard C, Erisman JW, Billen G, Bleeker A, Grenfelt P, van Grinsven H, Grizzetti B (eds) The European nitrogen assessment. Cambridge University Press, Cambridge, pp 62–81

Ondersteijn CJM (2003) Nutrient management strategies on Dutch dairy farms: an empirical analysis. PhD thesis, Wageningen University, The Netherlands

Ondersteijn CJM, Beldman ACG, Daatselaar CHG, Giesen GWJ, Huirne RBM (2002) The Dutch mineral accounting system and the European nitrate directive: implications for N and P management and farm performance. Agr Ecosyst Environ 92:283–296

Ondersteijn CJM, Beldman ACG, Daatselaar CHG, Giesena GWJ, Huirne RBM (2003) Farm structure or farm management: effective ways to reduce nutrient surpluses on dairy farms and their financial impacts. Livest Prod Sci 84:171–181

Pedersen SM, Bizik J, Costa LD (2005) Potato production in Europe: a gross margin analysis. Department of Economics, University of Copenhagen, Frederiksberg

Powell JM, Rotz CA, Weaver (2009) DM Nitrogen use efficiency in dairy production. In: Grignani C, Acutis M, Zavattaro L, Bechini L, Bertora C, Marino Gallina P, Sacco D (eds) Proceedings of the 16th Nitrogen Workshop-Connecting different scales of nitrogen use in agriculture. Università degli Studi di Torino, Turin, pp 241–242

Powell JM, Gourley CJP, Rotz CA, Weaver DM (2010) Nitrogen use efficiency: a potential performance indicator and policy tool for dairy farms. Environ Sci Pol 13:217–228

Ribaudo M, Delgado J, Hansen L, Livingston M, Mosheim R, Williamson J (2011) Nitrogen in agricultural systems: implications for conservation policy. Economic Research Report ERR-127. U.S. Dept. of Agriculture, US

Rotz CA (2004) Management to reduce nitrogen losses in animal production. J Anim Sci 82 (E-Suppl):E119–E137

Rotz CA, Taube F, Russelle MP, Oenema J, Sanderson MA, Wachendorf M (2005) Whole-farm perspectives of nutrient flows in grassland agriculture. Crop Sci 45:2139–2159

Rougoor CW, Dijkhuizen AA, Huirne RBM, Mandersloot F, Schukken YH (1997) Relationships between technical, economic and environmental results on dairy farms: an explanatory study. Livest Prod Sci 47:235–244

Schepers JS, Raun WR (eds) (2008) Nitrogen in agricultural systems. ASA, CSSA, SSSA Agronomy Monograph 49, Chicago

Schoumans OF, Rulkens WH, Oenema O, Ehlert PAI (2010) Phosphorus recovery from animal manure; Technical opportunities and agro-economical perspectives. Alterra-report 2158, Wageningen University, Wageningen, The Netherlands

Schröder JJ (2005a) Revisiting the agronomic benefits of manure: a correct assessment and exploitation of its fertilizer value spares the environment. Bioresour Technol 92(2):253–261

Schröder JJ (2005b) Manure as a suitable component of precise nitrogen nutrition. Proceedings 574, International Fertiliser Society

Schröder JJ, Neeteson JJ (2008) Nutrient management regulations in the Netherlands. Geoderma 144:418–425

Sheriff G (2005) Efficient waste? Why farmers over-apply nutrients and the implications for policy design. Rev Agric Econ 27:542–557

Tavella E, Scavenius IMK, Pedersen SM (2011) Report on cost structure and economic profitability of selected precision farming systems. http://www.futurefarm.eu/node/263. Accessed Oct 2011

Van den Ham A, van den Berkmortel N, Reijs J, Doornwaard G, Hoogendam K, Daatselaar C (2010) Nutrient management and economy on dairy farms (in Dutch). LEI report 09-066. The Hague, The Netherlands

Van Dijk W, Prins H, de Haan MHA, Evers AG, Smit AL, Bos JFFP (2007) Economic consequences at farm level of application limits for fertilizers and manures in 2006–2009 for dairy farming and arable farming and horticulture (in Dutch). Rapport PPO-report no. 365. Praktijkonderzoek Plant and Omgeving. Wageningen, The Netherlands

Van Dijk W, Burgers S, ten Berge HFM, van Dam AM, van Geel WCA, van der Schoot JR (2008) Effects of sub-optimal nitrogen fertilization on marketable yields and nitrogen uptake of arable and horticultural crops (in Dutch). Praktijkonderzoek Plant & Omgeving AGV. PPO-report nr. 366, Lelystad, The Netherlands

Chapter 3
Economics of Low Nitrogen Feeding Strategies

Ad M. van Vuuren, Carlos Pineiro, Klaas W. van der Hoek, and Oene Oenema

Abstract Livestock retains typically between 10 and 40 % of the protein-nitrogen in the animal feed in milk, egg and/or meat, depending also on animal productivity and management. The remaining 60–90 % of the nitrogen (N) is excreted in urine and faeces, and contributes to the emissions of ammonia (NH_3) and nitrous oxide (N_2O) into the atmosphere and to nitrate leaching to groundwater and surface waters. Low-N feeding strategies can help to minimize these environmental effects. Here, we discuss the economic cost of such low-N feeding strategies.

Low-N feeding strategies commonly include a shift in concentrate feed from high-protein to low-protein feed ingredients. An important prerequisite for such strategies is to maintain animal performance. Therefore a possible deficiency in essential amino acids is compensated by including synthetic amino acids.

For pigs, strategies to reduce N excretion may result in a decrease (up to € 2 per kg of NH_3 reduced) or in an increase (up to € 6 per kg of NH_3 reduced) in production costs, depending on the market prices of low-protein feed ingredients and synthetic amino acids. Costs were much higher (up to € 62) when no synthetic amino acids but standard feed ingredients were used to adjust the feed for the amino acid requirements. For poultry, no actual data were found in literature to compare the economic effects of low N feeding strategies in broilers and laying hens. For dairy cattle, a reduction in N excretion through low-N feeding strategies may result

A.M. van Vuuren
Wageningen UR, Livestock Research, P.O. Box 338, 6700 AH Wageningen, The Netherlands
e-mail: ad.vanvuuren@wur.nl

C. Pineiro
PigCHAMP Pro Europa S.L., C/Santa Catalina, 10, Segovia 40003, Spain
e-mail: carlos.pineiro@pigchamp-pro.com

K.W. van der Hoek
RIVM, Antonie van Leeuwenhoeklaan 9, 3721 MA Bilthoven, The Netherlands
e-mail: Klaas.van.der.Hoek@rivm.nl

O. Oenema (⊠)
Environmental Sciences Group, Alterra, Wageningen University, P.O. Box 47, NL-6700 AA Wageningen, The Netherlands
e-mail: oene.oenema@wur.nl

© Springer Science+Business Media Dordrecht 2015
S. Reis et al. (eds.), *Costs of Ammonia Abatement and the Climate Co-Benefits*,
DOI 10.1007/978-94-017-9722-1_3

in a profit of € 1.40 per kg of NH_3 reduced or in extra costs of € 6 per kg of NH_3 reduced.

Costs of low-N feeding strategies in near future will be influenced by the competing demands for low-N biomass by the growing livestock sector and the growing biofuel sector. Regulatory measures related to animal welfare may also have effect on the demand for low-N feed. As a consequence, it is uncertain whether costs for low-N feeding strategies will increase or decrease in the near future.

Keywords Ammonia emissions • Low nitrogen feeding • Nitrogen management • Agriculture • Animal husbandry

3.1 Introduction

For maintenance of body functions and production, livestock requires water, energy (carbohydrates), protein, essential fatty acids, vitamins and minerals. The requirements for protein in the diet may vary from 12 % to more than 20 %, depending on animal species and animal productivity. Protein consists of various amino acids which animals require in specific amounts for proper functioning. An important element in protein is nitrogen (N); about 16 % of the protein consists of N.

Dietary N is partly retained by the animal in meat, milk, eggs and offspring. However, a large percentage of the substances consumed in the feed are excreted again via faeces and urine – typically anywhere from 60 % to over 90 % of the N and the mineral nutrients present in the feed, depending on animal species, feed composition, and management. Faecal N is mainly organically bound, whereas urinary N is mainly urea N (in poultry: uric acid). Due the abundance of urease in the environment, urea is almost instantaneously hydrolysed into ammonium.

Livestock manures are main sources of ammonia (NH_3) and nitrous oxide (N_2O) in the atmosphere and of nitrate (NO_3^-) in groundwater and surface waters. The NH_3 and N_2O may be emitted from the manures in animal housing systems and manure storages as well as following the application of the manure to land. A direct linear relationship between the input of dietary N and N excretion in urine has been reported for pigs (Bracher and Spring 2010) and dairy cows (Kebreab et al. 2002). Also, a strong positive relationship has been reported for the amounts of N excretion in urine and NH_3 emissions (Monteny et al. 2002).

Due to safety margins implemented by the feeding industries, the dietary N supply is often in excess of what is required by the animals. Reducing dietary N input will reduce the environmental impact of N excretion by livestock animals (Oenema et al. 2008), but can also affect animal performance and health (Baker 2009; Bauchart-Thevret et al. 2009; Wu and Satter 2000). The implementation of low-N feeding strategies in practice depends also on the availability of low-N feed ingredients and on the cost of such low-N feeding strategies.

Here, we discuss the economic effects of measures to reduce the excretion of urea N or the inhibition of NH_3 volatilisation. Unfortunately, literature data on this topic is scarce, in part because relevant data are within the feeding industries and kept confidential. Inquiries within this sector were answered in general terms and provided no actual data.

3.1.1 Benefits of Low Nitrogen Feeding Strategies

Low-N feeding strategies to reduce NH_3 emissions from animal houses have significant benefits compared with end of pipe techniques (such as air scrubbers) used in modern poultry and pig housing systems. While end of pipe techniques clean the exhausted ventilation air (Melse et al. 2009) low N feeding strategies result in low NH_3 concentrations inside the animal house. Air quality inside animal houses impacts animals as well as farmers. In animal houses with higher NH_3 levels of the inside air performance of broilers is lower and disease susceptibility is increased (Beker et al. 2004; Ritz et al. 2004). In comparison with animal houses with dirty air quality weaning pigs grew faster in animal houses with clean air quality (Lee et al. 2005). Studies from Sweden and Canada show that air quality inside pig houses also affects the health of farmers (Donham et al. 1989; Charavaryamath and Singh 2006). It should be noted that air quality inside the animal house not only refers to NH_3 concentrations but often also to the dust content.

Moreover, feeding protein-rich diets cost feed energy. In grass feeding trials with dairy cattle it has been shown that the removal of the consumed excess N by the animal metabolism costs additional energy. So adjusting the protein intake to the advised feed requirements can save feed imports (Bruinenberg et al. 2002).

3.1.2 Feed Protein Costs

Low-N feeding strategies usually include a shift in concentrate feed from high-protein to low-protein feed ingredients. A possible deficiency in first-limiting amino acids, often methionine and lysine, is hereby compensated by including synthetic amino acids, which are available on the market.

Compared to soybean meal (ca. 500 g of crude protein/kg of dry matter) ingredients with a lower crude protein content are usually cheaper. However, if costs are expressed per kg of crude protein, low-protein ingredients are usually more expensive than soybean meal (Fig. 3.1). Thus, reducing the protein concentration in feeds by including low-protein ingredients results in higher feeding costs. This is due to a higher price of crude protein from low-protein feed ingredients and to extra costs for including synthetic amino acids.

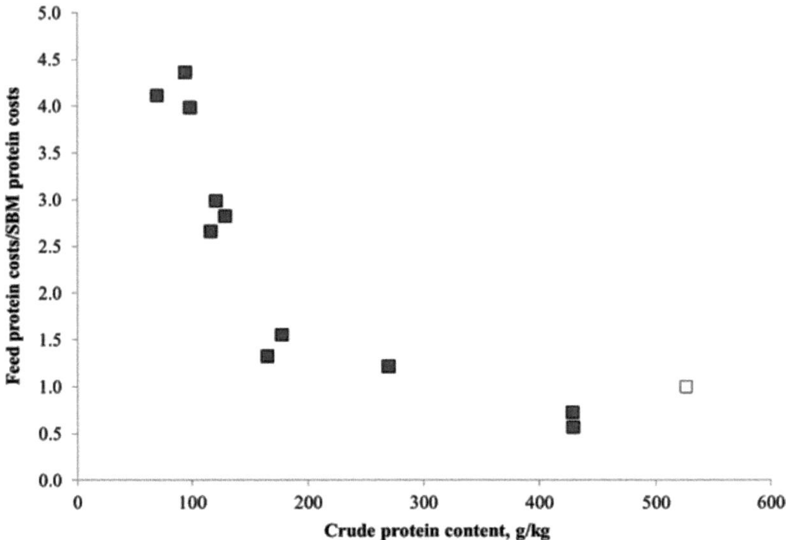

Fig. 3.1 Costs of protein of feed ingredients relative to costs of soybean meal (SBM) protein. Costs based on market prices in August 2011. Feed ingredients are barley, citrus pulp, maize gluten feed, maize, palm kernel expeller, rapeseed meal, soybean hulls, soybean meal (□), sugar beet pulp, sunflower meal, wheat, wheat middlings

3.2 Low-Nitrogen Feeding Strategies in Pigs

3.2.1 Introduction

Feeding strategies to lower ammonia emissions from pig production units are focussed on:

1. reducing urea concentration and excretion
2. reducing ammonia production and volatilization during storage and application.

3.2.2 Reducing Urea Concentration and Excretion

Two feeding strategies have been tested:

1. Reducing the total urinary and faecal N excretion
2. Shifting N excretion in urine towards N excretion in faeces.

Table 3.1 Indicative crude protein levels in BAT-feeds for poultry and pigs (European Commission 2003)

Species	Phase	Crude protein[a], g/kg feed
Broiler	Starter	200–220
	Grower	190–210
	Finisher	180–200
Turkey	<4 weeks	240–270
	5–8 weeks	220–240
	9–12 weeks	190–210
	13+ weeks	160–190
	16+ weeks	140–170
Layer	18–40 weeks	155–165
	40+ weeks	145–155
Weaner piglet	<10 kg	190–210
	10–25 kg	175–195
Fattening pig	25–50 kg	150–170
	50–110 kg	140–150
Sow	Gestation	130–150
	Lactation	160–170

[a]With adequately balanced and optimal digestible amino acid supply

3.2.2.1 Reducing Total N Excretion

Reducing crude protein intake has a large impact on total N excretion. An important prerequisite for such a strategy is to maintain performance (piglet production, growth) not only because this determines farmer's profitability, but also because a reduced performance will result in a relative reduction in N retention and consequently in a relative increase in N excretion.

Phase feeding or stage feeding has now largely been accepted for growing pigs. This strategy recognises the change in required energy-to-protein ratio during the growing period. Indicative crude protein levels in feeds considered as Best Available Technique for Intensive Rearing of Poultry and Pigs are presented in Table 3.1.

An increase in the energy-to-protein ratio can be achieved by the exchange of high-protein feed materials, such as soybean meal, by carbohydrate sources. The costs of such an exchange depend on the costs of high-protein sources and carbohydrate sources such as grains. Soybean meal is often 10 % more expensive relative to grains, and exchanging soybean meal by grains will then reduce feeding costs. However, grains are also used for human food and for fuel (ethanol production). The increased demand of grains for food, feed and fuel will increase the market prize of grains and grain prices may become higher than that of soybean meal.

Besides, ethanol production from grains yields rest products that are relatively high in protein concentrations. Products such as distiller's grains from maize and wheat may become important feed ingredients in future, competing with soybean meal. Protein in dried distiller's grains with solubles (DDGS) is approximate 80 %

of the price of protein in soybean meal. However, diets with high proportions of DDGS will be deficient in lysine, threonine and tryptophan.

3.2.2.2 Shifting N Excretion

Dietary fibre from plant cell walls will not be digested in the small intestine, because mammals do not synthesise and secrete cellulases. Dietary fibre entering the large intestine will be partly degraded by the micro flora yielding volatile fatty acids and microbial protein. Volatile fatty acids will be absorbed by the animal through the intestinal wall, whereas microbial protein will be secreted in the faeces. If microbial protein synthesis exceeds protein degradation in the large intestine, urea will be secreted into the intestinal lumen and incorporated in microbial protein. This will reduce blood urea concentrations and hence urinary urea excretion.

3.2.3 Reducing Ammonia Production and Volatilization

Strategies to reduce NH_3 production and volatilization are mainly focussed on reducing urinary pH. Supplementing the diet with benzoic acid, reducing the dietary cation-anion difference by selection of feed material or addition of anions (such as chloride or sulphate) are the main methods to acidify urine by feeding strategies.

3.2.4 Cost Calculation in the Netherlands

Aarnink et al. (2010) calculated the impact and costs of various feeding strategies to reduce NH_3 emissions from houses for growing–finishing pigs. Conclusions of this study are summarised in Table 3.2. The authors concluded that a reduction of dietary protein to a level of 135–140 g/kg of feed and a reduction of the dietary cation-anion difference are effective feeding strategies at relatively low costs.

3.2.5 Cost Calculation in Spain

Piñeiro et al. (2010) calculated costs but also accounted for possible changes in performance. Extra costs for phase-feeding or low-protein diets are presented in Table 3.3. In a favourable market, the feeding of low-protein diets will improve the farmer's profit per pig and place.

Table 3.2 Environmental impact and costs of feeding strategies to reduce ammonia emissions for pig houses

Strategy	Ref. value	Change	Unit	Change NH_3 emission (%)	Costs per pig place[a] (€)	Costs per 10 % NH_3 reduction (€)
Dietary protein	165	−15	g/kg of feed	−15	−2.07	−1.38
	165	−30		−30	5.91	1.97
Benzoic acid	0	10	g/kg of feed	−16	9.94[b]	6.21
Exchange $CaCO_3$ by $CaCl_2$ (or $CaSO_4$)	0	3	g Ca/kg of feed	−24	4.38	1.83
	0	6		−35	8.77	2.50
Dietary CAD[c]	320	−100	mEq/kg feed DM	−7		
Fermentable carbohydrates	180	50	g/kg of feed	−6		
	180	100		−12		

From Aarnink et al. (2010)

[a]Standardised for a farm with 4,200 pig places and assuming a consumption of 247 kg of feed per pig produced

[b]Growth-promoting effect of benzoic acid is not taken into account

[c]CAD = cation-anion difference

Table 3.3 Effective unit cost for feeding strategies (Piñeiro et al. 2010)

Technique	NH_3 reduction %	kg/place	Extra costs €/place	€/kg pig	€/kg NH_3 reduction
Phase feeding	10	0.32	1.52	0.0052	4.8
Low-N diet, unfavourable market	30	0.95	1.61	0.0053	1.7
Low-N, favourable market	30	0.95	−1.92	−0.0064	−2.0

3.2.6 Cost Calculation in Finland

Niemi et al. (2010) compared the value of a multi-phase feeding system for growing swine against a two-phase feeding system and the effects of changes in the prices of pork, grains and soybean meal. Changing from a two-phase to a multi-phase feeding system increased the annual return per pig place by € 1.60 (€ 1.35–€ 1.88). For an average Finnish pig farmer this would yield a 3 % improvement of agricultural income. The effect of various market changes are presented in Table 3.4. Multi-phase feeding is slightly less vulnerable to market changes than two-phase feeding.

Table 3.4 Net economic benefit (€ per pig place) of changing from a two-phase to a multi-phase feeding system and the impact on market changes (Niemi et al. 2010)

Scenario	Benefit multi-phase system	Income per pig place as affected by market changes, €	
		Two-phase	Multi-phase
Benchmark	27	0	0
10 % higher pig meat price	22	724	719
10 % higher piglet price	31	−358	−353
10 % higher barley price	28	−182	−180
10 % higher soybean meal price	25	−59	−57

3.2.7 Cost Calculation in Switzerland

Bracher and Spring (2010) calculated the feed costs using available feed ingredients to reduce the N level of feeds. A reduction in protein content of feed mixtures for fattening pigs had no effect on feed costs per se. Higher feed costs were predicted when requirements for specific essential amino acids were taken into account. Below 150 g of crude protein per kg of feed low feed costs could not be maintained when accounting for the requirements of essential amino acids, including isoleucine. Under these conditions feed costs increased by about 30 % when the crude protein level was decreased from 150 to 130 g per kg of feed.

Bracher and Spring (2010) also calculated costs of reducing the average crude protein concentration from 160 to 140 g per kg of feed in a two- and three-phase feeding system. They estimated an increase in feeding costs of 5–15 % for such a reduction. Assuming a reduction in NH_3-emission of 0.8 g per day when decreasing the crude protein concentration from 160 to 130 g per kg of feed (Hayes et al. 2004), a feed intake of 2.0 kg per day per growing pig and a feed price of 25 € per 100 kg, the calculations of Bracher and Spring (2010) would imply that the reduction of 1 kg of NH_3 costs € 62. This estimate is more than 30 time higher than estimates of Aarnink et al. (2010; Table 2.2), which seems attributable to the different approaches. While Bracher and Spring (2010) did not include synthetic amino acids, Aarnink et al. (2010) used commercial diets including synthetic amino acids.

3.3 Low Nitrogen Feeding Strategies in Poultry

3.3.1 Reducing Crude Protein Concentration

Similar to pig diets, reductions in protein concentration in the diet are possible, but requirements for the supply of essential amino acids have to be taken into account. Thus, a reduction in protein concentration by exchanging e.g. soybean meal by grains is possible and profitable. However, as in pigs, the actual reduction in dietary protein depends on the availability, price and metabolic efficiency of synthetic amino acids. Farrell (2005) fed diets with different protein concentration to laying hens from 25 to 45 weeks of age. Reducing the crude protein concentration from 172 to 139 g/kg with additional supplementation of glutamic acid reduced N excretion from 62 to 44 g per kg egg production.

Veens et al. (2009) plead for redefining the maintenance requirements for amino acids for laying hens. At high ingredient prices, methionine and tryptophan were economical in diets with crude protein concentrations between 190 and 210 g/kg. For a further reduction of dietary crude protein for laying hens, besides lysine, methionine and tryptophan, also threonine and valine will be required.

No actual data were found in literature to compare the economic effects of low N feeding strategies in broilers and laying hens.

3.3.2 Reducing Ammonia Volatilisation

Litter dry matter has a large impact on the volatilisation of NH_3 in poultry houses. Since 1st of January 2012, battery housing of laying hens is forbidden within the EU and, consequently, the effect of litter dry matter on NH_3 emissions has become more important. Emissions of NH_3 from litter can be reduced by maintaining high dry matter content, a low pH or low temperature, which minimize the degradation rate of organic nitrogen and thus the NH_3 volatilization (Groot-Koerkamp and Elzing 1996). Dry matter content and pH can be influenced by dietary strategies.

3.4 Low Nitrogen Feeding Strategies in Dairy Cattle

3.4.1 Introduction

A general feeding strategy for dairy cattle is to reduce total N intake by replacing high-N grass (silage) for low-N maize (silage) or other grain cereals in their diet. A lower N intake will reduce N excretion, which is monitored by controlling the urea concentration in milk. As mentioned previously, costs of such an exchange depend on the costs of high-protein sources and carbohydrate sources such as grains. Soybean meal is often more expensive relative to grains, and exchanging soybean meal by grains will then reduce feeding costs. However, grains are also used for human food and for fuel (ethanol production). The increased demand for grains as food and fuel will also increase the market prize of grains and grain prices may be higher than that for soybean meal. In October 2010, market prices for protein sources soybean meal and rapeseed meal were higher than prices for grains and therefore, low-protein supplements for dairy cows are less expensive than high-protein supplements. In dairy cattle feeding, products such as distillers grains from the ethanol production from maize and wheat are already commonly used feed materials. Such products can be fed in a dry form, but farms within a relatively small distance of ethanol plants are feeding distillers grains in a wet form.

Only at relative low protein concentrations and with a limited number of feed ingredients supplementing extra amino acids may be required. Lysine may be limiting in high maize-starch diets; methionine in diets with soybean as main protein source; histidine may be limited in diets with a large proportion of ruminal-degradable protein such as (unwilted) grass silage. For dairy cows amino acid supplements should be protected against ruminal degradation, which results in a relatively high price of such products.

3.4.2 Partial Replacement of Fresh Grass

Reducing the intake of high-protein fresh grass during the growing season by reducing grazing time per day and feeding a low-protein diet indoors is one of the feeding strategies to reduce N intake in dairy cattle. Partial replacement of herbage by maize silage during the growing season reduced N intake and improved the utilisation of ingested N without detrimental effects on milk performance. Silage maize is seen as an effective supplement in an herbage-based diet to reduce N excretion (Valk 1994; Valk et al. 2000).

Possible effects of feeding strategies on N excretion and economic profit were calculated by Mandersloot (1992). Feeding strategies were unrestricted grazing and no maize silage (U0); unrestricted grazing with 3 kg of maize silage dry matter at milkings (U3); restricted (daytime) grazing with 3 kg of maize silage dry matter fed indoors (night-time) (R3); restricted grazing with 6 kg of maize silage dry matter fed indoors (R6). It was assumed that feeding maize silage reduced grass intake by 1.2 (U3 and R3) or 3.6 (R6) kg of dry matter and that restricted grazing reduced grass intake by 10 %.

From the results summarised in Table 3.5 it can be concluded that restricted grazing reduced N excess by almost 20 % and reduced profitability by 1–5 %. Intensity of farming had no significant impact. The reduction in profitability was mainly due to the extra labour costs of harvesting residual grass and application of extra manure from the housed animals in the restricted grazing strategy.

Another strategy to reduce N intake in dairy cattle is decreasing the protein content of the herbage. This can be achieved through lowering the N fertilization rate and through harvesting herbage at an older physiological stage, i.e., at higher herbage yield level. Here, a compromise has to be found, because increasing maturity will decrease the feeding value and digestibility of the grass. Also, grazing losses may increase when herbage yield on offer is high. A modest decrease in fertiliser input does not affect intake and animal performance (Valk et al. 2000).

Table 3.5 Effects of grazing strategy and feeding maize silage during the growing season of grassland on N excess and farm profit, in per cent (Mandersloot 1992)

	Intensity							
	10,000 kg of milk/ha				15,000 kg of milk/ha			
	Unrestricted grazing		Restricted grazing		Unrestricted grazing		Restricted grazing	
	Maize silage intake (kg DM)				Maize silage intake (DM)			
	0	3	3	6	0	3	3	6
N excess	100	97	82	81	100	97	82	82
Profit	100	99	98	95	100	99	98	96

3.4.3 Combined Feeding and Other Management Strategies (De Marke)

The effects of environmental measures on an experimental low-input farm in the Netherlands ("De Marke") were reported by De Haan (2001). The main findings for combinations of feeding strategies are summarized in Table 3.6. Changing grazing management in which dairy cows were housed indoors during two periods (afternoon and night) reduced N surplus by 9 % compared to the basal situation. Extra costs amounted to € 0.12 per kg reduction in N surplus of the farm. Including more maize products in the diet raised these extra costs by € 0.90 (extra maize in summer period) to more than € 2 (feeding corn and low emission housing) per kg reduction of total N surplus. Farmer's income increased for siesta feeding and better accounting for DVE (intestine digestible protein) requirements, but decreased for strategies that included extra maize products in the diet.

From 1998, grazing time for the siesta grazing strategy at De Marke was reduced from 6 h/day to 4.5 h/day. This led to extra costs, resulting in a 10 % reduction in farmer's income (De Haan and ter Veer 2004). In the study of De Haan (2001), corn was fed to each of two production groups up to the average energy requirement of each group.

From 2000, lactating cows are housed as one group and silage maize is supplemented to cows individually, using a multifeeder. Using a multifeeder in combination with a partly exchange of maize by triticale production, reduced farmers income by 20 %. In that study, no data have been reported on N balances.

3.4.4 Wisconsin Simulation Study

Rotz et al. (1999) used the dairy farm model "DAFOSYM" to simulate the effects of reduced N feeding strategies for a 60-cow dairy farm. In this simulation study, improved N utilization was implemented by using two different protein sources: (1) a mix consisting of 50 % heat-treated soybean meal, 15 % blood meal and 25 % swine meat and bone meal and (2) 100 % roasted soybeans. Improving the protein

Table 3.6 Effect of feeding strategies on farm[a] economic results (De Haan 2001)

	Strategy (sequential implementation)					
	Base	+ siesta grazing[b]	+ DVE req.	+ more maize in summer[c]	+ shorter grazing period[d]	+ maize ear silage[e]
N surplus, kg/ha	242	221	219	155	142	132
N reduction, %	0	9	10	36	41	45
Extra costs, €	0	142	554	4,290	9,348	14,445
Extra costs, €/kg N reduction	–	0.12	0.44	0.90	1.70	2.39
Extra costs, €/kg milk	–	0.00	0.00	0.01	0.01	0.02
Extra farmers income[f], €/kg N red	–	1.44	1.51	−0.60	−1.49	−2.45

[a]Experimental farm "De Marke", situation 1998: 78 dairy cows and 55 ha of land area (grass and maize)
[b]+ less young stock + crop rotation
[c]+ previous strategies + catch crop under maize + reduced N fertilizer + reduced P fertilizer
[d]+ previous strategies + growing more maize
[e]+ previous strategies + low emission housing
[f]Net farm income + labour costs

supply reduced N import by more than 10 % (Table 3.7). Although the total export of N in milk, feed and animal sale was 500 kg lower compared to the original situation (soybean meal supplementation only), the net return increased absolutely as well as per kg exported N. However, it should be realised that meal of animal origin is not allowed in cattle feed within the EU and that the profitability estimations depends largely to the assumed prices (in this study: $ 250/tonne of soybean meal dry matter (DM), $ 120/tonne of maize grain DM and $ 330/tonne of protein mix DM).

3.4.5 Studies from Sweden

Data about N retention in feed, milk, faeces and urine were extracted from experiments with different cattle rations with high and low protein levels (Swensson 2003). The daily N output in the urine could be predicted by an equation based on the sum of AAT and PBV in the diet and the daily N output in the milk. AAT and PBV are abbreviations for amino acids absorbed in the rumen and protein balance in the rumen, respectively. Prediction of the daily NH_3 emission was more exact from the protein content of the diet than from the daily N excretion with the urine. This was explained by the effect of the different rations on the volumes of the excreted urine (Swensson 2003).

Table 3.7 Effect of protein supplement on nitrogen balance, costs and profit of a 60-cow dairy farm[a] (Rotz et al. 1999)

| | Supplement | | | |
	Soybean	Rumen protected	Mixture[b]	Roasted soybean
Milk yield, kg	10,000	10,000	10,000	10,000
N input, kg	19,946	17,354	17,306	17,784
N input, kg/kg milk	1.99	1.74	1.73	1.78
N input, relative to soybean, %	100	87	87	89
N export, kg	8,284	7,768	7,772	8,074
N export, relative to soybean, %	100	94	94	97
N loss, kg/kg milk	0.64	0.49	0.49	0.50
N loss, relative to soybean, %	100	77	76	79
Cost, 1,000$	199	193	193	195
Cost, relative to soybean, %	100	97	97	98
Cost, $/kg N exported	24.0	24.9	24.9	24.1
Cost, relative to soybean, %	100	104	104	100
Net return, $/kg N exported	−0.10	0.38	0.41	0.36

[a]60 mature cows, 52 replacement heifers on 70 ha of cropland under average weather conditions
[b]50 % heat-treated soybean meal, 15 % blood meal and 25 % swine meat and bone meal

The objective of the EU LIFE Ammonia project that ran from 1999 to 2003, was to demonstrate possibilities to decrease ammonia emissions from dairy farms (Sannö 2003). The project compared three dietary treatments, using the amount of applied N fertilizers to adjust the protein content of the grass-clover forage. The authors concluded that when undegradable and degradable proteins in diets are formulated properly, dietary protein concentrations of 160–170 g per kg dry matter can be used for cows in early lactation in commercial herds to improve N utilization without causing a simultaneous decrease in milk yield (Nadeau et al. 2007).

3.4.6 Concluding Remarks

In this chapter technical options to lower N excretion and NH_3 emissions on dairy farms by using feed management were discussed. Dairy farms differ widely in feed management; from pure grassland based systems to integrated grassland-arable production systems. Not all farms have the same possibilities to lower the N excretion. For example, pure grassland based systems do not always have the possibility to grow silage maize. The difference in reference situation between farms and countries is another difficulty in calculating the associated costs.

Grazing management offers possibilities to reduce N excretion by decreasing the amount of time cattle spent in the pasture. Feeding cattle indoor offers more possibilities to adjust the ration then when cattle are grazing outdoor. Although N excretion can be lowered this way, NH_3 emissions are increasing unless animal manure is applied with low emission techniques. Methane emissions from manure management will also increase with more housing because of the transfer of excreta from the pasture to the animal house.

In a recent assessment for emission reduction potential two options for feed management in dairy farming were mentioned: better adjusting and fine tuning to feeding requirement tables and a higher share of maize in the ration. It was estimated that reducing the annual N excretion from 138 kg per dairy cow in 2009 to 120 kg in 2020 and reducing the average milk urea concentration from 228 mg/kg of milk to 200 mg in 2020 will reduce NH_3 emission in the Netherlands by 3–5 kton. The costs were estimated at € 6–€ 10 per kg of NH_3 reduced (Koelemeijer et al. 2010).

3.5   Future Outlook

Global livestock production systems are rapidly evolving, due to the increasing demands for food by the increasing global population (e.g., Steinfeld et al. 2010). To fulfil these demands, feed production will have to increase drastically, perhaps by 50–100 % during the next 30–50 years (Bruinsma 2003). At the same time, there is an increasing demand for biomass for biofuel production. There is also increasing competition on the global markets for dairy products, meat and egg, through the effects of globalization and technology developments. Margins for farmers tend to go down, and farmers have to reduce cost of production. The larger farms managed by better educated farmers will survive, although some small-scale farms may continue to produce niche products with added value (Mazoyer and Roudart 2006).

Consumers, the retail sector and the processing industry increasingly set standards for agricultural production systems and processes. So far, consumers, retailers and processing industry have not set targets for protein levels in the diet based on concerns about NH_3 emissions. This cannot be excluded for the near future, as concerns about the environmental impacts of intensive livestock production systems increase. Currently, a big dairy processing industry in Western Europe is exploring the possibility of setting emission targets for greenhouse gas and NH_3 emissions from dairy farms (Frans Aarts, Wageningen University, personal communication, 2011). Also, governmental policies interfere with agricultural production processes, including livestock systems; farmers have to comply with regulations for animal welfare and emissions standards.

How do these developments affect animal nutrition and in particular the protein content in the animal feed? How will cost of low-protein diets evolve? There are no clear answers to these questions yet. Likely, the above developments have diverse effects on protein animal feeding, some will tend to decrease protein content, whereas others may increase protein content. Environmental policies, standard setting by processing industry and retail sectors, and pressure on the markets for feeds will likely lead to a lowering of the protein content. For example, the protein content of grass silage in The Netherlands has decreased by on average 2 g per kg per year between 1997 and 2009, mainly due to the effects of environmental policies; farmers increased the utilization of N from animal manure and decreased the fertilizer N input (Reijneveld et al. 2014; van Vuuren and Chilibroste 2013). In addition, protein-rich herbages have been replaced partly by low-protein fodder maize. This has been shown up in a drastic decrease in the milk-urea content during the last decade and also in a drastic decrease in the N excretion via manure per kg milk and beef produced.

Poultry production in the major poultry producing areas in the world (US, Brazil, EU, China, South-east Asia) is increasingly organized by a few large feed-millers and processing industries. These companies define the diets and set the standards tightly. They will be able to introduce new technologies, including the supplementation of amino acids rather easily, when economically attractive and/or demanded by environmental policies. The up-scaling in the production of synthetic amino acids and other supplements (vitamins, minerals) will likely make these amino acids and other supplements cheaper. This trend of increasing influence by few transnational corporations will continue in the near future (UNCTAD 2009).

On the other hand, increasing demand for carbohydrates for the biofuel industry (ethanol, biogas), the availability of high protein by-products from the biofuel industry, animal welfare regulations and economic urge for high production rates may lead to increasing N contents of the animal feed and, consequently, increasing N excretions per unit of milk, meat and egg produced.

The area of C-4 crops, like maize, is also likely to increase in future, because these crops have a higher production potential and higher water and nutrient use efficiency than C-3 crops like most of the herbages grown in the northern hemisphere. As a consequence, the cost of roughage production may go down, and the protein content of the animal feed too. Genetic engineering of crops may also contribute to increasing crop yields and to a lowering of the protein content of the animal feed.

Indeed, changes will be in the order of the day in the livestock production of tomorrow (Anthony 1939).

References

Aarnink AJA, Smits MCJ, Vermeij I (2010) Reduction of ammonia emission from houses for growing-finishing pigs by combined measures [in Dutch with English summary]. Report 366, Wageningen UR Livestock Research, Lelystad, The Netherlands

Anthony EL (1939) The effect of a changing economic situation. J Anim Sci 1939(1):44–48

Baker DH (2009) Advances in protein–amino acid nutrition of poultry. Amino Acids 37:29–41

Bauchart-Thevret C, Stoll B, Chacko S, Burrin DG (2009) Sulfur amino acid deficiency upregulates intestinal growth in neonatal pigs' methionine cycle activity and suppresses epithelial growth in neonatal pigs. Am J Physiol Endocrinol Metab 296:E1239–E1250

Beker A, Vanhooser SL, Swartzlander JH, Teeter RG (2004) Atmospheric ammonia concentration effects on broiler growth and performance. J Appl Poultry Res 13:5–9

Bracher A, Spring P (2010) Möglichkeiten zur Reduktion der Ammoniakemissionen durch Fütterungsmassnahmen bei Schweinen. Report Schweizerische Hochschule für Landwirtschaft SHL, Zollikofen, Switzerland

Bruinenberg MH, van der Honing Y, Agnew RE, Yan T, van Vuuren AM, Valk H (2002) Energy metabolism of dairy cows fed on grass. Livest Prod Sci 75:117–128

Bruinsma J (ed) (2003) World agriculture: towards 2015/30, an FAO perspective. Earthscan, London

Charavaryamath C, Singh B (2006) Pulmonary effects of exposure to pig barn air. J Occup Med Toxicol 1:10. doi:10.1186/1745-6673-1-10

De Haan MHA (2001) Economics of environmental measures on experimental farm 'De Marke'. Neth J Agric Sci 49:179

De Haan MHA, ter Veer DT (2004) Environmental strategies De Marke after 2008. Report 48, Animal Sciences Group, Lelystad, The Netherlands

Donham K, Haglind P, Peterson Y, Rylander R, Belin L (1989) Environmental and health studies of farm workers in Swedish swine confinement buildings. Br J Ind Med 46(1):31–37

European Commission (2003) Integrated pollution prevention and control. Reference document on Best Available Techniques for Intensive Rearing of Poultry and Pigs (BREF). http://eippcb.jrc. ec.europa.eu/reference/. Accessed 13 Aug 2014

Farrell DJ (2005) Matching poultry production with available feed resources: issues and constraints. Worlds Poult Sci J 61:299

Groot-Koerkamp PWG, Elzing A (1996) Degradation of nitrogenous components in and volatilization of ammonia from litter in aviary housing systems for laying hens. Trans ASABE 39:211

Hayes ET, Leek ABG, Curran TP, Dodd VA, Carton OT, Beattie VE, O'Doherty JV (2004) The influence of diet crude protein level on odour and ammonia emissions from finishing pig houses. Bioresour Technol 91:309–315

Kebreab E, France J, Mills JA, Allison R, Dijkstra J (2002) A dynamic model of N metabolism in the lactating dairy cow and an assessment of impact of N excretion on the environment. J Anim Sci 80:248–259

Koelemeijer R, van der Hoek D, de Haan B, Noordijk E, Buijsman E, Aben J, van Jaarsveld H, Hammingh P, van Tol S, Velders G, de Vries W, Wieringa K, Reinhard S, Linderhof V, Michels R, Helming J, Oudendag D, Schouten A, van Staalduinen L (2010) Investigation of additional measures for the programmed nitrogen approach. An investigation of environmental and economic effects, PBL-publication no. 500215001. Planbureau voor de Leefomgeving, The Hague

Lee C, Giles LR, Bryden WL, Downing JL, Owens PC, Kirby AC, Wynn PC (2005) Performance and endocrine responses of group housed weaner pigs exposed to the air quality of a commercial environment. Livest Prod Sci 93:255–262

Mandersloot F (1992) Farm economic consequences of reducing nitrogen losses at dairy farms. Report 138, Proefstation voor de Rundveehouderij, Schapenhouderij en Paardenhouderij, Lelystad, The Netherlands

Mazoyer M, Roudart L (2006) A history of world agriculture. Earthscan, London

Melse RW, Ogink NWM, Rulkens WH (2009) Overview of European and Netherlands' regulations on airborne emissions from intensive livestock production with a focus on the application of air scrubbers. Biosyst Eng 104:289–298

Monteny GJ, Smits MCJ, van Duinkerken G, Mollenhorst H, de Boer IJM (2002) Prediction of ammonia emission from dairy barns using feed characteristics Part II: Relation between urinary urea concentration and ammonia emission. J Dairy Sci 85:3389–3394

Nadeau E, Englund JE, Gustafsson AH (2007) Nitrogen efficiency of dairy cows as affected by diet and milk yield. Livest Sci 111:45–56

Niemi JK, Sevón-Aimonen M-J, Pietola K, Stalder KJ (2010) The value of precision feeding technologies for grow-finish swine. Livest Sci 129:13

Oenema O, Bannink A, Sommer SG, Van Groenigen JW, Velthof GL (2008) Gaseous nitrogen emissions from livestock farming systems. In: Hatfield JL, Follett RF (eds) Nitrogen in the environment: sources, problems, and management, 2nd edn. Elsevier Publishers, Amsterdam, pp 395–441

Piñeiro C, Montalvo G, Herrero M, Bigeriego M (2010) Influence of measures for the control of ammonia emissions in the profitability of pig farms in Spain. TFRN-5, 25 October 2010, Paris, France

Reijneveld JA, Abbink GW, Termorshuizen AJ, Oenema O (2014) Relationships between soil fertility, herbage quality and manure composition on grassland-based dairy farms. Eur J Agron 56:9–18

Ritz CW, Fairchild BD, Lacy MP (2004) Implications of ammonia production and emissions from commercial poultry facilities: a review. J Appl Poultry Res 13:684–692

Rotz CA, Satter LD, Mertens DR, Muck RE (1999) Feeding strategy, nitrogen cycling, and profitability of dairy farms. J Dairy Sci 82:2841

Sannö J-O (2003) LIFE ammonia. Towards a sustainable milk production – reducing on-farm ammonia losses. Final report. http://www.slu.se/Documents/externwebben/vh-fak/husdjurens-miljo-och-halsa/life_final_technical_report.pdf. Accessed 15 Aug 2014

Steinfeld H, Mooney H, Schneider F, Neville LE (eds) (2010) Livestock in a changing landscape: drivers, consequences and responses. Island Press, Washington, DC

Swensson C (2003) Relationship between content of crude protein in rations for dairy cows, N in urine and ammonia release. Livest Prod Sci 84:125–133

UNCTAD (2009) World investment report. Transnational Corporations, Agricultural Production and Development, United Nations Publications, Geneva

Valk H (1994) Effects of partial replacement of herbage by maize silage on N-utilization and milk production of dairy cows. Livest Prod Sci 40:241–250

Valk H, Leusink-Kappers IE, van Vuuren AM (2000) Effect of reducing nitrogen fertilizer on grassland on grass intake, digestibility and milk production of dairy cows. Livest Prod Sci 63:27–38

Van Vuuren AM, Chilibroste P (2013) Challenges in the nutrition and management of herbivores in the temperate zone. Animal 7(s1):19–28

Veens T, Namkung H, Leeson S (2009) Limits to protein in layer diets relative to mitigating ammonia emission. J Avian Biol 2:143

Wu Z, Satter LD (2000) Milk production during the complete lactation of dairy cows fed diets containing different amounts of protein. J Dairy Sci 83:1042–1051

Chapter 4
Ammonia Abatement by Animal Housing Techniques

Gema Montalvo, Carlos Pineiro, Mariano Herrero, Manuel Bigeriego, and Wil Prins

Abstract This chapter provides information on the additional costs incurred by farmers implementing measures to abate ammonia emissions from livestock housing systems, from values derived in a Spanish case study. The examples provide information on the extra costs to implement a technique, in comparison with a reference system. Calculations take into account the economic lifetime of the investment, deducting grants and including changes in performance. The information comes from an original study prepared at the request of the Spanish Ministry of Agriculture and the Environment (MAGRAMA, Comparison of ammonia emissions when using different techniques after application of slurry in Spain: summary of results 2004–2006. International conference on ammonia in agriculture: policy, science, control and implementation. Wageningen, The Netherlands, 2007). Information is provided for a variety of suitable methods for each livestock phase (Gestating Sows, Lactating Sows, Weaners and Growers-finishers). Following detailed information on each technique and its costs, a summary of the costs and emission reduction costs for this Spanish example is provided. Most effective techniques (ranked by both emissions reduction potential and maximising profitability) are listed with a discussion of complicating factors and potential caveats regarding implementation.

Keywords Ammonia emissions • Animal housing • Emission abatement • Agriculture • Animal husbandry

G. Montalvo (✉) • C. Pineiro
PigCHAMP Pro Europa S.L., C/Santa Catalina, 10, 40003 Segovia, Spain
e-mail: gema.montalvo@pigchamp-pro.com; carlos.pineiro@pigchamp-pro.com

M. Herrero
Feaspor, Ctra. Soria, no 11, 40196 La Lastrilla, Segovia, Spain
e-mail: feaspor@feaspor.es

M. Bigeriego
Spanish Ministry of Agriculture and Environment, C/Almagro, 33, 28010 Madrid, Spain
e-mail: mbigerie@magrama.es

W. Prins
Ministry of Infrastructure and the Environment, Directorate for Climate, Air and Noise, The Hague, The Netherlands

© Springer Science+Business Media Dordrecht 2015

S. Reis et al. (eds.), *Costs of Ammonia Abatement and the Climate Co-Benefits*,
DOI 10.1007/978-94-017-9722-1_4

4.1　Introduction

The purpose of this chapter is to provide information on the additional costs incurred by farmers implementing measures to abate ammonia emissions from livestock housing systems. The case study presented here is from the Spanish pig sector, and therefore the calculations and methods employed are specific to that sector and region, however, standard concepts and a transparent methodology have been used, to allow conversion for local conditions.

The examples provide information on the extra costs to implement a technique, in comparison with a reference system. In each case, the reference system is described, followed by the abatement technique. Calculations were carried out according to the methodology set out in the IPPC Reference Document on Best Available Techniques for Intensive Rearing of Poultry and Pigs (European Commission 2003). This takes into account the economic lifetime of the investment, deducting grants and including changes in performance. The information comes from an original study prepared at the request of the Spanish Ministry of Agriculture and the Environment (MAGRAMA 2007).

Considerable advice and assistance has been provided by a number of technical experts, machinery and building manufacturers and suppliers, and by farmers and contractors during the course of the preparation of these costings, so we acknowledge their help with thanks.

4.2　Methodology

4.2.1　Unit Cost

The calculation of unit cost requires a clear understanding of:

- The proposed techniques to be introduced to reduce emissions
- The whole range of systems of production and management that are found on affected farms in the country concerned.
- The impact that the introduction of the technique will have on particular farm production and management systems in both physical and financial terms.

These techniques relate mainly to the housing category, however when appropriate other parts of the farm system (such as manure storage or slurry application) are referred to if they are a by-product of the technique. Unit costs are the annual increase in costs that a typical farmer will bear as a result of introducing a technique per unit of production. The general approach to the calculation of unit costs is as follows:

Table 4.1 'Units' used for assessing costs

Category		Units/Year		Details
Housing (and feed)		Per head		Building capacity
		Per kg NH$_3$		
		Per kg pig produced		
Manure and slurry storage		Per m^3 or tonnes		Liquid slurry (including dilution) and solid manure (including bedding)
Manure and slurry treatment		Per kg NH$_3$		
Manure and slurry application		Per kg pig produced		

- Define the physical and husbandry changes resulting from implementation of the abatement technique based on a thorough understanding of current farming systems.
- For each technique, identify those areas where costs or performance changes will be associated with the introduction of that technique.
- In all cases, only those costs directly associated with the techniques should be considered.
- As the assessment of costs is at farm level, any grants that are available should be deducted from expenditure.

The category that techniques fall into will determine the physical units that are used to define the population or quantities of manure, are used in subsequent calculations. The relationship can be seen from Table 4.1.

Unit costs should be calculated according to the general approach described below:

- Current cost should be used for all calculations
- Capital expenditure, after deducting any grants, should be annualised over the economic life of the investment.
- Annual running costs should be added to the annualised cost of capital
- Changes in performance have a cost and should be taken into account

This total sum is divided by the annual throughput to determine the 'unit cost'. Annual cost is expressed by the units shown in Table 4.1.

For pig production, costs are also expressed as € per pig produced, extending the work developed in other studies (Pineiro et al. 2005). The following assumptions are used in the calculations in this chapter:

- Equivalences:

    - 1 sow productive place produces 20.00 pigs marketed/year
    - 1 gestating place produces 26.60 pigs marketed/year
    - 1 lactating place produces 80.00 pigs marketed/year
    - 1 nursery place produces 5.79 pigs marketed/year

- 1 grower place produces 2.94 pigs marketed/year
- 1 pig marketed produces 1.25 m^3 of slurry

Pig marketed $=$ 100 kg body weight

4.2.2 Calculating the Annualisation of Capital Expenditure

The calculation of the annual cost *(Ann_C)* of capital items has been based on the following relationship:

$$(1)\ Ann\_C = C \times Dep \quad (2)\ Dep = \left[\frac{r(1+r)^n}{(1+r)^n - 1}\right]$$

Where: *Dep = Depreciation factor*
$C=$ Capital cost (€)
$r=$ Interest rate (%)
$n=$ Write off period (years)

Write off periods according to investment type are given in Table 4.2 and the subsequent depreciation factors, using an amortisation rate of 5 % ($r=0.05$,) (the current rate of interest commonly incurred by farmers seeking medium term loans).

The write-off periods reflect the economic life over which the investment should be considered rather than operational life. Annual costs for repairs used in this study are based on the estimations carried out by Nix (2003).

4.3 Results

Results are presented by phase (i.e. Gestating sows, lactating sows, etc), then by technique.

4.3.1 Gestating Sows

4.3.1.1 Sow Production Assumptions

- Sow production: 20 pigs per sow
- Pig marketed: 100 kg
- Gestating sows: 75 % of sows

Production per place (sow) $=$ 2,000 kg
Production per gestating place $=$ 2,667 kg

Table 4.2 Capital write-off periods

Investment type	Write off period (Years)
New buildings	20
Equipment installed in buildings	10
Machinery	6

(a)   Partial slat and reduced pit

Reference system: Total slat and deep pit (more than 60 cm)
Proposed system: Pit reduction (50 % width)

Technical Description

- Building capacity: 230 places
- Building surface: 460 m^2
- Floor: 50 % concrete slat
- Pit: rectangular section on average 60 cm deep

Depreciation Assumptions

- Pit life: 10 years
- Repairs: 1 % investment costs
- Interest rate: 5 % annual)

(b) Littered system

Reference system: Total slat and deep pit (more than 60 cm).
Proposed system: Change the total slat for solid floor, and add straw. It is necessary
to build a dunghill (60 m^2) to store waste litter until the field application.

Technical Description

- Building capacity: 30 places
- Building surface: 42 m^2
- Floor: 100 % slat

Depreciation Assumptions

- Concrete solid floor life: 10 years
- Dunghill life: 15 years
- Repairs (for both): 1 % investment costs
- Interest rate: 5 %

Other Assumptions

- Straw cost: 0.04–0.10 €/kg
- Straw need: 2.92 kg per sow and week
- Ratio occupation building: 85 %
- Manpower: 2 h per week (manual); 0.78 h per week (mechanized)
- Manpower cost: 15 € per hour (manual); 30 € per hour (mechanized with tractor
  and fuel)

Table 4.3 Cost calculations for partial slat and reduced pit (gestating sows)

		Capital cost (€)	Unit cost (€/unit)	Cost per kg (€/kg pig)
New building[a]				
Investment cost				
	Disassembly and assembly facilities	3,770		
	Pit reformation	1,386		
	Concrete for pits	4,819		
Total investment cost		**9,975**		
Annual cost				
	Amortization		5.27	
	Repairs		0.41	
Total			**5.68**	**0.0021**
Existing building[b]				
Annual cost *(120 % of reference costs)*			**6.83**	**0.0030**

[a]In the case of new buildings, the cost of adding a pit is the same as it would be for the reference system

[b]In the case of existing buildings it is advisable to consider that it may cost 20 % more than for the reference system (as it depends on the conditions of the existing structure)

- Slurry produced per sow: 2.50 m^3 per place per year
- Slurry application cost: 0.70 €/m^3
- Pit need: 25.2 m^3
- Pit construction cost: 10 €/m^3

(c) Frequent manure removal

Reference system: Total slat and deep pit (more than 60 cm). Pit is emptied when is full.

Proposed system: To empty the pit weekly.

These costs are the same for other production phases (except for lactating sows where it is not a suitable technique due to low slurry flow) – therefore they are not detailed further in other sections – but are included in the final summary table as the ammonia reductions do vary by phase.

4.3.2 Lactating Sows

4.3.2.1 Sow Production Assumptions

- Sow production: 20 pigs per sow
- Pig marketed: 100 kg
- Lactating sows: 25 % of productive sows

Table 4.4 Cost calculations for littered system (gestating sows)

	Capital cost (€)	Unit cost (€/unit)	Cost per kg (€/kg pig)
New buildings			
Investment cost			
Dunghill construction	3,412		
Annual cost			
Dunghill amortization		10.96	
Dunghill repairs		1.14	
Straw		5.16–12.90	
Manpower (mechanized)		40.56	
Pit cost[*]		−8.40	
Slurry application[*]		−1.75	
Total		**47.61–55.35**	**0.0179–0.0208**
Existing building[a]			
Investment cost			
Floor substitution	1,120		
Dunghill construction	3,412		
Annual cost			
Floor amortization		4.83	
Dunghill amortization		10.96	
Floor repairs		0.37	
Dunghill repairs		1.14	
Straw		5.16–12.90	
Manpower (manual)		52	
Slurry application[*]		−1.75	
Total		**72.71–80.45**	**0.0273–0.0302**

[a]Negative values indicate a saving of costs, in the case of a new building, or offset costs due to re-use of slurry on the farm

Table 4.5 Costs of frequent manure removal (gestating sows)

	Capital cost (€)	Unit cost (€/unit)	Cost per kg (€/kg pig)
Investment cost			
Cost of frequent manure removal for gestating sows	0	0	0
Annual cost			
Cost of frequent manure removal for gestating sows	0	0	0

Production per place (sow) = 2,000 kg
Production per lactating place = 8,000 kg

(a) Combination of a water and manure channel

Table 4.6 Cost of combination of a water and manure channel (lactating sows)

	Capital cost (€)	Unit cost (€/unit)	Cost per kg (€/kg pig)
***New buildings*[a]**			
Investment cost			
Construction	2,640		
Annual cost			
Amoritization		2.85	
Repairs		0.44	
Total cost, new building		**3.29**	**0.0004**
Existing buildings[a]			
Investment cost			
Construction costs	13,440		
Annual cost			
Amortization		14.50	
Repairs		2.24	
Total cost, existing building			
Favourable conditions[a]		**16.74**	**0.0021**
Less favourable conditions[a]		**20.09**	**0.0025**

[a]As the existing conditions can be less favourable than paper based examples, we have included 'favourable conditions' and 'less favourable conditions' (favourable conditions +20 %)

Reference system: Total slat and deep pit (more than 50 cm).
Proposed system: The manure pit is split up into a wide water channel at the front and a small manure channel at the back.

Technical Description

- Building capacity: 120 places
- Building surface: 684 m^2
- Floor: 100 % slat
- New pit with two channels (water channel and manure channel)

Depreciation Assumptions

- Pit life: 10 years
- Repairs: 2 % investment costs
- Interest rate: 5 % annual

Other Assumptions

- m^2 brick partition: 20 €/m^2
- Disassembly and assembly facilities (12 places): 1,080 €

(b) Manure pan underneath

Reference system: Total slat and deep pit (more than 50 cm).
Proposed system: A board with a very smooth surface is placed under the slatted floor.

Table 4.7 Cost of manure pan underneath slatted floor (lactating sows)

		Capital cost (€)	Unit cost (€/unit)	Cost per kg (€/kg pig)
New buildings[a]				
Investment cost		14,070		
	Construction			
Annual cost				
	Amoritization		15.18	
	Repairs		2.34	
Total cost, new building			**17.52**	**0.0022**
Existing buildings[a]				
Investment cost				
	Construction costs			
Annual cost				
	Amortization		26.84	
	Repairs		4.14	
Total cost, existing building				
Favourable conditions[a]			**30.98**	**0.0039**
Less favourable conditions[a]			**37.18**	**0.0046**

[a]As the existing conditions can be less favourable than paper based examples, we have included a 'favourable conditions' and 'less favourable conditions' (favourable conditions +20 %)

Technical Description

- Building capacity: 120 places
- Building surface: 684 m^2
- Floor: 100 % slat

Depreciation Assumptions

- Pit life: 10 years
- Repairs: 2 % investment costs
- Interest rate: 5 % annual

Other Assumptions

- m^2 brick partition: 20 €/m^2
- Disassembly and assembly facilities (12 places): 1,080 €
- Concrete + PVC sheet (for pit): 1,407 €/12 places

As noted previously, frequent slurry removal is not a suitable option for lactating sows, due to slow slurry flow.

4.3.3 Weaners

4.3.3.1 Production Assumptions

- Occupation days: 63
- Number of rotations per year: 365/63 = 5.79 rotations per place per year
- Pig marketed: 100 kg

Production per place: 579 kg pig per place per year

(a) Manure channel with sloped floor

Reference system: Total slat and deep pit (more than 65 cm).
Proposed system: A sloped floor to separate faeces and urine is placed under the slated floor.

Technical Description

- Building capacity: 1,320 places
- Number of rooms: 11
- Building surface: 500 m^2
- Floor: 100 % slat
- Pit: Sloped floor

Depreciation Assumptions

- Pit life: 10 years
- Repairs: 2 % investment costs
- Interest rate: 5 % annual

Other Assumptions

- m^2 brick partition: 20 €/m^2
- PVC sheet (for pit): 8.35 €/m^2
- Disassembly and assembly facilities (per room): 275–1,375 € (function of installation characteristic and material of panels and slat).
- Pit cost for new installations (with sloped floor): 0 € to +30 %
- Pit cost for existing installations (with sloped floor): 6,960.36 € (in this building)

(b) Partial Slat

Reference system: Total slat and deep pit (more than 70 cm).
Proposed system: To install a slated floor (2/3 slat floor and 1/3 solid floor).

Technical Description

- Building capacity: 1,320 places
- Building surface: 500 m^2
- Floor: 75 % slat
- Pit: rectangular section. 65 cm average deep

Table 4.8 Cost of manure channel with sloped floor (weaners)

		Capital cost (€)[*]	Unit cost (€/unit)	Cost per kg (€/kg pig)[a]
New buildings				
Investment cost				
	Construction	0–2,088		
Annual cost		0–0.2		
	Amoritization			
	Repairs	0–0.03		
Total cost, New building		0–0.23		0–0.0004
Existing buildings (*with collapsible PVC pens*)				
Investment cost				
	Construction costs	11,258		
Annual cost				
	Amortization		1.10	
	Repairs		0.17	
Total cost, existing building			**1.27**	**0.0022**
Existing buildings (*with fixed metallic pens*)				
Investment cost				
	Construction costs	23,558		
Annual cost				
	Amortization		2.31	
	Repairs		0.36	
Total			**2.67**	**0.0046**

[a]Where two values are shown, they indicate a minimum and maximum

Depreciation Assumptions

- Solid floor life: 10 years
- Repairs: 2 % investment costs
- Interest rate: 5 % annual

Other Assumptions

- Disassembly and assembly facilities (per room 120 places): 275–1,375 € (depending on whether the system has PVC or fixed metal pens)
- m^2 brick partition: 20 €/m^2
- Blind PVC Slats cost: 16 €/m^2

For new installations, the cost of installing partially-slatted floors for houses for transition piglets is approximately the same as for the reference system.

(c) Frequent manure removal

As detailed in Sect. 4.3.1, there is no cost to this technique.

Table 4.9 Cost of partial slat (weaners)

		Capital cost (€)	Unit cost (€/unit)	Cost per kg (€/kg pig)
Existing buildings *(PVC pens)*				
Investment cost				
	Construction	7,777		
Annual cost				
	Amoritization		0.76	
	Repairs		0.12	
Total cost, existing buildings (PVC Pens)			**0.88**	**0.001**
Existing buildings *(fixed metallic pens)*				
Investment cost				
	Construction costs	19,877		
Annual cost				
	Amortization		1.95	
	Repairs		0.30	
Total cost, existing building (fixed metallic pens)			**2.25**	**0.0026**

4.3.4   Growers-Finishers

4.3.4.1   Production Assumptions

- Occupation days: 124
- Number of rotations per year: 365/124 = 2.94 rotations per place per year
- Pig marketed: 100 kg

Production per place: 294 kg pig per place per year

(a)   Partial slat

Reference system: Total slat and deep pit (more than 70 cm).
Proposed system: To install a slated floor (2/3 slat floor and 1/3 solid floor).

Technical Description

- Building capacity: 1,440 places
- Building surface: 1,450 m^2
- Floor: 75 % slat
- Pit: rectangular section. 70 cm average deep

Depreciation Assumptions

- Solid floor life: 10 years
- Repairs: 2 % investment costs
- Interest rate: 5 % annual

Table 4.10 Cost of partial slat (growers-finishers)

		Capital cost (€)	Unit cost (€/unit)	Cost per kg (€/kg pig)
New buildings[a]				
Investment cost				
	Construction	0		
Annual cost				
	Amoritization	0		
	Repairs	0		
Total cost, new building		**0**		**0**
Existing buildings[b]				
Investment cost				
	Construction costs	34,820		
Annual cost				
	Amortization		3.13	
	Repairs		0.48	
Total cost, existing building				
Favourable conditions[b]			**3.61**	**0.0123**
Less favourable conditions[b]			**4.33**	**0.0147**

[a]For a new building, the reduced pit cost is the same than the reference system
[b]For existing building, it is advisable to consider a security economic margin of 20 % (the existing conditions can be less favourable than the example)

Other Assumptions

- Disassembly and assembly facilities (per room 120 places): 2,540 €
- m^2 brick partition: 20 €/m^2
- Concrete solid floor: 15 €/m^2

(b) Littered

Reference system: Total slat and deep pit (more than 60 cm).
Proposed system: Change the total slat for solid floor, and add straw. It is necessary to build a dunghill (120 m^2) to store waste litter until the field application.

Technical Description

- Building capacity: 1,440 places (10 rooms)
- Building surface: 1,450 m^2
- Floor: 100 % slat
- Pit: rectangular section. 70 cm average deep

Depreciation Assumptions

- Concrete solid floor life: 10 years
- Dunghill life: 15 years

Table 4.11 Cost of littered system (growers-finishers)

		Capital cost (€)	Unit cost (€/unit)	Cost per kg (€/kg pig)
New buildings				
Investment cost				
	Dunghill construction	6,825		
Annual cost				
	Dunghill amortization		0.46	
	Dunghill repairs		0.05	
	Straw		3.72–9.28	
	Manpower		24.00	
	Pit cost[a]		−6.32	
	Slurry application[a]		−1.75	
Total			**20.16–25.72**	**0.0686–0.0875**
Existing building[a]				
Investment cost				
	Floor substitution	41,600		
	Dunghill construction	6,825		
Annual cost				
	Floor amortization		3.74	
	Dunghill amortization		0.46	
	Floor repairs		0.29	
	Dunghill repairs		0.05	
	Straw		3.72–9.28	
	Manpower (manual)		30.00	
	Slurry application[a]		−1.75	
Total			**36.51–42.07**	**0.1242–0.1431**

[a]Negative values indicate a saving of costs, in the case of a new building, or offset costs due to re-use of slurry on the farm

- Repairs (for both): 1 % investment costs
- Interest rate: 5 %

Other Assumptions

- Straw cost: 0.04–0.10 €/kg
- Straw need: 2.2 kg per place and week
- Ratio occupation building: 85 %
- Manpower: 2 h per place per week (manual); 0.8 h per place per week (mechanized)
- Manpower cost: 15 € per hour (manual); 30 € per hour (mechanized with tractor and fuel)
- Slurry produced per sow: 2.50 m^3 per place per year
- Slurry application cost: 0.70 €/m^3
- Pit need: 910 m^3
- Pit construction cost: 10 €/m^3

Table 4.12 Cost of manure channel (growers-finishers)

		Capital cost (€)[a]	Unit cost (€/unit)	Cost per kg (€/kg pig)[a]
New buildings				
Investment cost				
	Construction	0–7,037		
Annual cost				
	Amoritization	0–0.63		
	Repairs	0–0.10		
Total cost, new building		**0–0.73**		**0–0.0025**
Existing buildings[b]				
Investment cost				
	Construction costs	62,142		
Annual cost				
	Amortization		5.59	
	Repairs		0.86	
Total cost, existing building				
Favourable conditions[b]			**6.45**	**0.0219**
Less favourable conditions[b]			**7.74**	**0.0263**

[a]Where two values are shown, this indicates the minimum and maximum values
[b]For existing building, it is advisable to consider a security economic margin of 20 % (the existing conditions can be less favourable than the example)

- Installation cost per room (120 pigs), 2,540 €
- Concrete slats cost: 15 €/m^3

(c) Manure channel

Reference system: Total slat and deep pit (more than 65 cm).
Proposed system: A sloped floor to separate faeces and urine is placed under the slated floor.

Technical Description

- Building capacity: 1,440 places
- Building surface: 1,450 m^2
- Floor: 100 % slat
- Pit: Sloped floor

Depreciation Assumptions

- Pit life: 10 years
- Repairs: 2 % investment costs
- Interest rate: 5 % annual

Other Assumptions

- m^2 brick partition: 20 €/m^2
- PVC sheet (for pit): 8.35 €/m^2
- Installation cost per room (120 pigs), 2,540 €
- Pit cost for new installations (with sloped floor): 0 € to +30 %
- Pit cost for existing installations (with sloped floor): 23,458 € (in this building)

(d) Frequent removal of manure

As noted in Sect. 4.3.1, there is no cost to the frequent removal of manure technique.

4.4 Summary and Conclusions

4.4.1 Overview of Extra Costs

A summary of the calculated costs is included in Table 4.14, along with the predicted reduction in Ammonia by adaptation (ammonia emissions for reference techniques is provided in Table 4.13). This is then used to calculate the extra cost per (in Euro) per kg of NH_3 saved by adaptation.

The extra costs incurred for housing strategies for ammonia abatement have a very wide range, from 0 to 212.8 €/kg NH_3 abated. The GAINS model estimates the costs of abatement strategies for housing in Spain at 27.6 €/kg NH_3 (Klimont and Winiwarter 2011), which is within the same order of magnitude as many of the values in this chapter. Pineiro et al. (2005), therefore concluded that the systems were roughly equivalent.

It is clear from Table 4.14 that littered systems (especially in the case of gestating sows) are not only very costly, but are the least effective in terms of cost efficiency for ammonia abated (212.8 €/kg NH_3). As average annual profitability from 2003 to 2009 in the Spanish pig sector was 0.0494 €/kg pig, (MAGRAMA 2010) installing a littered system into an existing building would decrease the profitability by 60 %. Therefore the costs for this technique are prohibitive. However it is also clear that the technique of frequent manure removal in the gestating, weaners and growers-finishers phase can be a cost-free emission reduction strategy of 25–30 % depending on the system (if no extra labour is required to be bought in for the work to take place) and is suitable for new and existing building systems. According to this study the addition of partial slats to growers-finishers and weaner housing in new buildings is potentially without (extra) cost, which would make it the best 'construction' option for reducing ammonia in housing, it does however make cleaning more difficult.

Between the two extremes described already there are then a range of costs and systems, which when the profitability of the pig sector is also taken into account, can make the best way forward for the farmer and for legislation, more complicated.

Table 4.13 Ammonia emissions from reference housing techniques, by livestock phase, from Spanish Inventory data

Livestock phase	Ammonia emissions from reference housing technique (kg/place)
Gestating sows	2.71
Lactating sows	3.73
Weaners	0.72
Growers-finishers	3.15

Looking at the values here (acknowledging that this is in isolation by technique and not within a full production chain setting), Table 4.15 shows the most efficient techniques by phase, considering (a) the lowest cost per unit NH_3 abated and (b) ranked by increasing cost per kg meat (which can be compared to profitability in the pig sector). The best options are the same in all cases except for that of weaners in existing buildings, where partial slats would be chosen over a manure channel with slope, if a decrease in profitability (per unit) were considered the most important.

It is also worth considering at this point, the proportion of ammonia emissions from livestock systems and from each phase in the pig sector in this example from Spain (Fig. 4.1). This would suggest that from an emissions perspective it is worth tackling emissions from housing, and that focussing on the growing-finisher rooms could be key, due to their proportion of total housing (in this example from a farrow to finish farm). According to the information in the study, this would suggest that using partial slats (in both the case of new and existing buildings), is the most cost effective measure in the pig sector in Spain.

Table 4.14 Summary of costs associated with ammonia reduction techniques in livestock housing, Pig Sector, Spain

Phase	Adaptation	New or Existing	Extra cost (€/place)	Extra cost (€/kg pig)	% Reduction in NH_3	NH_3 reduction (kg/place)	Extra cost (€/kg NH_3)
Gestating sows	Partial slat and reduced pit	N	5.68	0.0021	35	0.95	6
	Partial slat and reduced pit	E	6.83	0.003	35	0.95	7.2
	Littered system[a]	N	47.67–55.41	0.0179–0.0208	14	0.38	125.4–145.8
	Littered system[a]	E	72.72–80.46	0.0272–0.0302	14	0.38	191.4–211.7
	Frequent manure removal	-	0	0	25	0.68	0
Lactating sows	Combination water-manure channel[b]	N	3.29 (3.95)	0.0004 (0.0005)	52	1.94	1.7 (2)
	Combination water-manure channel[b]	E	16.74 (20.09)	0.0021 (0.0025)	52	1.94	8.6 (10.4)
	Manure pan underneath[b]	N	17.52 (21.04)	0.0022 (0.0026)	32	1.19	14.7 (17.7)
	Manure pan underneath[b]	E	30.98 (37.18)	0.0039 (0.0046)	32	1.19	26 (33.8)
Weaners	Manure channel with sloped floor[a]	N	0–0.23	0.0004	60	0.43	0–0.5
	Manure channel with sloped floor[c]	E	2.67	0.0046	60	0.43	6.2
	Manure channel with sloped floor[d]	E	1.28	0.0022	60	0.43	3.0
	Partial slat	N	0	0	25	0.18	0
	Partial slat[d]	E	0.88	0.001	25	0.18	4.9
	Partial slat[c]	E	2.25	0.0026	25	0.18	12.5
	Frequent manure removal	-	0	0	25	0.18	0

Growers-finishers	Partial slat	N	0	0	30	0.95	0
	Partial slat[b]	E	3.61 (4.33)	0.0123 (0.0148)	30	0.95	3.8 (6.9)
	Littered system	N	25.72	0.0875	20	0.63	40.7
	Littered system	E	42.07	0.1431	20	0.63	66.6
	Manure channel with sloped floor	N	0.73	0.0263	10	0.32	2.3
	Manure channel with sloped floor	E	6.45	0.0219	10	0.32	20.4
	Frequent manure removal	–	0	0	30	0.95	0

[a]Minimum and maximum values stated
[b]Less-favourable costs (favourable +20 %), listed in brackets
[c]Fixed metallic pens
[d]Collapsible PVC pens

Table 4.15 Most efficient techniques by phase, ranked by ammonia reduction and by maximising profitability (minimum and favourable conditions listed)

Phase	New or existing	Minimising cost by ammonia reduction (€/place)		Maximising profitability when considering a technique (€/kg pig)	
Gestating sows	N	Partial slat and reduced pit.	6	Partial slat and reduced pit.	0.0021
	E	Partial slat and reduced pit.	7.2	Partial slat and reduced pit.	0.003
Lactating sows	N	Combination water-manure channel.	1.7	Combination water-manure channel.	0.0004
	E	Combination water-manure channel.	8.6	Combination water-manure channel.	0.0021
Weaners	N	Manure channel with sloped floor.	0	Manure channel with sloped floor.	0
	E	Manure channel with sloped floor.. (collapsible pvc pens)	3	Manure channel with sloped floor.. (collapsible pvc pens)	0.0022
Growers-finishers	N	Partial slat.	0	Partial slat.	0
	E	Partial slat.	3.8	Partial slat.	0.0123

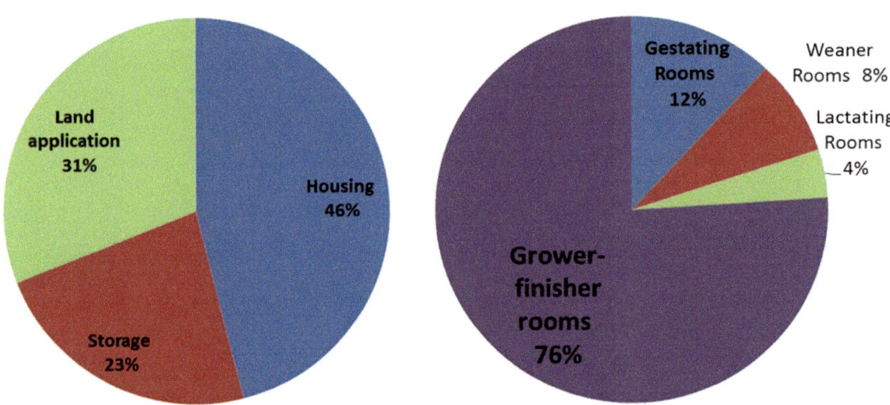

Fig. 4.1 Breakdown by livestock system (**a**) and by phase (**b**)

References

European Commission (2003) Integrated Pollution Prevention and Control (IPPC). Reference document on best available techniques for intensive rearing of poultry and pigs, July 2003

Klimont Z, Winiwarter W (2011) Integrated ammonia abatement – modelling of emission control potentials and costs in GAINS. IIASA Interim Report IR-11-027. IIASA, Laxenburg, Austria

MAGRAMA (2007) Comparison of ammonia emissions when using different techniques after application of slurry in Spain: summary of results 2004–2006. International conference on ammonia in agriculture: policy, science, control and implementation. Wageningen, The Netherlands

MAGRAMA (2010) Documento Guía Español para la asistencia técnica para la implantación de la Directiva IPPC en España para los sectores de cría intensiva de cerdos y aves de corral. http://www.magrama.gob.es/es/ganaderia/publicaciones/copia_de_default.aspx. Accessed 14 Aug 2014

Nix J (2003) Farm management pocketbook. Imperial College at Wye, 33rd edn. Agro Business Consultants Ltd, Melton Mowbray

Pineiro C, Montalvo G, Bigeriego M, Herrero M (2005) Emissions from European agriculture. Wageningen Academic Publishers, Wageningen

Chapter 5
Ammonia Abatement with Manure Storage and Processing Techniques

Andrew VanderZaag, Barbara Amon, Shabtai Bittman, and Tadeusz Kuczyński

Abstract In this chapter we examine techniques for reducing NH_3 emissions from stored manure, including: covers, storage design, and manure processing. Most techniques are for liquid manure stores, whereas there are few methods for solid manures and more research is needed. The cost effectiveness for reducing NH_3 emissions by each technique was estimated using installation and operating cost data from several sources and published data on baseline emissions and emission reductions of the techniques. A key uncertainty in these cost-efficacy estimates is the baseline emission used to calculate the quantity of N conserved. It is therefore important to obtain regional baseline emission factors (consistent with national emission inventories) to ensure accurate cost estimates.

Most category 1 and 2 techniques cost less than 10 €/kg of abated NH_3-N. Techniques costing less than 5 €/kg NH_3-N are available for mitigating emissions from every type of manure investigated (cattle slurry, cattle farm-yard manure (FYM), pig slurry tanks, pig slurry lagoons, pig FYM, and poultry manure). The most economical strategy was to allow cattle manure tanks to crust. Other highly economical techniques were floating materials, which abated NH_3 for much lower cost than structural covers.

The values of co-benefits were estimated for precipitation exclusion, N retention, reduced odour, and greenhouse gas mitigation. Combining these benefits

A. VanderZaag (✉)
Agriculture and Agri-Food Canada, 960 Carling Ave., Ottawa, ON K1A 0C6, Canada
e-mail: Andrew.VanderZaag@agr.gc.ca

B. Amon
Leibniz Institute for Agricultural Engineering Potsdam-Bornim e.V. (ATB),
Max-Eyth-Allee 100, 14469 Potsdam, Germany
e-mail: bamon@atb-potsdam.de

S. Bittman
Agriculture and Agri-Food Canada, 6947 Highway 7, PO Box 1000, Agassiz,
BC V0M 1A0, Canada
e-mail: shabtai.bittman@agr.gc.ca

T. Kuczyński
University of Zielona Góra, Zielona Góra, Lubuskie 65-516, Poland
e-mail: T.kuczynski@iis.uz.zgora.pl

© Springer Science+Business Media Dordrecht 2015
S. Reis et al. (eds.), *Costs of Ammonia Abatement and the Climate Co-Benefits*,
DOI 10.1007/978-94-017-9722-1_5

reduced the net costs substantially so we encourage additional efforts to thoroughly evaluate co-benefits in future analyses.

Keywords Ammonia emissions • Manure storage • Emission abatement • Agriculture • Manure spreading

5.1 Introduction

Gaseous ammonia (NH_3) emissions are undesirable for environmental quality and health. Chronic NH_3 deposition can exceed critical loads and damage sensitive ecosystems; acute effects arise from the direct toxicity of NH_3 exceeding critical levels for vulnerable species. Adverse human health effects stem from NH_3 acting as an aerosol precursor which leads to the development of fine particulate matter and smog. Furthermore, NH_3 emissions reflect a poor design from a farm-systems perspective. On-farm N is "given away" into the atmosphere while farmers purchase synthetic N that is industrially extracted from the atmosphere and transported long distances.

Agriculture is the dominant source of NH_3 emissions. In Canada, for instance, agriculture contributed 89 % (414 kt) of the NH_3 emitted from all sources in 2009 (Environment Canada 2011). Emissions from animal production are the dominant source as illustrated by UK data where animal production contributed 84 % of the agricultural total (Misselbrook et al. 2008). Within animal agriculture the manure storage contributes a modest amount (13 % in the UK example).

Although manure storage is not the largest NH_3 source it is a worthwhile target for emission mitigation strategies. The manure storage is a suitable stage for relatively intense mitigations since it is a point source (vs. grazing areas), is similar across broad regions (vs. manure application which is affected by crops, soil, topography, and climate) and changes do not affect animal productivity (vs. housing or diet strategies). Furthermore, it is one of the easiest abatement techniques to monitor (Webb et al. 2005). As discussed elsewhere in this book by Klimont and Winiwarter (2015), mitigations must address all farm management stages. There is no single "magic-bullet" solution (Wohl 1996) and without a complete strategy, N conserved at one stage will be lost downstream and no benefit gained.

Manure storages provide a valuable service by offering flexible manure application timing. Rather than spreading every day, the farmer may apply manure when it is best suited to the crop and avoid periods where damage (e.g., soil compaction) or environmental contamination might occur (e.g., rain, frozen soils). For most farms this also saves fuel and labor costs (Martin and Matthews 1983).

Several types of manure storage exist depending on farm management. The main division is between solid and liquid manures. Solid manures contain bedding materials and are common with poultry, beef cattle, smaller tie-stall dairies, and small swine operations. Solid manures are stored in outdoor heaps or stacks. Liquid manures (DM <5 %; ASABE 2013) and slurries (DM ~5–12 %) are common on larger farms and are becoming more prevalent as farm sizes increase. Types of liquid storages include outdoor tanks made of concrete or steel (above or below grade), and earthen basins. Containment below slatted floors are considered a part of housing, discussed earlier in the book by Montalvo et al. (2015). All storage systems are emptied at least once per year and differ from anaerobic lagoons which are treatment systems that operate for years before being emptied (ASABE 2013).

The focus of this chapter is to analyze mitigation strategies to NH_3 emissions from manure storage systems. The main components are: (i) the technical feasibility of the mitigation measures, (ii) evidence that NH_3 emissions are reduced in comparison to a baseline scenario, (iii) the cost of installation and operation, and (iv) assessment of abatement cost-effectiveness including co-benefits.

5.1.1 Mechanisms of NH_3 Emissions

A discussion of the driving forces that cause NH_3 emissions is helpful for understanding the rationale of mitigation strategies. Mechanistic models describing emissions from liquid manure have been developed (Ni 1999) and although there is more uncertainty about the processes in solid manure, similar mechanisms are involved (Sommer and Hutchings 2001). In brief, enzymatic breakdown of urea or uric acid and other organic biomass produces NH_3 which is equilibrium with NH_4^+. Acidic conditions shift the equilibrium to NH_4^+ and basic conditions to NH_3. Concentration gradients cause NH_3/NH_4^+ to diffuse towards the emitting surface. At the manure surface NH_3 overcomes gas- and liquid-phase resistances and diffuses into the air where it is moved by convection. Transfer within the liquid phase is temperature dependent and gas-phase transfer is dependent on temperature and air velocity. Therefore there are several targets for intervention: reduce the emitting surface area to volume ratio, reduce the surface air velocity, reduce the manure temperature, reduce the pH, or collect contaminated air and treat it.

5.2 Technical Description of Covers for Slurry Stores

Covering manure storages is a mitigation approach for use on both liquid and solid manures and combined with manure processing. There are a number of variations each suited to certain manure systems and with advantages and disadvantages.

5.2.1 Covers for Liquid Manures and Slurry

Many cover types have been used on liquid manures and slurries. In general, the types can be classified based on permeability, structure, and composition (Table 5.1). Impermeable covers tend to have longer lifespans and offer the

Table 5.1 Types of covers for liquid manure storage

Impermeable		Permeable	
Structural	Floating	Floating synthetic	Floating natural
Tent	Plastic film	Plastic fabrics	Straw
Lid (wood, concrete)	Negative air pressure	Plastic tiles	Crusts
Positive air pressure		Clay balls	Other materials
Storage bag		Other materials	

possibility of biogas collection if safety criteria are met. Structural covers are suited to concrete tanks or small earthen stores but may require modifications to existing tanks and tanks must be emptied before installation. Floating impermeable covers use the manure itself as support and require minimal modification to existing structures. Permeable covers are somewhat simpler because they float and do not require pumps to handle accumulations of gas or water. Synthetic permeable covers are more durable and reliable than natural ones but cost more. Manure agitation and handling is impeded by covers that float or are pressurized. Agitation and pumping poses a great risk of damaging floating covers so farm operations must be done carefully. Repairing a damaged cover is difficult or impossible and floating covers may plug pumps.

The most suitable cover depends on the facility. For instance, lagoons which have a relatively constant liquid level are well suited to floating covers (permeable or impermeable) that must be secured at the edges; whereas, slurry tanks which have a more variable liquid level are better suited to a solid roof or lid (which does not touch the manure) or a floating cover that can move up and down with the liquid level. Generally, floating covers are not suitable for dairy slurry with high solids content because vigorous agitation is required to break apart matted solids and this is extremely difficult to do beneath a floating cover. Straw is an exception to this generality as it can enhance crust development. Most cover types can be installed on existing manure stores. Concrete lids, however, should be part of the initial tank design and are not suitable for retrofits. Each cover type is discussed in the following sections and details are summarized in Table 5.2.

5.2.1.1 Impermeable Structural Covers

Most covers in this category have a headspace between the liquid and the cover. Gases accumulate in this space, which are potentially flammable (CH_4) or fatal (H_2S). Safety protocols mitigate but do not eliminate risks posed by these gases.

Tent Coverings

Tent structures are suitable for circular tanks (Fig. 5.1). A vertical pole made of resistant materials is installed at the centre of the tank to support the tent. In one

Table 5.2 Summary of covers for abating ammonia from storages and their operational aspects, co-benefits, advantages and disadvantages

Type	Suitable for	Existing stores	Liquid Level	Life (year)	Operation equip.	Access location	Risk[a]	Co-Benefits or negatives	Advantages	Disadvantages
Tent	Tank	Yes	Varied	>10	None	Single	Med.	Less volume[b]. Less odour.	Precipitation exclusion; access panel for agitation and pumping.	Support structure required; empty tank for installation.
Wood	Tank	Yes	Varied	>10	None	Single	Med.	Less volume[b] Less odour.	Reliable; precipitation exclusion.	Empty tank for installation.
Concrete	Concrete Tank	No	Varied	20–30	None	Single	Low	Less volume[b], odour.	Long-lived; precip. exclusion.	Initial cost; new installations
Pressure supported	Tank or Basin	Yes	Varied	?	Air pump, flare or filter	None (must deflate)	High	Less volume[b]. Less odour.	No structure; eliminates snow and rainwater; gas flared or biofiltered.	Electricity and pumps; wind damage; punctures; deactivate to pump or agitate manure.
Storage bag	Small pig farm	No	Varied	?	None	Single	Med.	Less volume[b]. Less odour.	Precipitation exclusion; simple installation.	Small size.
Plastic impermeable	Tank or Lagoon	Yes	Fixed, Varied	>8	Water removal	Single	High	Less volume[b]. Less odour.	No support structure; Rainwater collection prevents dilution.	Damage by storms, snow, ice.
Negative air pressure	Tank or Lagoon	Yes	Varied	>10	Pumps (gas, water), biofilter	Single	High	Less volume[b]. Less odour. Gas collection to remove CH_4.	No support structure; Prevents manure dilution; extra material allows liquid level to fluctuate.	Requires energy to remove water, collect gas, flare; damage by snow and ice; difficult to agitate/ pump.

(continued)

Table 5.2 (continued)

Type	Suitable for	Existing stores	Liquid Level	Life (year)	Operation equip.	Access location	Risk[a]	Co-Benefits or negatives	Advantages	Disadvantages
Plastic permeable	Lagoon	Yes	Fixed	>10	None	Single	Low	Less odour. More N_2O	No pumps required;	Snow may cause sinking; access port needed for agitation.
Plastic permeable	Tank	Yes	Varied	?	None	Single	High	Less odour. More N_2O	No structure; no pump	Potential damage by wind, snow, ice; access for agitation.
Plastic tiles	Swine or Digestate	Yes	Varied	25	None	Full	None	Less odour.	Easy installation; no maintenance; flexible; robust in storms, winter.	Initial cost.
Clay balls	Any-shape basin	Yes	Varied	>5	None	Full	Low	Less odour.	Self-floating; fits any shape; easy to install; farm disposal acceptable.	Initial cost; some material lost with agitation and pumping; wind effects; can clog pumps.
Straw	Lagoon or storage.	Y	Varied	<1	Chopper pump	Full	None	Less odour. More N_2O	Low cost; on-farm material.	Difficult to apply evenly; wind effects; sinking; special pump to avoid clogging.
Crust	Cattle slurry	Y	Varied	<1	None	Full	None	Less odour. More N_2O	No cost; naturally occurring.	Inconsistent thickness, durability, and efficacy.

[a]Risk is a qualitative scale reflecting the potential for catastrophic damage to the cover or adverse health effects from exposure to headspace gases
[b]Precipitation exclusion leads to lower stocking volume requirements. In dry areas the cover may reduce evaporation and cause increased manure volume

Fig. 5.1 Retrofitted tent cover in Poland (Photo: T. Kuczynski)

configuration, this centre pole supports beams or other material that connect to the top of the tank walls. This structure is covered by impermeable material such as PVC film that resists degradation from UV and rot. In another configuration, the centre pole directly supports the foil (e.g., PES) which is clamped to the top of the walls. Tent covers are commercially available for new or existing tanks (e.g., Erich Stallkamp ESTA GmbH, Germany).

The conical design sheds precipitation which increases manure capacity in wet regions. The tent is not airtight; therefore biogas will not pressurize the headspace. Manure handling and agitation equipment can be inserted into the tank through a removable section of the tent or via ports on the sides of the tank on new systems.

Lid (Wood or Concrete)

Solid horizontal lids made of wood or concrete are supported by the tank walls. Concrete lids should be part of the original concrete tank design and require ventilation to release biogas pressure. Properly designed concrete lids have a long lifespan of up to 50 years. Improper designs are prone to damage from gases in the headspace including corrosion and sagging.

Wooden lids (e.g., 20 mm thick; Sommer et al. 1993) are preferable for retrofitting an existing tank. Pressure vents are unnecessary unless the wooden lid has been designed to be gas-tight. Access ports to allow manure pumping and handling are required for any type of lid.

Positive Air Pressure

Positive air pressure covers are appealing because of their relatively low cost compared to lids and tents. In addition, they can be installed on a tank containing manure since no support column is inserted into the tank. These covers are made of flexible UV-stabilized material (e.g., poly vinyl chloride, PVC; ethylene propylene diene monomer, EPDM). Edges of the cover are tightly secured to the walls of a tank or into the earth surrounding a basin. Air is continuously blown under the cover by a fan which pressurizes the headspace and supports the cover. The result is a dome-shape that is wind-resistant and sheds precipitation.

Positive air pressure covers have been installed on concrete tanks (Zhang and Gaakeer 1998) and earthen lagoons (Funk et al. 2004a). Results of the former study were promising as the cover operated successfully for 3 years and manure removal was accomplished with the same equipment as before installation. The latter study, however, reported difficulties when the cover material tore and could not be repaired. Eventually the cover was removed.

Another drawback is that the cover deflates when electricity fails. Backup power reduces but does not eliminate this risk. Ropes installed across the surface of the liquid prevent the cover from falling into the liquid but it will still be susceptible to wind damage. The technology has potential but the risk of catastrophic damage must be eliminated before it can be recommended.

Storage Bag

These systems are large bags (or "bladders") made with flexible impermeable materials such as PVC coated polyester fabric. Commercial products for temporary use have a capacity of 200 m^3. Empty bags can be rolled and transported by a light-duty vehicle. When in use, the bags are placed on a flat surface and gradually filled with liquid manure. Vent tubes release excess biogas from the bag. More permanent systems are available with capacity up to 5,000 m^3 with electric agitation systems and multiple access ports; these systems are typically installed within a shallow earthen basin. Expected lifespan from the manufacturers is approximately 10 years with proper use.

5.2.1.2   Impermeable Floating Covers

Plastic Film

Plastic films made of impermeable materials like high density polyethylene float on the manure surface. The material is anchored to the sides of the tank or basin as shown for basins in Fig. 5.2. Precipitation accumulates on the surface of the cover. Weights and floats added to the cover cause precipitation to collect at sumps for removal. In dry regions water can be left on the surface to evaporate. Excluding precipitation is an important feature in wet regions because it increases the system's manure holding capacity. The weights also prevent lifting caused by gases trapped under the cover. Vents along the perimeter of the basin release excess pressure and vented gas can be directed to the atmosphere (lowest cost), a biofilter, or flared to destroy methane and

Fig. 5.2 Plastic covers on earthen basins in Poland (Photo: From archives Poldanor S.A.)

odour. The system must be properly designed to allow changes in the liquid level without tearing the cover. A design for concrete tanks by Geomembrane Technologies, Inc (Canada) (described in Bluteau et al. 2009) is fastened to the tank walls and supported by straps to prevent tearing when the liquid level is low. That design includes precipitation exclusion and has an access port for agitation.

Some designs suited for concrete tanks are not fastened to the walls. Instead a film of EPDM or PVC is stretched onto a floating frame (e.g. HDPE pipes and polystyrene; Bucon Industries BV, The Netherlands). This design allows the cover to freely move vertically with changes in the liquid level. However, care must be taken to avoid damage from strong winds. An Italian design for digested manure shown in Fig. 5.3 maintains alignment by a ring around a central pole and a tether on the perimeter. It also has a sump for rainwater collection.

Negative Air Pressure

Negative air pressure covers are made of flexible high density polyethylene membranes and can be installed on tanks or small earthen basins. Edges of the cover are securely anchored around the basin and a duct system is installed below the cover and connected to a fan that removes gases from below the cover and exhausts them to the atmosphere or biofilter. The vacuum serves to keep the cover tight to the surface which prevents wind damage. Although electricity is used continuously, periodic loss of electricity is not likely to cause damage.

This approach is not suited for large basins. Biogas bubbles at the centre of the basin take too long to reach the vacuum at the edges. Therefore the cover lifts and is susceptible to wind damage. Wave damage can also be an issue on large basins.

Like plastic films, precipitation accumulates on the cover and must be removed via a collection and pump system. Accessing the manure for agitation and handling poses a challenge and in some cases compressed air has been used for agitation with

Fig. 5.3 A floating cover developed within the EU Agro Biogas project designed by Ecomembrane S.R.L. The photos show it (**a**) being installed, and (**b**) in use on biogas effluent tanks (Photo: B. Amon)

the cover in place (Barry 2006). Winter maintenance and ice damage have occurred in Canada and the effective lifespan was estimated at 7 years or less (Barry 2006). Improved designs may substantially increase the life expectancy.

5.2.1.3 Permeable Synthetic Covers

Plastic Fabrics

Many designs for permeable synthetic covers have been tested and developed. Generally these designs involve a fabric and a foam layer large enough to cover the entire liquid surface. The materials must be resistant to degradation from rot, UV, and chemical influences. To prevent wind damage the materials are secured at the edge of the storage. Difficulty during agitation and pumping has been reported in early research (Bicudo et al. 2004) but recent designs include a removable section so that agitation and manure pumping equipment can be inserted (Layfield Environmental, Alberta, Canada). Two main categories of permeable synthetic covers are (i) covers designed for constant liquid levels (lagoons) and (ii) those designed for variable liquid levels (most slurry storages).

Changing liquid levels can cause the material to tear where it is fastened at the edges. Therefore in the cover can be designed with extra material to avoid tearing when the tank is emptied, or the cover can be fastened to the tank via pulleys or other tension devices that allow movement but provide support. Another alternative is to have a free-floating cover with a slightly smaller diameter than the tank (a permeable version of Fig. 5.3). This design allows vertical movement but risks sinking under heavy snow and ice, or wind damage at high liquid levels.

The simplest design is a layer of geotextile fabric (e.g., woven polypropylene filaments) which is typically between 0.3 and 2.4 mm thick (Clanton et al. 2001). Special material is required since normal geotextile is designed to be buried and therefore has minimal UV protection. However, buoyancy has been inconsistent and geotextile was ineffective at reducing NH_3 emissions (Clanton et al. 2001).

To address the buoyancy issue, companies (e.g., Baumgartner Environics, MN, USA; Huesker, Inc., NC, USA; Layfield Environmental, Alberta, Canada) have added a foam layer below the geotextile (VanderZaag et al. 2010a) or sandwiched between fabric layers (Zahn et al. 2001). The foam is typically between 20 and 50 mm thick and either open-cell (i.e., permeable) or closed-cell (i.e., impermeable) with perforations or spacings to allow gas exchange. An example of this material on a lagoon is shown in Fig. 5.4. This composite cover material has been quite effective at reducing NH_3 emissions (Miner et al. 2003; VanderZaag et al. 2010a, b).

Fig. 5.4 Floating permeable cover installed on an earthen lagoon containing swine manure (Photo: Baumgartner Environics, Inc.)

Installation requires expertise and is labor intensive. Typically the cover is assembled in a flat area adjacent to the slurry storage. During construction, pieces of cover material are bonded together into a single unit. Once assembled, floats are attached to the leading edge and the cover is pulled across the liquid. Edges are then anchored to tank or buried in a trench around an earthen storage.

The expected lifespan on a lagoon without removing the cover is 10 years or more. On tanks the lifespan is uncertain because of potential damage caused by wind or operations (e.g., aggressive agitation, excess change in liquid levels).

Plastic Tiles

Floating tiles made of plastic or polystyrene foam have been studied (e.g., De Bode 1991; Williams 2003) but were unsuccessful because of wind disturbance. Foam pellets encountered the same wind problems.

Great advances have been made recently in developing a hexagonal plastic tile system that is not disturbed by wind yet is extremely buoyant (Fig. 5.5). The product is commercially available under the name Hexa-Cover (Hexa-Cover ApS, Denmark). Tiles are made of recycled polypropylene, weigh 280 g each, and have a footprint ~330 cm^2. Tiles are delivered in 2 m^3 bags which are emptied onto the surface of a slurry tank or lagoon (with or without liquid in it). Geometry causes the tiles to naturally disperse over the liquid surface and eventually form a single tightly-spaced layer. Thus, installation takes very little time and requires no

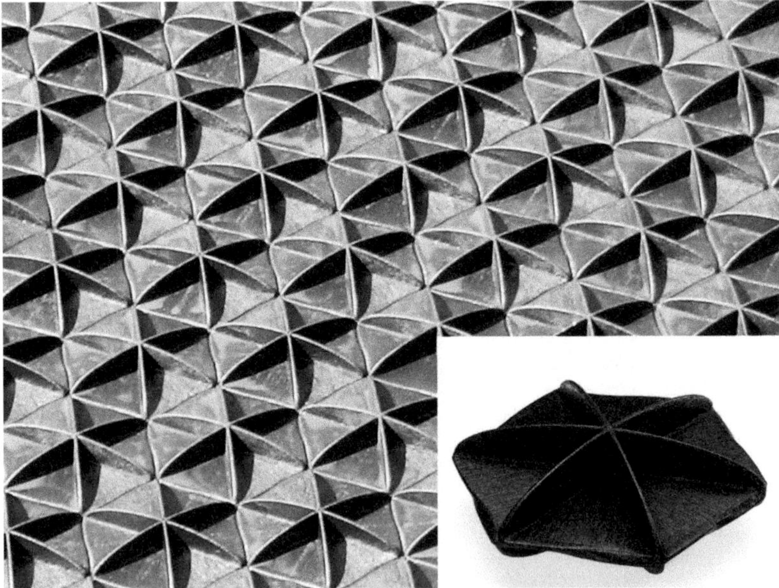

Fig. 5.5 Hexa-Cover plastic tiles floating on liquid manure with a single tile shown in the bottom right (Photo: Hexa-Cover ApS)

specialized expertise or equipment. Furthermore, the tiles are adaptable to changing management systems as they can be removed from one tank and placed on another, or more tiles can be added if a basin is expanded. Resale is also possible.

Hexagonal tiles are suitable for non-crusting slurries such as pig manure or anaerobic digestate and can be installed on tanks or lagoons of any dimension. In earthen storages with sloped berms tiles along the edge will simply lay on the ground when the liquid level declines and will float again when the liquid rises.

Small gaps between the tiles allow precipitation to pass through. Another important advantage compared to fabric covers is that the tiles are not damaged by storms, snow-loads, or layers of ice. Furthermore, the tiles require no maintenance and allow equipment access to any part of the tank at any time. Tiles remain at the surface during agitation and will not become entrained in slurry pumps so long as the tank is not emptied to the depth of the impeller.

Clay Balls

Clay balls, or Light Expanded Clay Aggregates (LECA) are small buoyant balls about 2 cm diameter that are commercially available under several trade-names (Leca®, Macrolite®; Fig. 5.6). Installation is simple—the balls are poured onto manure—and no modifications are required for the existing storage. The floating layer readily adapts to changing liquid levels and dimensions. As a general guideline, enough material must be added to create a 20 cm thick layer.

The expected lifespan of the balls is 10 years (Nicolai et al. 2002), however, in practice some balls are lost each year during slurry mixing and emptying, therefore the actual lifetime is reduced. Issues during agitation have also been reported where the balls clump together and plug pumping equipment (Funk et al. 2004b).

Fig. 5.6 Floating Leca® balls on a pair of lagoons

Other Materials

Other materials have been explored in research studies and may merit further research. Perlite is a heat-treated volcanic glass with low density. It is available under the name Pegülit® (ETH/OAM International Trading & Recycling GmbH, Germany). It is durable and will float after agitation (Hörnig et al. 1999).

Zeolite absorbs ammonia and is buoyant. It has been tested alone (Portejoie et al. 2003) and combined with other materials (Miner et al. 2003). A permeable synthetic cover with added zeolite is commercially available from Huesker, Inc.

Water and waste-water industries use hollow plastic balls (e.g., E-balls; Siemens Canada) but this has not been tested on liquid manure. Unlike clay balls, only a single layer is needed to cover the surface and the floatation properties are very good. Drawbacks are that a large portion of the liquid is exposed and rotation of the balls carries a film of liquid (and NH_4^+) to the surface.

Vegetable oil is buoyant and fits any store dimensions. However, it degrades and gives off foul odours (VanderZaag et al. 2008). Use of petroleum oil has potential negative impacts on land (Williams 2003). Recent improvements (i.e., RAPP, Juergens Environmental Control, USA) may make vegetable oil useful for manure under slatted floors (see the chapter on housing by Montalvo et al. 2015).

5.2.1.4 Permeable Natural Covers

Straw

Straw covers have been widely studied and are used on lagoons and earthen manure storages up to 0.8 ha to control odour emissions during the summer (PAMI 1996; Nicolai et al. 2006). A bale chopper is used to slice the straw and propel it onto the liquid. Evidently barley straw has the best durability compared to other cereals (Filson et al. 1996; PAMI 1996) and at least 15 cm thickness is needed for effective performance (Hörnig et al. 1999; Clanton et al. 2001).

Straw covers are short-lived, however, as they sink within 2–6 months and can be damaged by heavy rain. Strong wind is also detrimental as it pushes the material to the leeward side of the storage while exposing the upwind side (Sommer 1997). Thus it is difficult to ensure the entire surface is always covered. Since the material eventually sinks, agitation and pumping equipment must be modified to handle extra solids (e.g., a straw chopping blade; Zhang and Gaakeer 1998).

The short life of a straw cover is not overly problematic in regions with cold winters where the liquid surface is frozen for 4–5 months. Emissions of NH_3 from frozen liquid manure are relatively low anyway (VanderZaag et al. 2010b) so covers provide less benefit in winter. In cold climates it might be sufficient to apply straw once each year in spring, whereas warmer regions may require two or more applications to maintain year-round coverage.

Fig. 5.7 Dairy cattle slurry tank with a natural crust (Photo: T. Amon)

Crusts

Slurries with adequate dry matter and biogas production will form a natural crust (Fig. 5.7). It is comprised of fibrous feed material, manure solids, and bedding carried by gas bubbles to the surface where they coalesce. Crusts develop best where evaporation exceeds precipitation (Misselbrook et al. 2005) and generally form most readily on cattle slurry (Mannebeck 1985; Smith et al. 2004). A study in Ireland assumed that all cattle slurry formed a crust (Hyde et al. 2003); however, crusts will not form on farms that frequently agitate (Misselbrook et al. 2000).

Misselbrook et al. (2005) noted that crusts formed on slurries with $>1\,\%$ DM and crust formation began after 10–20 days, and stabilized after 40–60 days. Smith et al. (2004) observed that the nature of DM affects crust formation (e.g. more crusting with grass silage vs. maize-fed cattle). They also found "robust" crusts did not form without accumulating at least 25 cm of evaporative loss.

Although farmers cannot control crust development, for cattle manure stored in a tank it has been suggested that the loading method will influence crust formation (Rotz and Chianese 2009). Adding manure from the top of the tank causes disruption to the surface and decreases the crust. In contrast, adding manure via a pipe at the bottom of the tank leaves the crust undisturbed so it may establish a better seal and further reduce NH_3 losses. Emptying manure from beneath the crust might also help keep the crust in tact; however, we do not have data to confirm this.

Another role tank design can play in promoting crust development is decreasing the surface area to volume ratio. This design gives more manure to provide crust-forming materials for each unit of surface. In a series of tests, Smith et al. (2004) found that deeper tanks had the thickest and strongest crusts although it took longer for the crusts to form because of slow evaporation. In contrast, tanks with high surface area to volume ratios tended to form crusts quickly (perhaps due to more evaporation) but the crusts were not robust (Smith et al. 2004). It is unclear if the benefits of a robust crust outweigh the extra time it takes to develop.

Other Materials

Other farm materials have been investigated for their potential as floating covers. These include cornstalks, corn cobs, rice hulls, tree bark, wood chips, etc. (see Meyer and Converse 1982; VanderZaag et al. 2008; Yagüe et al. 2011). Advantages are low cost and the use of farm residues. Disadvantages are similar to straw.

5.2.2 Covers for Solid Manures

Solid manures are often stored in heaps which are prone to NH_3 emissions. One study reported that total N loss was 27 % during 7 days of storage (Schulze-Lammers et al. 1997). Although a convenient method for reducing these emissions has not been developed, two approaches have shown promise: covering and compaction.

Manure pile coverings can take the form of a roof with open walls or plastic sheeting (Chadwick 2005). Solid roofs exclude precipitation but allow gas exchange across the surface of the pile. This will produce a drier pile which may actually increase NH_3 loss because the surface remains porous and allows ammonia to diffuse out of the entire pile. In contrast, piles that are exposed to rainfall develop a surface crust that reduces NH_3 loss (ADAS 2004).

Therefore, effective covers must be tight to the pile surface. Plastic sheeting is sufficient and should be applied soon after the pile has been made. Chadwick (2005) combined compaction and a plastic cover on beef manure and reported NH_3 emissions were reduced more than 90 % during one experiment with high emissions (high N content of the manure). However, in another experiment the treatment did not have much effect on emissions since losses in the uncovered treatment were decreased by rainfall. A co-benefit of covering is a reduction in risk of N leaching and runoff in high rain areas/seasons and a reduction of odour and flies.

Controlling losses from solid manures is only justified if losses were limited prior to storage and if losses are also controlled during spreading. If most solid manure is spread on grass and left on the surface then any NH_3 reductions at the storage phase will be negated by higher losses in the field.

5.3 Manure Storage Design as an Abatement Strategy

Designing manure storages to minimize the surface-area to volume ratio will reduce NH_3 emissions by decreasing the emitting area and increasing the distance that NH_4+/NH_3 must diffuse before escaping to the atmosphere. A co-benefit of this approach is reduced cost to cover the storage and reduced rainwater accumulation in uncovered stores. Table 5.3 illustrates this point by showing four tank designs with equal volume but different dimensions. Tank 1 is shallow with a large radius and therefore has a large surface area. Consequently, Tank 1 has the highest potential for emitting NH_3 and the annual precipitation accumulation fills 20 % of the tank leaving less room for manure. In contrast, Tank 4 is twice as deep with a smaller radius and therefore has only half the surface area of Tank 1. As result, NH_3 emissions from Tank 4 are assumed to be reduced by 50 % and precipitation occupies a smaller fraction of the tank volume. If impermeable covers were installed on these tanks the precipitation benefit would not matter but the small-radius tanks would cost less to cover. There would also be less risk of wind damage but greater risk of damage due to movement with filling and emptying. With earthen storages, the same principles apply.

With increasing wall height the wall area increases, partially offsetting the reduced surface area. In the case of Tank 1 vs. Tank 4, the wall area increases by 41 % while the floor area decreases by 50 %. This will affect the cost of construction if walls are more expensive than the floor. As tank depth increases the manure head-pressure rises and will require thicker concrete walls, more excavation (to bury more of the tank below grade), or glass-lined steel walls. All of these factors raise the cost. For example, the low walls in Tank 1 could be built 20 cm thick but the walls in Tank 4 might require 40 cm thickness, therefore raising the required amount of concrete and steel reinforcement bars. Another factor is that shallow tanks are easy to agitate and empty with equipment placed over the wall; whereas, tall tanks are difficult and less safe to agitate from the top so they have valves in the side for agitation and pumping.

Design of earthen storages is also an option although depth may be limited by the soil profile and groundwater. Another possibility is to replace earthen storages with tanks that have a lower surface area to volume ratio. The net effect could be a 25–50 % reduction in emissions.

Table 5.3 Several tank designs with equal volume of 5,000 m^3 illustrating the relative surface area of each design and the impact on precipitation accumulation

Tank	Height (m)	Radius (m)	Wall area (m^2)	Surface area (m^2)	Relative surface Area (%)	Precipitation accumulation (m^3/y)	Volume as precipitation (%)
1	2.5	25.2	396	2,000	100 %	1,200	24 %
2	3.5	21.3	469	1,428	71 %	857	17 %
3	4.5	18.8	532	1,112	56 %	667	13 %
4	5.0	17.8	560	1,000	50 %	600	12 %

Calculations assume a cylindrical tank shape and net precipitation of 0.6 m per year

5.4   Low-Emission Manure Processing

Manure processing systems for either liquid or solid manure involve additional investment of time and money. Generally these systems are best suited to large farms that achieve an economy of scale and to help large farms deal with accumulations of phosphorus near the feeding facility. Other factors may also cause a farm to consider advanced processing, such as odour complaints from neighbours or market demand for processed manure.

5.4.1   Liquid

5.4.1.1   Solid-Liquid Separation

The solid and liquid manure fractions can be separated using many physical and/or chemical methods (e.g., decantation, filtration, screw press, decanting centrifuge flocculation) to achieve agronomic objectives. For instance, the liquid fraction tends to have a higher N: P ratio which is better suited to crop requirements. The solid fraction contains a higher concentration of P and organic matter which can be economically transported greater distances where there is a need for manure organic matter and P (Hjorth 2009). Despite the benefits, there has not been evidence that separation reduces overall NH_3 losses from farms. Rather, Amon et al. (2006) found total storage emissions of NH_3 were higher from the separated fractions (especially the solid fraction) compared to unaltered dairy slurry. This may be partially offset by lower emissions from the land-application stage for the liquid fraction (Balsari et al. 2008).

If storage time after separation is minimized (i.e., separation immediately before spreading) then the agronomic benefits of separation can be realized without impacting storage emissions; however, separation does not reduce emissions from the storage so it is not considered a mitigation strategy for storage emissions. In fact the manure liquid fraction tends to have a slightly higher pH and is unlikely to form a natural crust, so it would tend to increase NH_3 loss.

5.4.1.2   Separating Faeces and Urine

Much of the N in cattle or swine urine is present as urea which is rapidly dissociated to NH_4^+ by urease. However, as there is no urease enzyme in the urine itself, the combination of urine and faeces (which contains urease) on the housing floor enables rapid dissociation of the urea. By separating the faeces from the urine shortly after excretion the urea N can be conserved and NH_3 emissions avoided.

Unfortunately, it is extremely difficult to separate urine and faeces on a farm. Several separation techniques have been tested in lab or pilot-scale studies. Generally the techniques involve one or more aspects of conveyor belts, gutters, grooved floors, or other barn modifications which bring the urine to one storage tank and faeces to another (Ndegwa et al. 2008). Despite the potential, it does not appear that this technology is presently mature enough for widespread implementation (Vaddella et al. 2010) and there is added cost for building two storage tanks.

5.4.1.3 Anaerobic Digestion

Anaerobic digestion is a microbial conversion of organic matter into methane and carbon dioxide under anaerobic conditions. It is an appealing technology because the methane that is produced can be used to power a generator that yields electricity and heat. In addition, the digestion process reduces odours, pathogens, and can also recover bedding materials while retaining the nutrient value of the manure in the digested effluent. A common digester style is a continuously mixed circular concrete tank with a gas-tight cover and a heat-exchange system to maintain the desired temperature. Digesters are suited for manures of all animal types and for both slurry and farmyard manures (Massé et al. 1996; Bujoczek et al. 2000; Mata-Alvarez et al. 2000; Wilkie 2000).

There should be minimal NH_3 emissions directly from the digestion unit; however, the liquid effluent has to be stored in a tank before transporting to the field. The effluent is high in ammonia/ammonium and since it has a higher pH there is the potential for higher emissions of NH_3 compared to undigested manure and significantly higher emissions have been reported (Clemens et al. 2006; Balsari et al. 2011). Therefore, the effluent storage tanks must be covered to prevent NH_3 emissions (Balsari et al. 2011) and as such, anaerobic digestion is not a mitigation strategy for NH_3 emissions on its own. However, a digestate cover that collects biogas (such as the one developed by the EU Agro-biogas project shown in Fig. 5.3) provides additional energy production that may reduce the payback period for the cover to less than 1 year (Balsari et al. 2011).

5.4.1.4 Acidification

Lowering the pH of stored manure will reduce NH_3 emissions because of the shift in equilibrium from NH_3 towards NH_4^+. This strategy has been tried for decades (e.g., Stevens et al. 1992) but was not widely implemented due to unresolved issues of safety (exposure to strong acids) and inconvenience (manure foaming).

Recent developments have produced a new acidification technique that overcomes the previous difficulties (Kai et al. 2008). This commercially available technology (Infarm A/S, Denmark) involves pumping a portion of the slurry to a treatment tank where concentrated sulphuric acid is injected and aeration occurs. The acidified slurry (pH 6.3) is then sent to the manure storage tank. This system should be suitable for cattle or pig farms and should be compatible with any type of new or existing storage. Generally, for both pig and cattle slurry the amount of acid required to achieve a desired pH depends on the NH_4^+-N concentration in the slurry (Stevens et al. 1989).

Agronomic benefits include less field losses and more available N from the acidified manure (Kai et al. 2008); but, adding sulphur to manure could lead to a trade-off of increased production of odourous compounds, especially H_2S which is also toxic and a potential safety risk to operators.

5.4.2 Solid

5.4.2.1 Compaction

Compaction of the manure pile can be accomplished by driving a tractor over the pile in multiple passes such that the wheel tracks cover the entire surface. This process would have to be repeated after each new layer of fresh manure is added. The process increases the density of the pile and reduces pore space which slows gas exchange and NH_3 loss (Dewes 1996). In spite of the potential emission reduction, the quality of compacted manure is unproven. Before this can be a recommended practice, agronomic questions must be answered, such as: how does compaction impact the nutrient value of packed manure, can it be evenly applied to crops, and will it be incorporated so conserved NH_3 will not be lost in the field?

5.4.2.2 Composting

Composting is a treatment for solid manures by which organic material is biologically converted into stable humus under aerobic conditions. It is a beneficial soil amendment that improves physical properties of soil and the high temperatures in the compost pile destroy pathogens and weed seeds (not always consistently).

Composting manure requires balancing the C/N ratio and aerating the pile either by turning or by aeration (passive or forced). Aeration tends to increase NH_3 losses compared to non-composted static piles and active aeration has greater losses than passive aeration (Amon et al. 1998; Hao and Chang 2001). Co-benefits of composting include reduced N_2O and CH_4 emissions, no NH_3 emissions after composting (Amon et al. 1998) and considerably lower hauling costs.

To counteract the increased NH_3 emissions plastic covers can be used. However, it has been shown that a porous tarpaulin allowed significant NH_3 loss (Sommer 2001). Impermeable materials that prevent gas exchange are preferable and can reduce emissions by about 90 % compared to an uncovered pile (Karlsson 1996a, b).

Cover systems for composting are commercially available. Some are designed to fit over concrete compost bunkers. One such device combines a large tarpaulin with a spool that allows the cover to be deployed for coverage or retracted for adding or turning the compost (Curry Industries, Canada). A design for outdoor piles uses an impermeable UV-resistant fabric to covers the compost pile which contains systems for forced aeration and collecting exhaust gases (ECS Inc, USA).

Another alternative is to completely enclose composting facilities ("in vessel") and pass the exhaust air through a biofilter or ammonia scrubber (GICOM B.V., Netherlands). This approach is more common in municipal composting for odour control but may be too costly for farm use.

5.5 Effects on NH3 Emissions and Co-benefits

5.5.1 Baseline Emissions

To estimate emission reductions with an abatement technique, in absolute terms, we must know the baseline emission rate. For liquid manure, baseline emissions are taken for a manure storage type without any cover or surface crust. Baseline emission values derived from all available studies for various cattle, pig, and poultry manure stores have been compiled in Table 5.4. To enable cost comparisons, data were converted from NH_3 to NH_3-N using the coefficient 14/17 and adjusted to yearly values. These estimates assume that the stores always contain some manure and therefore emit ammonia every day of the year (Misselbrook et al. 2000; Misselbrook et al. 2007).

The baseline data will not represent local variability since emissions depend on several factors including regional climate, animal diets, etc. For example, based on data reported by McGinn et al. (2008) and Flesch et al. (2009) we estimate a summer emission rate in western Canada of 2.0 kg NH_3-N m^{-2} $year^{-1}$ for uncrusted cattle slurry, which is higher than the annual value we used (1.4; Table 5.4). For German conditions, Döhler et al. (2011) used reference emissions

Table 5.4 Mean baseline flux data summarized from several studies

Manure source	Storage type	Mean flux (kg NH_3-N m^{-2} y^{-1})	References
Cattle FYM		0.9	Misselbrook et al. (2000)
Cattle Slurry (crusted)	Tank or Lagoon	0.8	Misselbrook et al. (2000)
			Misselbrook et al. (2007)
			Hyde et al. (2003)
			Döhler et al. (2011)
Cattle Slurry (uncrusted)	Tank or Lagoon	1.4	Misselbrook et al. (2000)
			Misselbrook et al. (2007)
Pig Slurry	Tank	2.9	Misselbrook et al. (2000)
			Misselbrook et al. (2007)
			Hyde et al. (2003)
Pig Slurry	Lagoon	0.9	*Ibid.*
Pig FYM	–	1.8	Misselbrook et al. (2000)
Poultry (solid)	–	2.6	Misselbrook et al. (2000)
			Hyde et al. (2003)

of 4.8 kg NH_3-N m^{-2} $year^{-1}$ for pig slurry, which is much higher than the UK values. Reference emissions from warmer countries would tend to be even higher.

There is limited published information on baseline emissions. This adds uncertainty to cost-efficacy estimates since baseline emissions are used directly to calculate the quantity of N, in kg, conserved by a mitigation strategy. Any technology that reduces emissions will be more cost-effective for a high than a low emission source. It is therefore important to obtain regional baseline emission factors to ensure accurate cost estimates. Baseline values also need to be consistent with national emission inventories. Based on Table 5.4 the most cost effective techniques will target pig and poultry manure rather than cattle manures.

5.5.2   Emission Reductions

Abatement strategies for NH_3 emissions from stores were assigned to three categories based on practicality and data to demonstrate their emission reductions. Cost was not a factor.

- **Category 1**: practical for use and research has quantitatively demonstrated their efficacy at least at an experimental scale.
- **Category 2**: promising but currently lack adequate research on their abatement efficacy or it will always be difficult to quantify their abatement efficacy. These can be used as part of an abatement strategy if the local situation permits.
- **Category 3**: ineffective for NH_3 mitigation or impractical. The following strategies are in this category: positive air pressure covers, vegetable oil, waste petroleum oil, solid-liquid separation, separating faeces from urine, anaerobic digestion, and compaction.

Experimental data on emission reductions relative to uncovered baseline for various cover types are presented in Table 5.5. Maximum and minimum values were selected based on the highest and lowest emission reduction value across several studies. A conservative estimate value was then assigned that was deemed to be achievable in practice over the life of the cover. Some of these values are consistent with the draft guidance document for preventing and abating ammonia emissions from agricultural sources (Table 17 of WGSR-48); however, there are several differences: (i) Storage bag has been assigned to Category 2 and given a lower NH_3 emission reduction (80 % vs. 100 %) due to a lack of published supporting data, (ii) floating covers have been disaggregated into several types and assigned a higher emission reduction (80 % vs 60 %) though they are shown here as Category 2 because they are not suitable for all manure storage types, (iii) Hexacover has been assigned to Category 1 however it has been given an emission reduction of 60 % (lower than the minimum reported) until additional research can confirm its efficacy is indeed ca. 90 %; (iv) low technology floating covers have been assigned higher emission reductions than in the guidance document based on

Table 5.5 Effect of cover type on NH_3 emissions reduction relative to uncovered baseline

Type	NH$_3$ emission reduction (%) Est.	Min	Max.	Cat	References and comments
Tight lid					***May not be suitable on existing stores.***
Tent	80	70	99	1	Menzi et al. (1997) cited in Reidy and Menzi (2007); Hornig et al. (1998) (tent covered lagoon); De Bode 1991: (Pilot slurry tanks)
Lid (wood)	80	80	95	1	Nicolai et al. (2002)
Lid (concrete)	80	80	95	1	New Stores only. Menzi et al. in Reidy and Menzi, 2007
Storage bag	80	?	?	2	Reduction potential assumed to equal that of impermeable plastic covers. Data are scarce.
Plastic cover solid manure	50	0	90	2	Depends on how quickly the pile is covered and variable emissions from the reference pile. Chadwick (2005)
Floating covers					***Management and other factors may limit use***
Impermeable plastic	80	59	95	2	Durability in winter is a concern. Zhang and Gaakeer (1998); Nicolai et al. (2002); Portejoie et al., 2003; Scotford and Williams (2001)
Negative air pressure	80	0	95	2	English and Fleming (2006); Funk et al. 2004b
Permeable synthetic	60	45	95	2	Devries et al. (1980); Karlsson (1996a, b); Hörnig et al. (1999); Nicolai et al. (2002); Miner et al. (2003); VanderZaag et al. (2008)
Hexacover	60	90	98	1	DLG (2005) on liquid pig manure covered 98 % of surface. Need field tests. Wind effect uncertain.
Low tech floating cover					***Unsuitable if materials cause management problems***
Clay balls	65	65	95	1	Nicolai et al. (2002); Sommer et al. (1993); Bundy et al. (1997); Williams (2003)
Straw	50	25	85	2	Depends on thickness. Uncertain durability and coverage in field. Nicolai et al. (2002); Berg et al. (2006); VanderZaag et al. (2008; 2009)
Natural cover					***Not suitable where slurry is spread frequently***
Crust	40	20	90	2	Variable thickness and duration. Nicolai et al. (2002); VanderZaag et al. (2008); Misselbrook et al. (2007); Bicudo et al. (2004); De Bode (1991); Williams (2003); Misselbrook et al. (2005)
Tank design					***New construction only. Subject to local regulation.***
Low SA/V of store	40	0	50	2	No measurements.
Additives					
Acidifying	80	70	95	1	Aneja et al. (2001); LefCourt and Meisinger (2001); Kai et al. (2008)

available data, (v) low surface-area manure tanks have been deemed Category 2 because there are no measurements to verify its emission reduction.

Emission reductions should consider the full life of the technology in a field setting. For example, a straw cover will reduce emissions better in a research setting with minimal wind and no precipitation compared to a farm setting where high winds and heavy rain disrupts the material. Similarly, with most cover types larger emissions will occur during manure handling when covers are "pulled back" or access ports are opened. Effectiveness also depends on how well the technology is designed, implemented and maintained. These factors are not typically evaluated in laboratory studies. Therefore, there is potentially a large difference between experimental results and on-farm results.

5.6 Cost of Implementation

5.6.1 Methodology

Implementation cost was calculated for all category 1 and 2 techniques. The cost analysis included two components: capital and operating costs. Cost data were obtained from multiple sources including published literature, companies, and personal communication regarding actual installations. Prices were converted to Euros using current exchange rates (1 CAD = 0.72 €; 1 USD = 0.68 €; 1 £ = 1.11€). Data from the literature were adjusted to constant 2011 Euros by assuming 2 % annual inflation and using the publication date as the base year. For example, 1.00 € in 2000 would be 1.24 € in 2011. We did not attempt to quantify opportunity costs—in other words, how this investment influenced other farm investments—but we suggest this and assessing other indirect costs and benefits for future work.

Capital costs were limited to the additional costs for the abatement technique; the cost of the manure storage itself was not included. For capital costs we strove to obtain estimates that include the cost of both materials and installation, which is important because of the wide range of installation costs for various techniques and regions. Capital costs were amortized over the lifespan of the project with a constant interest rate of 7 % and an annual payment schedule using the equation $C_{an} = C \times [(1+i)^L \times i]/[(1+i)^L - 1]$, where C_{an} is the annualized cost of the project (€), C is the initial capital cost (€), i is the constant interest rate (decimal), and L is the lifetime of the project (year). This was implemented using the PMT function in Excel®. We did not account for taxes, support-payments, or lending fees.

Operating costs include consumables, maintenance, insurance, and the additional effort required to manage manure. Operating costs can be highly variable for certain technologies. In particular, the effects of manure agitation and extreme weather on maintenance and longevity of the technology are difficult to predict. Since on-farm maintenance data were scarce we calculated the maintenance as a fixed percentage of the capital costs per year (Webb et al. 2006). However, we adjusted the rate between zero and 2 % depending on the technology. The zero rate was given to techniques that require no maintenance, no insurance, and allow full access to the manure. Two percent was applied to techniques requiring maintenance, insurance, pumps, flares, and restricting access to the manure.

5.6.2 Results for Manure Storage Covers

Results of the cost analysis of liquid manure storage covers are summarized in Table 5.6, below. It is evident that there is a large range of costs among the cover types. Besides natural crust, which has no cost, straw covers and other floating permeable covers were the least expensive options. Hexacover was also among the least costly options. On the other hand, covers made of concrete, wood, and tents were among the most expensive. Storage bags were the most expensive; however, this technology incorporates the cost of both a cover and storage.

A large variation of costs also exists within cover types. Most cover types have a range of costs that is larger than the mean. This is probably indicative of several factors, including: incomplete accounting for installation costs, variation in regional labor cost, shipping, commodity price fluctuations during the past decade, and economies of scale. Another source of variability is the non-standardized designs of most covers. For example, the thickness of straw or Leca used is varied and there are many design variations for permeable covers and tents.

In spite of the variability within and among covers, the progression of costs follows a logical trend. The least expensive options are natural, then floating

Table 5.6 Annualized costs for various liquid manure covers from multiple data sources

| | Annualized Cost | | | |
| | By area ($€\ m^{-2}\ y^{-1}$) | By volume ($€\ m^{-3}\ y^{-1}$) | Range by volume ($€\ m^{-3}\ y^{-1}$) | |
Cover type	Mean	Mean	Min	Max
Natural Crust	0	0	–	–
Straw[a]	1.64	0.47	0.22	0.72
Floating Permeable	1.84	0.53	0.13	0.89
Hexacover	3.00	0.86	0.74	0.98
Clay balls	3.89	1.11	0.45	2.34
Floating Impermeable	5.38	1.54	0.14	2.92
Wood	6.31	1.80	1.52	2.09
Tent	9.30	2.66	1.32	4.64
Concrete	10.85	3.10	1.92	4.27
Storage Bag	22.64	6.47	4.31	8.63

The reported costs include both capital and maintenance costs following the methodology described in the text. For all materials a storage depth of 3.5 m was used to convert between unit surface area and cubic meter of manure storage capacity
[a]Assuming that two applications of straw were required each year

permeable materials, then impermeable floating materials, and finally structural covers. Thus, we may be confident in ranking the techniques.

Cover costs are affected by the interest rate used for amortization and long-lived materials are the most sensitive. To illustrate this we recalculated the costs with a 5 % interest rate (instead of 7 %) and the cost of concrete covers (30 year lifetime) decreased by 19 % while the cost of Hexacovers (25 year lifetime) fell 17 %. Costs of most covers fell between 5 and 9 % and the cost of straw was unaffected. Therefore in a low interest rate economy the long-lived covers become proportionally more economical. Importantly, rankings based on mean cost did not change with a drop in interest rate from 7 to 5 % from those shown in Table 5.6.

Details of the analysis for each technology are described below:

Concrete Lids Cost data were obtained from Canadian and European sources (English and Fleming 2006; Dux et al. 2005; Raaflaub and Menzi 2011). All sources estimated an economic life of 30 years and we used a low maintenance rate of 0.25 % of the initial cost. There was a large range of estimates for the initial cost, ranging from 73 to 180 € m^{-2}.

Wooden Lids Cost data were obtained from Canadian and European sources (English and Fleming 2006; Raaflaub and Menzi 2011). Both sources estimated a 10 year lifespan and the initial cost estimates ranged from 33 to 48 € m^{-2}.

Tent Covers Cost estimates were obtained from several sources in North America, EU, and the UK (Dux et al. 2005; English and Fleming 2006; Webb et al. 2006; Raaflaub and Menzi 2011; personal communication T. Kuczynski). There was a large range of economic lifespan estimates, between 10 and 20 years. Capital cost and installation estimates ranged from 37–100 € m^{-2}. Maintenance was estimated at 2 % of capital cost per year.

Storage Bags Cost data for storage bags were obtained from North American manufacturers. A range of expected lifespans between 5 and 20 years were reported. We used 10 years as the economic life in the calculations and maintenance charges of 1 % for bags without agitation and 2 % with agitation. Variability in the cost data arise from options such as portability and internal agitation systems. Note that unlike other technologies, cost of storage bags includes the cost of the store. Assuming a concrete manure tank has a 30 year lifespan and costs about 20 € m^{-3} for a 500 m^3 tank, the annual charge is about 2 € m^{-3} year^{-1}. Deducting this "storage" value from the cost of the storage bag (Table 5.7) leaves an annual cost of about 6.50 € m^{-3} year^{-1} for the "cover" value. Thus, storage bags remain the most expensive cover option. This ignores other benefits to storage bags such as flexibility and portability.

Floating Impermeable Covers Cost data were obtained from North American and European sources (Powers 2004; Dux et al. 2005; English and Fleming 2006; Raaflaub and Menzi 2011; personal communication T. Kuczynski, J. Baumgartner, and Geomembrane Technologies Inc.). A range of materials and designs were included in this category including HDPE, HDPE/foam composites, and negative

Table 5.7 Abatement cost for category 1 and 2 techniques

Abatement measure	Baseline emissions kg NH$_3$-N m^{-2} y^{-1}	Emission reduction (%)	Cost ($€$ m^{-3} y^{-1})	Cost of abatement $€$ (kg NH$_3$-N)$^{-1}$
Tent	Cattle: 1.4	80	2.66	8.31
	Pig tank: 2.9			4.01
Wood lid	Cattle: 1.4	80	1.80	5.63
	Pig tank: 2.9			2.72
Concrete lid	Cattle: 1.4	80	3.10	9.69
	Pig tank: 2.9			4.68
Storage bag	Pig tank: 2.9	80	6.47	9.79
Floating impermeable, negative air pressure	Cattle: 1.4	80	1.54	4.81
	Pig tank: 2.9			2.32
	Pig lagoon: 0.9			7.49
Floating permeable synthetic covers	Cattle: 1.4	60	0.53	2.21
	Pig tank: 2.9			1.07
	Pig lagoon: 0.9			3.44
Hexacover	Cattle: 1.4	60	0.86	3.58
	Pig tank: 2.9			1.73
	Pig lagoon: 0.9			5.57
Clay balls	Cattle: 1.4	65	1.11	4.27
	Pig tank: 2.9			2.06
	Pig lagoon: 0.9			6.64
Straw	Cattle: 1.4	50	0.47	2.35
	Pig tank: 2.9			1.13
	Pig lagoon: 0.9			3.66
Natural crust	Cattle: 1.4	40	0.00	0.00
Covered farmyard manure piles	Cattle FYM: 0.9	50	1.10	4.89
	Pig FYM: 1.8			2.44
	Poultry: 2.6			1.69
Covered composting	Cattle FYM: 0.9	50	2.50	11.11
	Pig FYM: 1.8		5.00	11.11
	Poultry: 2.6		5.00	7.69
Low Surface Area to Volume Design	Cattle: 1.4	40	0.13	0.81
	Pig tank: 2.9		0.13	0.39
	Pig lagoon: 0.9		1.50	14.58
Acidification	Pig tank: 2.9	80	10.82	16.32
	Pig lagoon: 0.9			52.60

Data were converted between area and volume assuming a depth of 3.5 m for tanks, basins, and lagoons, and FYM is stacked 2 m

air pressure covers fitted with gas flares and rainwater collection pumps. The lifespan estimated by all sources was consistently between 8 and 10 years. The range of initial cost estimates ranged from 1.70 $€$ m^{-2} for a simple cover without rainwater collection to 63 $€$ m^{-2} for a more advanced design. Maintenance was estimated at 2 % per year since rainwater collection and other pumps are required.

Floating Permeable Covers on Lagoons Cost data were obtained from North American and UK sources (Powers 2004; English and Fleming 2006; Webb et al. 2006; personal communication J. Baumgartner). Materials ranged from simple geotextile to composites of geotextile with permeable foam. The estimated lifespan was 5 years for geotextile and 10 years for composites. In all cases these were installed on lagoons with relatively constant liquid levels. Webb et al. (2006) calculated the annualized cost for floating covers was nearly 50 % higher on slurry lagoons compared to tanks. Capital cost estimates ranged from less than 2 € m^{-2} for a simple geotextile to about 20 € m^{-2} for a composite. Maintenance was estimated at 1 % per year.

Hexacover Cost data were obtained from the manufacturer in Denmark (personal communication S. Madsen of Hexa-Cover ApS). We used an economic life of 25 years based on the manufacturer's data. Because installation can be done easily by the farmer, we used zero installation costs and zero maintenance charges. The cost of materials ranged from 30 to 40 € m^{-2} and we used 35 € m^{-2} in calculations. We did not account for shipping costs.

Clay Balls Cost estimates were obtained for Leca® and Macrolite® clay balls from several sources in North America and the EU (Powers 2004; Dux et al. 2005; English and Fleming 2006; Raaflaub and Menzi 2011; personal communication T. Kuczynski). Several factors contributed to the large range of annualized cost values: (i) wide range of economic lifespan estimates from 5 to 10 years; (ii) capital costs ranged from about 14–27 € m^{-2}, which is partly due to variation in application rate (between 7 and 20 cm thick) and partly due to shipping costs. Maintenance was estimated at 1 % per year for the 5 year lifespan and 2 % per year for the 10 year lifespan to account for losses and solids management.

Straw Covers Cost estimates from 0.50 to 1.50 € m^{-2} were obtained from North American sources (Powers 2004; English and Fleming 2006). Since straw is short-lived we assumed two applications per year were necessary to provide coverage comparable to other techniques. Therefore, the cost was annualized by doubling the cost for a single application. We added a 2 % maintenance charge for handling additional solids during agitation and land spreading.

Natural Crust Natural crusts were considered zero cost.

5.6.3 Results for Other Techniques

Low SA/V Tank Design The cost of building a new tank with lower surface area to volume ratio is difficult to generalize. Based on Canadian data we assume a concrete tank costs three to four times more than an earthen basin of the same volume (ignoring the cost of land used). Further assuming both systems have a 30 year lifetime and equal maintenance costs, the added cost for building a concrete tank instead of an earthen basin would be ca. 1.50 € m^{-3} year^{-1}. We cannot give a simple estimate for the cost of changing from building a low to a high wall concrete tank, but we will make a sample calculation for two tanks with 5,000 m^3 capacity to provide a comparison. In this example the high surface area tank has 25 m radius, 2.5 m wall height, and 20 cm wall thickness. The low surface area tank has 18 m radius, 5 m wall height, and 40 cm wall thickness. In this case the taller, low surface-area tank costs an additional 0.13 € m^{-3} $year^{-1}$ using a 30 year economic life, 1 % maintenance, and local building costs.

Acidification Cost data for acidification for a pig farm were taken from Kai et al. (2008), who provide an annual cost of 60 € per 500 kg livestock unit. Assuming a 500 kg livestock unit produces 6 m^3 of manure each year (average of boars and sows, ASABE 2010). Acidification is expensive for reducing manure storage emissions; however, it also reduces emissions from the animal house and land application.

Covered FYM and Covered Compost The cost of covering solid manure and compost piles with plastic sheeting was assumed to be similar. For a cost estimate we used the value given by Webb et al. (2006) where the total annual cost for covering farmyard manure stores was calculated as 1.10 € m^{-3} $year^{-1}$.

Determining the cost of composting is complex because there are many approaches depending on land cost, co-substrates, labour, and machinery. Based on information summarized by Fabian et al. (1993), costs for a loader-turned windrow system were on the order of 3–5 USD per ton which is about 1–1.5 € m^{-3} in 2011 for solid dairy manure (assuming a bulk density of 267 kg m^{-3}; Landry et al. 2002) and 3–5 € m^{-3} for solid poultry or pig manure (assuming a bulk density of 1,000 kg m^{-3}; Landry et al. 2002). Costs for windrow composting with specialized turning machines are generally lower than turning with a tractor loader; conversely, in-vessel composting costs were an order of magnitude greater. Clearly there is a wide range of costs, but in all cases the cost of composting at least doubles the cost of covering the manure. The cost of composting plus covering dairy manure is around 2.5 € m^{-3} $year^{-1}$ and for pig and poultry around 5 € m^{-3} $year^{-1}$.

5.7 Cost-Effectiveness of Abatement Strategies

The cost of abating NH_3 emissions was calculated for each technique in categories 1 and 2 using the results of Sects. 5.5 and 5.6. Results show a wide range in the abatement costs per unit of NH_3 emission reduction (Table 5.7, Fig. 5.8).

As a general ranking, the most economical techniques for liquid manures were: various floating permeable covers followed by floating impermeable material, then structural covers, tank design, and acidification. Covering solid manures was quite economical for pig, cattle, and poultry. Not surprisingly, the most economical technique was to allow cattle manure to form a crust. Next lowest in cost were several technologies for pig manure that cost between 1 and 2 €/kg of abated NH_3-N loss: covering pig FYM, which has high baseline emissions, and three types of

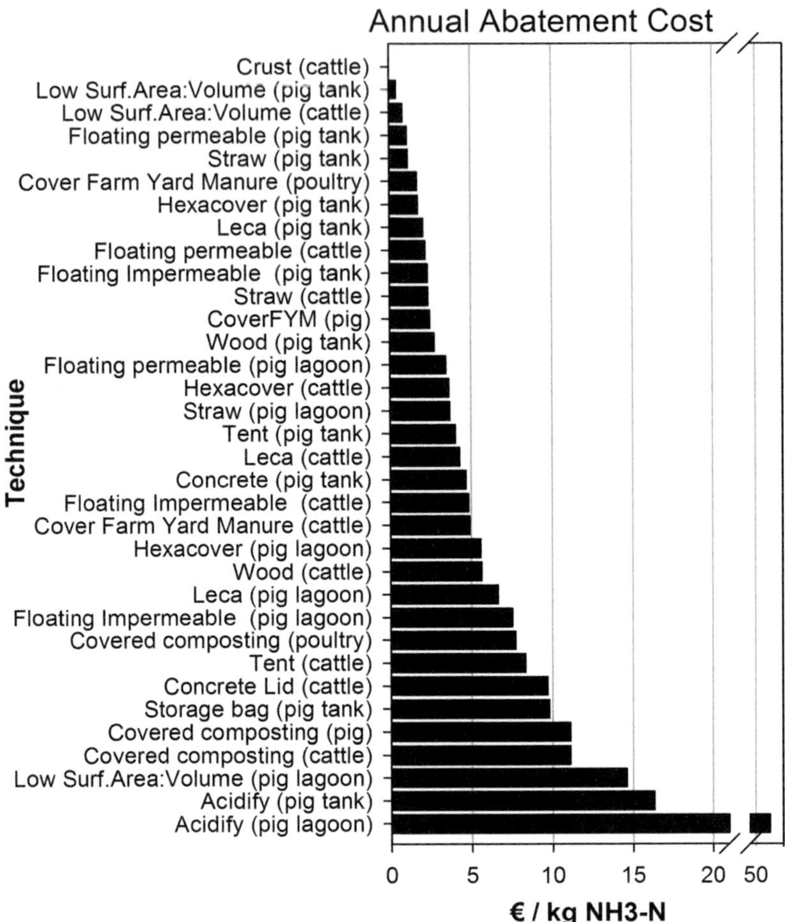

Fig. 5.8 Abatement cost for category 1 and 2 techniques in ascending order

floating covers on pig tanks (straw, Hexacover, and permeable synthetic). The most expensive techniques were above 10 €/kg of abated NH_3-N loss and included acidification and reduced surface area designs for pig manure lagoons.

At 5 €/kg of abated NH_3-N loss there are several options available for cattle slurry, pig tanks, and pig lagoons. This threshold would also include covering for solid manure from pigs, cattle, and poultry.

Abatement costs are very sensitive to baseline emissions. For this reason it is important that baseline emissions for making calculations are the same values used in National Emission Inventories. Covering a tank or FYM has the same cost whether it contains pig or cattle manure and therefore it will be more cost effective to cover high emitting manures. The fact that four of the five most economical strategies were for pig manure tanks illustrates this. Despite its importance, there are limited published baseline emission data from various storage systems. Clearly more baseline data will greatly reduce uncertainty of abatement cost estimates.

5.7.1 Value of Co-benefits

It is important to recognize that improved manure storage systems achieve co-benefits in addition to ammonia abatement. Co-benefits include precipitation exclusion, reduced odours, retained nitrogen, and reduced methane emissions. Incorporating these aspects into the cost-benefit analysis is challenging. Some elements like precipitation exclusion, retained N, and reduced odours are more beneficial in some regions than others (e.g., wet vs. dry regions; N deficit vs. surplus; remote vs. highly populated). Furthermore, as with many environmental externalities, odour and methane emissions are challenging to assign a price. Despite these challenges, we attempt to incorporate these co-benefits for cover techniques on liquid manure systems to illustrate how important they are in the cost-benefit analysis and to emphasize that additional work is needed to improve these estimates. Methods used to estimate these values are described in Sect. 5.7.2.

Precipitation exclusion is a tangible benefit that can be immediately obtained by farmers in wet regions in the form of reduced store capacity and lower manure transport costs. According to our simple calculations the value of this co-benefit was about $0.40 €$ m^{-3} $year^{-1}$ for all impermeable covers. An added benefit of excluding rain water is improving the certainty of storage requirements. Uncertainty means that uncovered stores need additional capacity for wet years and also present the risk of needing to empty prematurely in years of extremely high rainfall. These co-benefits are difficult to quantify and depend on the risks associated with untimely manure application. Deducting this cost savings from the values shown in Table 5.6 makes all impermeable covers less expensive and reduces the average cost of floating impermeable covers close to the cost of expanded clay balls.

Nitrogen retention is another benefit that can be immediately obtained with any technique that abates NH_3 loss. Using our simple calculations the annual value of this benefit ranges from 0.03 to $0.18 €$ m^{-3} $year^{-1}$ with the greatest benefits accruing to covers with large percent emission reduction installed on pig manure tanks.

Reduced emissions of greenhouse gases and odour are less tangible in that the benefits do not currently accrue to the farmer. At the time of writing, the carbon trading market is unstable and carbon offset protocols are not available in all jurisdictions. Moreover, the quantity of offsets may be too small to market by individual farmers so an aggregator would often be needed. Nevertheless, our simple estimate gives a carbon credit value ranging from $0.29 €$ m^{-3} $year^{-1}$ for covers with low methane abatement potential to $2.33 €$ m^{-3} $year^{-1}$ for covers with gas collection and flaring. However, most covers do not abate methane.

The value of odour abatement is very site-specific and is influenced by proximity of neighbours, bylaws, risk of litigation, etc. Therefore the value of abatement is highly uncertain but significant, ranging from 1.50 to $2.41 €$ m^{-3} $year^{-1}$. The largest benefits accrued to impermeable covers.

Combining multiple co-benefits has a substantial effect on the cost/benefit analysis of manure storage covers. For example, on pig manure storages, combining the tangible benefits of precipitation exclusion and N retention lowered the mean cost for all covers (excluding storage bags) to $2.50 \, €\, m^{-3} \, year^{-1}$ or less. Combining all co-benefits narrowed the spread of costs among cover types and reduced the mean cost for each cover-type below zero (excluding storage bags). After accounting for co-benefits in our scenario, most covers returned a small profit rather than incurring an expense.

This analysis is intended to illustrate the potential value of co-benefits including some that are difficult to quantify. These benefits are location specific and may not accrue directly to the farmers who bear the costs. Therefore the numbers presented here should be used with caution.

5.7.2 Methods Used for Co-benefit Analysis

Precipitation Exclusion By diverting precipitation, impermeable covers reduce manure dilution. In turn, this decreases the cost of transporting the manure to the fields. Based on the work of Ghafoori et al. (2007), we used a value of $4 \, €\, m^{-3}$ for the cost of transporting manure by truck over short distances. In the analysis we then assumed a wet region has 0.5 m of net precipitation (i.e., precipitation − evaporation) that would not need to be transported if an impermeable cover was installed. Using these figures, each square meter of tank surface would avoid $0.5 \, m^3$ of precipitation, which equates to $1.44 \, €\, m^{-2} \, year^{-1}$. Using a storage depth of 3.5 m would convert to annual savings of $0.41 \, €/m^3$ of capacity. An additional benefit is that a smaller tank is needed to store a farm's manure (irregular rainfall requires surplus capacity to insure against wet years); however, we did not account for this effect. If considered, these factors will proportionately add benefit to impermeable covers for new store construction.

Retained Nitrogen Reducing NH_3 emissions leads to more N retained on the farm. This N has an economic value that we assumed was equal to the cost of purchasing synthetic N fertilizer, which is a fluctuating price. In this case we used a value of $1 \, €/kg$ N. The mass of retained N for each technique was calculated based on information in Table 5.7.

Reduced GHG Emissions Many NH_3 abatement techniques also reduce methane emissions which are the dominant GHG emitted by liquid manure storages on a CO_2-equivalent basis (CO_2e). We ignored N_2O and NH_3 (an indirect GHG). Methane emission reductions were assigned as follows: 10 % to impermeable covers, 10 % to straw covers, and 80 % to systems with gas collection and flaring. For illustrative purposes we assumed a baseline CH_4 emission rate of $5 \, m^3\text{-}CH_4/m^3\text{-}$ manure for all liquid systems (cf. Massé et al. 2003), a storage depth of 3.5 m, methane was converted to mass at $0.667 \, kg \, m^{-3}$ and to CO_2e using a global warming potential of 25 (IPCC 2007). A "carbon credit" value of $10 \, €/ton$ of CO_2e ($0.01 \, €/kg$) was used.

Reduced Odour Emissions Odour emission reductions were assumed to equal NH_3 emission reductions on a percentage basis (Table 5.7). We used emissions of volatile organic compounds (VOC) as a proxy for odour, and a baseline VOC emission rate was taken as 250 µg m^{-2} s^{-1} for all liquid manure systems based on data from a swine manure storage (Bicudo et al. 2002). The economic value of VOC emissions was estimated as 1.35 €/kg-VOC based on an economic study of environmental externalities of landfills and waste incineration (European Commission 2000). More specific quantification of the benefit of manure odour reductions is needed.

5.8 Conclusions

In this chapter we reviewed several technologies that have been proposed for reducing emissions from stored manure. Most of the available techniques are for liquid manure stores, whereas there are few methods currently suited for solid manures so clearly research is needed. For liquid manures the technical feasibility is an important aspect that depends on proportion of manure solids (for crusts), changes in liquid levels, and local weather. Some strategies were relegated as Category 3 because of technical limitations and inadequate evidence that they can reduce ammonia emissions.

The majority of category 1 and 2 techniques cost less than 10 €/kg of abated NH_3-N loss. Techniques costing less than 5 €/kg NH_3-N are available for mitigating emissions from every type of manure investigated (cattle slurry, cattle farm-yard manure, pig slurry tanks, pig slurry lagoons, pig FYM, and poultry manure). The most economical strategy was to allow cattle manure tanks to crust. Other highly economical techniques were floating materials, which abated NH_3 for much lower cost than structural covers, but the latter are more attractive for new structures since they will save needed storage capacity.

While the value of co-benefits are difficult to quantify and generalize, we estimated the value of both tangible (precipitation exclusion and N retention) and intangible (reduced odour and GHGs). Including the value of tangible co-benefits for covers installed in a wet region on pig manure tanks substantially reduced the cost for all cover types, especially impermeable covers. Combining tangible and intangible benefits reduced the mean cost for each cover type to less than zero, but this is contextual and must be treated with caution. Our assessment suggests the importance of co-benefits for cost-benefit analysis and we encourage additional efforts to thoroughly evaluate these factors.

Acknowledgements The authors would like to acknowledge the input from numerous people in the manure covering and processing industry for their comments and discussions. The assistance of Dr. J. Chilima in reviewing the literature is also appreciated.

References

ADAS Gleadthorpe Research Centre (2004) Management techniques to minimise ammonia emissions from solid manures. Final report DEFRA Project code WA0716. http://randd. defra.gov.uk/Document.aspx?Document=WA0716_6885_FRP.doc. Accessed 1 Oct 2014

Amon B, Amon T, Boxberger J, Pöllinger A (1998) Emissions of NH_3, N_2O, and CH_4 from composted and anaerobically stored farmyard manure. In: Proceedings of the RAMIRAN conference. pp 209–216. http://www.ramiran.net/doc98/FIN-POST/AMON-BAR.pdf. Accessed 1 Oct 2014

Amon B, Kryvoruchko V, Moitzi G, Amon T (2006) Greenhouse gas and ammonia emission abatement by slurry treatment. Int Congr Ser 1293:295–298

Aneja VP, Roelle PA, Murray GC, Sutherland J, Erisman JW, Fowler D, Asman WAH, Patni N (2001) Atmospheric nitrogen compounds II: emissions, transport, transformation, deposition and assessment. Atmos Environ 35:1903–1911

ASABE (American Society of Agricultural and Biological Engineers) 2010 Manure production and characteristics. ASAE D384.2 Mar 2005 (R2010), ASABE, 2950 Niles Road, St. Joseph, MI 49085-9659. Available from www.ANSI.org, http://webstore.ansi.org/RecordDetail.aspx? sku=ASAE+D384.2+MAR2005+(R2010). Accessed 1 Oct 2014

ASABE (American Society of Agricultural and Biological Engineers) (2013) Uniform terminology for rural waste management. ASAE S292.5 Oct 1994 (R2013), ASABE, 2950 Niles Road, St. Joseph, MI 49085-9659. http://webstore.ansi.org/RecordDetail.aspx?sku=ASAE+S292.5 +OCT1994+(R2013). Accessed 1 Oct 2014

Balsari P, Dinnuccio E, Santoro E, Gioelli F (2008) Ammonia emissions from rough cattle slurry and from derived solid and liquid fractions applied to alfalfa pasture. Aus J Exp Agric 48:198–201

Balsari P, Dinnuccio E, Gioelli F (2011) Coverage of digestate storage tanks: why it is necessary and technical solutions. In: Progress in biogas, Stuttgart-Hohenheim, Germany

Barry D (2006) Experience with the negative air pressure cover at Arkell Research Station. Canadian Pork Council, University of Guelph, Guelph

Berg W, Brunsch R, Pazsiczki I (2006) Greenhouse gas emissions from covered slurry compared with uncovered during storage. Agr Ecosyst Environ 112(2–3):129–134

Bicudo J, Schmidt D, Powers W, Zahn J, Tengman C, Clanton C, Jacobson L (2002) Odor and VOC emissions from swine manure storages. In: Odor and toxic air emissions specialty conference. Water Environment Federation, Albuquerque, USA

Bicudo JR, Clanton CJ, Schmidt DR, Powers W, Jacobson LD, Tengman CL (2004) Geotextile covers to reduce odor and gas emissions from swine manure storage ponds. Appl Eng Agric 20 (1):65–75

Bluteau M, Becze L, Demopoulos GP (2009) The dissolution of scorodite in gypsum-saturated waters: evidence of Ca-Fe-AsO4 mineral formation and its impact on Arsenic retention. Hydrometallurgy 97:221–227

Bujoczek G, Oleszkiewicz J, Sparling R, Cenkowski S (2000) High solid anaerobic digestion of chicken manure. J Agric Eng Res 76:51–60

Bundy DS, Li X, Zhu J, Hoff F (1997) Malodour abatement by different covering materials. In: Proceedings of the international symposium on ammonia and odour control from animal production facilities. CIGR and EurAgEng Publication, Rosmalen, pp 413–420

Chadwick DR (2005) Emissions of ammonia, nitrous oxide and methane from cattle manure heaps: effect of compaction and covering. Atmos Environ 39:787–799

Clanton CJ, Schmidt DR, Jacobson LD, Nicolai RE, Goodrich PR, Janni KA (1999) Swine manure storage covers for odour control. Appl Eng Agric 15(5):567–572

Clanton CJ, Schmidt DR, Nicolai RE, Jacobson LD, Goodrich PR, Janni KA, Bicudo JR (2001) Geotextile fabric-straw manure storage covers for odor, hydrogen sulfide, and ammonia control. Appl Eng Agric 17(6):849–858

Clemens J, Trimborn M, Weiland P, Amon B (2006) Mitigation of greenhouse gas emissions by anaerobic digestion of cattle slurry. Agr Ecosyst Environ 112(2–3):171–177

De Bode MJC (1991) Odour and ammonia emissions from manure storage. In: Nielsen VC, Voorburg JH, L'Hermite P (eds) Odour and ammonia emissions from livestock farming. Elsevier Applied Science, London, pp 59–66

DeVries H, Stevenson R, Hayes R, Turnbull JE, Clayton RE (1980) Experiences with floating covers for manure storages. Presented at the 1980 CSAE-SCGR 60th annual AIC conference, paper no. 80–218. ASAE, University of Alberta, Edmonton, AB

Dewes T (1996) Effect of pH, temperature, amount of litter and storage density on ammonia emissions from stable manure. J Agric Sci 127:501–509

DLG (German Agricultural Society Test Center) (2005) DLG test report 5451F – Hexa-cover ApS emission reduction of odour and ammonia. http://www.dlg-test.de/pbdocs/5451F_e.pdf. Accessed 1 Oct 2014.

Döhler H, Eurich-Menden B, Rößler R, Vandré R and Wulf S (2011). UN ECE-Convention on long-range transboundary air pollution – Task force on reactive nitrogen: Systematic cost-benefit analysis of reduction measures for ammonia emissions in agriculture for national cost estimates. Federal environment agency (Umweltbundesamt) project no. (FKZ) 312 01 287. Available online at: http://www.uba.de/uba-info-medien/4207.html

Dux D, Van Canaegem L, Steiner B, Kaufmann R (2005) FAT-Bericht Nr. 642, Forschungsanstalt Agroscope Tanikon, Switzerland

English S, Fleming R (2006) Liquid manure storage covers. Final report. Prepared for Ontario Pork. University of Guelph, Rodgetown Campus, Ridgetown, Ontario, Canada

Environment Canada (2011) National pollutant release inventory – ammonia (NH_3) emissions for Canada 2009. http://unfccc.int/national_reports/annex_i_ghg_inventories/national_invento ries_submissions/items/8108.php. Accessed 1 Oct 2014

European Commission, DG Environment (2000) A study on the economic valuation of environmental externalities from landfill disposal and incineration of waste. Final main report, October 2000. http://ec.europa.eu/environment/waste/studies/pdf/econ_eva_landfill_report.pdf. Accessed 14 Aug 2014

Fabian E, Richard T, Kay D, Allee D, Regenstein J (1993) Agricultural composting: a feasibility study for New York farms. Cornell University Feasibility Study. http://compost.css.cornell. edu/feas.study.html. Accessed 14 Aug 2014

Filson H, Henley WT, Reding BE, Sand D (1996) Natural and biological odour control methods for swine earthen manure storage areas. In: Proceedings of international conference on air pollution from agricultural operation, 7–9 February, Kansas City, MO. Midwest Plan Service, Iowa State University, Ames

Flesch TK, Harper LA, Powell JM, Wilson JD (2009) Use of dispersion analysis for measurement of ammonia emissions from Wisconsin dairy farms. Agric Eng 52:253–265

Funk TL, Hussey R, Zhang Y, Ellis M (2004a) Synthetic covers for emissions control from earthen embanked swine lagoons, part 1: positive pressure lagoon cover. Appl Eng Agric 20(2):233–238

Funk TL, Mutlu A, Zhang Y, Ellis M (2004b) Synthetic covers for emissions control from earthen embanked swine lagoons part II: Negative pressure lagoon cover. Appl Eng Agric 20(2):239–244

Ghafoori E, Flynn P, Feddes J (2007) Pipeline vs. truck transport of beef cattle manure. Biomass Bioenerg 31:168–175

Hao X, Chang C (2001) Gaseous NO, N2O, and NH_3 loss during cattle feedlot manure composting. Phyton-Ann Rei Bot A 41:81–93

Hjorth M (2009) Flocculation and solid-liquid separation of animal slurry; fundamentals, control and application. Ph D thesis, Aarhus University, Denmark

Hörnig G, Turk M, Wanka U (1999) Slurry covers to reduce ammonia emission and odour nuisance. J Agric Eng Res 73:151–157

Hyde B, Carton O, O'Toole P, Misselbrook T (2003) A new inventory of ammonia emissions from Irish agriculture. Atmos Environ 37(1):55–62

IPCC (2007) Climate change 2007: the physical science basis. Contribution of Working Group I to the Fourth Assessment Report of the Intergovernmental Panel on Climate Change Solomon S, Qin D, Manning M, Chen Z, Marquis M, Averyt KB, Tignor M, Miller HL (eds). Cambridge University Press, Cambridge/New York

Kai P, Pedersen P, Jensen JE, Hansen MN, Sommer SG (2008) A whole-farm assessment of the efficacy of slurry acidification in reducing ammonia emissions. Eur J Agron 28:148–154

Karlsson S (1996a) Åtgärder för att minska ammoniakemissionerna vid lagring av stallgödsel. JTI rapport Lantbruk & Industri nr 228. JTI – Swedish Institute of Agricultural and Environmental Engineering, Uppsala, Sweden

Karlsson S (1996b) Measures to reduce ammonia losses from storage containers for liquid manure. Paper 96E-013 presented at AgEng 96, Madrid, Spain

Klimont Z, Winiwarter W (2015) Estimating costs and potential for reduction of ammonia emissions from agriculture in the GAINS model. In: Reis S, Howard CM, Sutton MA (eds) Costs of ammonia abatement and the climate co-benefits. Springer, Amsterdam

Landry H, Laguë C, Roberge M, M Alam (2002) Physical and flow properties of solid and semi-solid manure as related to the design of handling and land application equipment. In: Canadian Society of Agricultural Engineering annual meetings, Paper No. 02–214. http://www.engr.usask.ca/societies/csae/PapersAIC2002/CSAE02-214.pdf. Accessed 14 Aug 2014

Lefcourt AM, Meisinger JJ (2001) Effect of adding alum or zeolite to dairy slurry on ammonia volatilization and chemical composition. J Dairy Sci 84:1814–1821

Mannebeck H (1985) Covering manure storage tanks to control odour. In: Nielson VC, Voorsburg JH, L'Hermite P (eds) Odour prevention and control of organic sludge and livestock farming. Elsevier Applied Publishers, London

Martin RO, Matthews DL (1983) Comparison of alternative manure management systems: effect on the environment, total energy requirement, nutrient conservation, contribution to corn silage production and economics. Agway Inc., USEPA, R.S. Kerr Environmental Research Laboratory, Ada, OK, 74820, Syracuse

Massé D, Patni N, Droste R, Kennedy K (1996) Operation strategies for psychrophilic anaerobic digestion of swine manure slurry in sequencing batch reactors. Can J Civ Eng 23:1285–1294

Massé D, Croteau F, Patni N, Masse L (2003) Methane emissions from dairy cow and swine manure slurries stored at 10 °C and 15 °C. Can Biosyst Eng 45:6.1–6.6

Mata-Alvarez J, Macé S, Llabrés P (2000) Anaerobic digestion of organic solid wastes. An overview of research achievements and perspectives. Bioresour Technol 74:3–16

McGinn SM, Coates T, Flesch TK, Crenna B (2008) Ammonia emissions from dairy cow manure stored in a lagoon over summer. Can J Soil Sci 88(4):611–615

Meyer DJ, Converse JC (1982) Controlling swine manure odors using artificial floating scums. Trans ASABE 25(6):1691–1695

Miner JR, Humenik FJ, Rice JM, Rashash DMC, Williams CM, Robarge W, Harris DB, Sheffield R (2003) Evaluation of a permeable 5 cm thick, polyethylene foam lagoon cover. Trans ASABE 46(5):421–1426

Misselbrook TH, Van Der Weerden TJ, Pain BF, Jarvis SC, Chambers BJ, Smith KA, Phillips VR, Demmers TGM (2000) Ammonia emission factors for U.K. agriculture. Atmos Environ 34:871–880

Misselbrook TH, Nicholson FA, Chambers BJ (2005) Predicting ammonia losses following the application of livestock manure to land. Bioresour Technol 96:159–168

Misselbrook TH, Chadwick DR, Chambers BJ, Smith KA, Williams JR, Demmers TGM (2007) Inventory of ammonia emissions from UK agriculture 2006. Defra contract report AC0102, London, UK

Misselbrook TH, Chadwick DR, Chambers BJ, Smith KA, Williams JR (2008) Inventory of ammonia emissions from UK agriculture 2007. Defra contract report AC0112, London, UK

Montalvo G, Pineiro C, Herrero M, Bigeriego M (2015) Ammonia abatement by animal housing techniques. In: Reis S, Howard CM, Sutton MA (eds) Costs of ammonia abatement and the climate co-benefits. Springer, Amsterdam

Ndegwa PM, Hristov AN, Arogo J, Sheffield RE (2008) A review of ammonia emission mitigation techniques for concentrated animal feeding operations. Biosyst Eng 100:453–469

Ni J (1999) Mechanistic models of ammonia release from liquid manure: a review. J Agric Eng Res 72(1):1–17

Nicolai RE, Pohl S, Schmidt D (2002) Covers for manure storage units. South Dakota State University, Brookings

Nicolai RE, Clanton CJ, Janni KA, Malzer GL (2006) Ammonia removal during biofiltration as affected by inlet air temperature and media moisture content. Trans ASABE 49(4):1125–1138

PAMI (1996) Research update 698. Hog lagoon odour control-A treatment using floating straw. In Int. Conf. on air pollution from agricultural operations, 375–378, Ames, Iowa: Mid-West Plan Service, Iowa State University, USA

Portejoie S, Martinez J, Guiziou F, Coste CM (2003) Effect of covering pig slurry stores on the ammonia emission processes. Bioresour Technol 87(3):199–207

Powers W (2004) Practices to reduce ammonia emissions from livestock operations. Iowa State University, Ames

Raaflaub M, Menzi M (2011) Wirtschaftliche Aspekte von Massnahmen zur Reduktion von Ammoniakemissionen. Schweizerische Hochschule fur Landwirtschaft SHL/Bundesamt fur Umwelt, unpublished

Reidy B, Menzi H (2007) Assessment of the ammonia abatement potential of different geographical regions and altitudinal zones based on a large-scale farm and manure management survey. Biosyst Eng 97:520–531

Rotz CA, Chianese DS (2009) The dairy greenhouse gas model reference manual, version 1.2. Pasture and watershed management research unit, Agricultural Research Service – USDA. http://www.ars.usda.gov/Main/docs.htm?docid=21345 Accessed 1 Oct 2014

Schulze-Lammers P, Römer G, Boeker P (1997) Amount and limitation of ammonia emission from stored solid manure. In: Voermans JAM, Monteney GJ (eds) Ammonia and odour control from animal production facilities. Elsevier, Amsterdam, pp 43–48

Scotford IM, Williams AG (2001) Practicalities, costs and effectiveness of a floating plastic cover to reduce ammonia emissions from a pig slurry lagoon. J Agric Eng Res 80(3):273–281

Smith DR, Moore PA, Haggard BE Jr, Maxwell CV, Daniel TC, vanDevander K, Davis ME (2004) Effect of aluminum chloride and dietary phytase on relative ammonia losses from swine manure. J Anim Sci 82:605–611

Sommer S (1997) Ammonia volatilization from farm tanks containing anaerobically digested animal slurry. Atmos Environ 31:863–868

Sommer SG (2001) Effect of composting on nutrient loss and nitrogen availability of cattle deep litter. Eur J Agron 14:123–133

Sommer SG, Hutchings NJ (2001) Ammonia emission from field applied manure and its reduction - invited paper. Eur J Agron 15:1–15

Sommer SG, Christensen BT, Nielsen NE, Schjorring JK (1993) Ammonia volatilization during storage of cattle and pig slurry: effect of surface cover. J Agric Sci 121:63–71

Stevens R, Laughlin R, Frost J (1989) Effect of acidification with sulphuric acid on the volatilization of ammonia from cow and pig slurries. J Agric Sci 113:389–395

Stevens RJ, Laughlin RJ, Frost JP (1992) Effects of separation, dilution, washing and acidification on ammonia volatilization from surface applied cattle slurry. J Agric Sci 1119:383–389

Vaddella V, Ndegwa P, Joo H, Ullman J (2010) Impact of separating dairy cattle excretions on ammonia emissions. J Environ Qual 39:1807–1812

VanderZaag AC, Gordon RJ, Glass V, Jamieson RC (2008) Floating covers to reduce gas emissions from liquid manure storages: a review. Appl Eng Agric 24(5):657–671

VanderZaag AC, Gordon RJ, Jamieson RC, Burton DL, Stratton GW (2010a) Permeable synthetic covers for controlling emissions from liquid dairy manure. Appl Eng Agric 26(2):287–297

VanderZaag AC, Gordon RJ, Jamieson RC, Burton DL, Stratton GW (2010b) Gas emissions from straw covered liquid dairy manure during summer storage and autumn agitation. Trans ASABE 52(2).599–608

Webb J, Menzi H, Pain BF, Misselbrook TH, Dammgen U, Hendriks H, Dohler H (2005) Managing ammonia emissions from livestock production in Europe. Environ Pollut 135 (3):399–406

Webb J, Ryan M, Anthony S, Brewer A, Laws J, Aller M, Misselbrook T (2006) Cost-effective means of reducing ammonia emission from UK agriculture using NARSES model. Atmos Environ 40:7222–7233

Wilkie AC (2000) Anaerobic digestion: holistic bioprocessing of animal manures. In: Proceedings of the animal residuals management conference. Water Environment Federation, Virginia, USA, pp 1–12

Williams A (2003) Floating covers to reduce ammonia emissions from slurry. In: Proceedings of the international symposium on gaseous and odour emissions from animal production facilities. Horsens, Denmark, pp 283–291

Wohl J (1996) A literature review of the economics of manure management options in the lower Fraser valley. Prepared for the British Columbia Ministry of Environment, Land and Parks, Canada

Yagüe M, Guillén M, Quílez D (2011) Effect of covers on swine slurry nitrogen conservation during storage in Mediterranean conditions. Nutr Cycl Agroecosyst 90:121–132

Zahn JA, Tung AE, Roberts BA, Hatfield JL (2001) Abatement of ammonia and hydrogen sulfide emissions from a swine lagoon using a polymer biocover. J Air Waste Manage Assoc 51 (4):562–573

Zhang Y, Gaakeer W (1998) An inflatable cover for a concrete manure storage in a Swine facility. Appl Eng Agric 14(5):557–561

Chapter 6
Cost of Ammonia Emission Abatement from Manure Spreading and Fertilizer Application

J. Webb, John Morgan, and Brian Pain

Abstract Abatement of emissions following the application of manures to land has been identified as a priority to reduce emissions of ammonia (NH_3). However, the conventional method of spreading slurry, surface broadcasting by splash plate applicator, is rapid and inexpensive while application techniques which reduce emissions of NH_3 impose an additional cost on the farmer. We critically reviewed the methodology used to estimate the additional costs of spreading livestock manures to land by the use of reduced NH_3 emission (RAE) techniques. The input values used to calculate costs were as follows: purchase price of tractors and RAE spreading machines; depreciation, of both the tractor and the spreader; interest rates, on loan or expended capital; fuel consumption; repairs; labour costs. This approach is consistent with the method to estimate the cost of abatement techniques given in the BREF guidance document. As a result of revising the calculations with updated input costs the additional cost of spreading using reduced-emission spreading equipment was from £0.52 to £0.65 per m^3 slurry applied by RAE techniques. The costs of spreading manure arise mainly from labour (27 %), Fuel (23 %), spreader costs (10 % from repairs and maintenance and 12 % from depreciation) and tractor costs (11 % from repairs and maintenance and 10 % from depreciation). Since most of the labour and fuel costs incurred during the spreading of slurry arise from travelling from the slurry store to the field, the reduced work rates of RAE machines, and greater fuel requirements for pulling injectors, incur only moderate additional costs. These cost estimates were within the range of additional costs reported by commercial farmers who had adopted RAE techniques in a pilot study. Estimates of the additional costs of applying livestock manures by RAE techniques will inevitably vary due to differences among farms and contracting operations with respect to the volumes of manure to be spread, differences in fuel and labour costs and in depreciation and interest rates. Nevertheless, we conclude that the estimates of £0.65 m^{-3} for slurry and £0.54 t^{-1} for manures spread using farm equipment are broadly reliable. We also conclude that

J. Webb (✉)
Ricardo-AEA, Gemini Building, Harwell, Didcot OX11 0QR, UK
e-mail: j.webb@ricardo-aea.com

J. Morgan • B. Pain
Creedy Associates, Great Gutton Farm, Shobrooke, Crediton, Devon EX17 1DJ, UK
e-mail: john.morgan@creedyassociates.com; brian.pain@creedyassociates.com

© Springer Science+Business Media Dordrecht 2015

S. Reis et al. (eds.), *Costs of Ammonia Abatement and the Climate Co-Benefits*,
DOI 10.1007/978-94-017-9722-1_6

the costs of RAE techniques for manures applied by contractor will be substantially less than for manures applied by farm machinery.

Keywords Ammonia emissions • Emission abatement • Agriculture • Manure spreading • Fertiliser application

6.1 Introduction

Abatement of emissions following the application of manures to land has been identified as a priority in the development of national and international approaches to reduce emissions of ammonia (NH_3) (e.g. Webb et al. 2005). However, the conventional method of spreading slurry, surface broadcasting by splash plate (**SP**) applicator, is rapid and inexpensive (Bittman et al. 1999) while application techniques which reduce emissions of NH_3 impose an additional cost on the farmer. These costs arise either through the greater capital costs of reduced application techniques and the increased labour costs due to the reduced work rate or through the additional charges levied by contractors for spreading manures using reduced-NH_3 emission (RAE) methods. Initial estimates of the additional costs of using RAE methods in the UK ranged from £1.44 m^{-3} of slurry applied by the trailing hose (TH) applicator to £2.84 m^{-3} for open slot injection (OSI) (Webb et al. 2006). The additional costs of applying slurry by RAE spreading techniques were recently estimated to be £0.52 m^{-3} for all three types of spreader, TH, OSI and the trailing shoe (TS) (Webb et al. 2010). This cost estimate was within the range of actual costs on commercial farms which adopted RAE methods reported by Anon. (2002), which ranged from £0.42 to £2.11 m^{-3}. In this paper we have updated the approach used in the earlier review by Webb et al. (2010) to assess the costs of RAE spreading techniques and how robust the findings were and report to the TFRN. The results of the 2009 review of the effectiveness of RAE techniques and their impacts on emissions of nitrous oxide (N_2O) and on crop recovery of manure N, were published in Webb et al. (2010).

6.2 The Basis of the Methodology

One of the foundations of the methodology was that with most machine-based operations on farm, the running cost of the tractor pulling the implement, is a considerable element of the overall cost.

Moreover, in order to accurately assess the additional costs of RAE spreading techniques all costs must be estimated, including those that will also be incurred from spreading using a conventional splash plate machine. Thus the additional costs of using RAE methods will be the total costs of spreading manures using RAE methods minus the total costs of spreading using the splash plate.

Hence, the annual running costs of a suitable tractor were reviewed alongside those associated with RAE spreading equipment such as trailing hose (TH), trailing shoe (TS) and open slot injection (OSI) machines as well as cultivation equipment required to incorporate solid manures.

The input values used to calculate costs in the spreading review exercise carried out in 2009 were as follows:

- Purchase price of machines.
- Depreciation, of both the tractor and the spreader.
- Interest rates, on loan or expended capital.
- Fuel consumption.
- Repairs.
- Labour costs.

Depreciation is an estimate of the decreasing value of an asset over time. In theory, this deprecation, if set aside in a reserve fund, should be sufficient to replace the asset as and when the time comes.

The approach used is consistent with the method to estimate the cost of abatement techniques given on p 329 of the BREF guidance document (Anon. 2003). Following that guidance, unit costs should be calculated as below:

- Current costs should be used for all calculations.
- Capital expenditure should be annualized over the economic life of the investment.
- Annual running costs should be included.
- Take into account changes in performance.
- The total sum is divided by the annual throughput to determine the 'unit cost'

6.2.1 Depreciation

Depreciation occurs for three reasons, obsolescence, gradual deterioration with age and wear and tear as a result of use. While the first two reasons are time-related and are largely age-related, wear and tear is a result of use and is directly linked to the hours a machine is worked during a year. If the first two reasons predominate then depreciation tends to be more of a fixed cost. If the machine does many hours the wear and tear associated with this use becomes the key element of its depreciation over time and in effect makes deprecation more of a variable cost (rises proportionately as hours worked increases) than a fixed one.

While the rate of depreciation is not directly linked to hours worked there is a close link between the two. As a consequence the more hours a machine works the less the depreciation per hour, and hence per m^3 of slurry spread. This reducing rate of depreciation will to some extent be countered by the likely increase in repairs and maintenance cost as the hours worked increases.

The cost estimates assumed a 6 year write-off for the machines with a final sale price of 10 % of the purchase price.

6.2.2   Interest on Capital

The capital invested in the ownership of a machine either results in lost interest, if the purchase was made with saved money, or actual interest payments if capital is borrowed. This lost or extra interest payment should be taken into account when working out the cost of running a machine. When working out the interest payment associated with a machine the average interest paid over the life of the machine is used and is therefore calculated on the average capital invested. Typically interest is calculated on half the initial capital cost on the basis that the machine is being written off over the time and the depreciation money is being invested and earning interest in preparation for the machine's replacement. The interest rate charged on this capital will inevitably depend on the personal situation of the machine's owner and the interest rates at the time. Whether interest is being lost, as a result of money from reserves being used to finance a machine, or paid as a result of borrowed finance, the rates paid in the UK are to a large extent linked to the Bank of England base rate.

6.2.3   Repairs

While it is tempting to assume that repair costs rise or fall directly in proportion to the hours the machine works like other variable costs, certain repair costs are in fact related to the machine's age. Examples of such fixed costs would be battery life. Nevertheless, repair and maintenance costs while not directly linked to workrate are in practice very closely associated with the amount of work undertaken and for practical reasons repair costs tend to be linked to hours worked. Repair costs for tractors are typically assumed to be in the region of 8 % of the initial capital purchase price of the machine. Repair costs associated with spreaders, tends to be more directly linked to the number of hours worked.

Difference between splash plate machines and RAE spreaders.

The key differences when calculating the direct differences in cost between RAE spreaders and splash plates were:

- Slower work rate of RAE spreaders and therefore greater tractor costs per unit of slurry spread.
- Smaller repair costs associated with splash plate machinery due to less soil/ machine contact than the TS and OSI machines and less moving parts.
- Greater purchase costs of RAE spreaders.

These differences have been used to calculate the spreading costs *vs.* a splash plate tanker.

6.2.4 *Labour*

The relatively slower work rate of the RAE spreaders will inevitably increase the cost per hour of using this equipment. However, as indicated in Sect. 6.2.4, the amount of time actually spent in field spreading is only a small proportion of the total time required to load and transport tankers to and from the field where manure is to be spread.

6.2.4.1 Cost of Rapid Incorporation of Solid Manures

It might be argued that there are no inherent extra costs associated with rapid incorporation of solid manures in most circumstances. Incorporation of manures will usually occur at some stage after spreading. The key issue is the time between spreading and incorporation. In the UK manures are often spread to arable land over the winter in the period between harvest of cereals and planting of spring-sown crops such as sugarbeet and potatoes. Normal practice is for the manures to remain on the soil surface until cultivation prior to planting in March or April. Hence if manures are to be incorporated soon after spreading this is likely to introduce an additional cultivation, since in the interval between ploughing subsequent soil settlement and weed growth might require another cultivation before seedbed establishment. Even for manures applied in late summer or early autumn, shortly before planting autumn-sown crops, there is an issue with respect to logistics. Many farmers find it difficult to have spreading and incorporation machinery on site at the appropriate time. For a small farming operation rapid incorporation relies on the farmer regularly swapping machines on his tractor which can waste significant quantities of time and slow down work rates. For larger operations liaison with farm workers and contractors is needed to ensure required machines are on site at the same time. In addition, work rates of the incorporation and spreading machinery need to be matched to keep all units working at optimum speeds. Matching work rates is not simple particularly bearing in mind the different travelling times between different blocks of land for spreaders and different work rates of cultivation machinery based on soil type, machinery size etc.

6.2.5 *Final Note*

As a general rule farmers rarely calculate reliable costs of machinery operation and therefore generating information from this source alone is difficult and potentially inaccurate. Contractors will in many cases attempt to calculate tractor and machine running costs and will use these to set rates charged to clients. Where possible these contractor costs have been used to verify the costs of running machines. The results of the Defra Pilot Farms study (Laws et al. 2003) have been examined in an attempt to put the results of the 2009 review into context.

6.3 The Values Used for Inputs, Including the Basis for Any Assumptions and the Uncertainties in the Estimates

As indicated above, the input values used to calculate costs in the spreading review exercise carried out in 2009 were as follows:

- Purchase price of machines.
- Depreciation, of both the spreader and the tractor.

    - straight line (between purchase price and resale/scrap price).
    - diminishing balance, reduces by a fixed % each year.

- Interest, on loan or foregone on expended capital.

    - in the review we used 4.5 % to reflect the current economic situation.
    - 7 % has been used in previous calculations.

- Fuel consumption.
- Repairs.

    - will depend upon hours worked, important consideration in estimating additional costs associated with the tractor.

- Labour costs.

6.3.1 Purchase Price of Machines

Our assessment did not indicate any consistent differences in the purchase price of RAE spreading machines (see Sect. 6.4.4), and we used an average price of £28,000. The tractor was priced at £51,000.

6.3.2 Depreciation

Typical depreciation rates were estimated to be in the region of 15 % of the initial purchase price. Initial annual depreciation would be greater than this but late in the machine's life rates would decrease. Using an initial capital cost of a tanker plus

RAE spreader of £28,000 based on information obtained from suppliers over winter 2008/2009, depreciation would typically be £4,200/year. It was assumed that the machine is sold after 6 years of use at 10 % of its initial purchase price.

6.3.3   Repair and Maintenance

Machines that require direct soil to machine contact such as injectors will inevitably have greater repair and maintenance costs that those that do not. Repair costs of individual farm machines over a long enough time to come up with genuine averages are very rarely kept. Budgeting estimates of 7 % of initial purchase price are quoted for the first 200 h of use with an additional 2 % of purchase cost per 100 h above this 200 h base for machines with limited soil to machine contact. Annual repair costs in the region of 13 % of the initial purchase price should therefore be budgeted on machines that do 500+ hours. Assuming an initial capital cost of a tanker plus band spreader of £28,000 repairs and maintenance costs would therefore typically be £3,640/year.

6.3.4   Labour

Hourly rates of £9–10/h are typical for tractor drivers and foreman/supervisors respectively in the UK. The relatively slower work rate of the RAE spreaders will inevitably increase the cost per hour of using this equipment. Interestingly while there is a difference in the output of the machinery in the field, when spreading with a tanker-based system, most of the time taken is a result of the travel to and from the field. Even with slower application rates of RAE spreaders compared with surface spreading, the overall increase in work rate is relatively small. Table 6.1, taken from a DARD technical note 'Alternative Spreading Systems' (Frost and Mulholland 2004), shows the difference in number of tanker loads over an 8 h period between two different spreading systems with system B being based around a slower RAE spreader.

Table 6.1 The number of tanker loads over an 8 h period applied by two different spreading systems with system B being based around a slower RAE spreader

	Travel time to field from slurry store	Time to spread slurry in field	Travel time to slurry store from field	Time to fill tanker at slurry store	Efficiency	Tanker loads per 8 h
A	4 min	4 min	4 min	4 min	80 %	24
B	4 min	6 min	4 min	4 min	80 %	22
Difference A–B		+50 %				−9 %

6.4 Making It Clear Exactly How the Numbers Are Derived

The above approaches produced the estimates of annual costs given in Table 6.2. Brief summaries of the basis for each estimate are also provided. All costs were estimated for both systems in order to calculate the difference, rather than trying to identify the differences and cost those.

The large proportion of costs attributed to fuel is consistent with that reported in farming literature.[1]

Spreader depreciation and interest payments are the running costs directly related to the purchase price of a spreader. These account for just 13.6 % of the

Table 6.2 Components of the estimates of the total costs of spreading made by Webb et al. (2010), based on machine costing £28,000

	Assumption 1	Assumption 2	Assumption 3	Annual cost (£)	% of total
Tractor depreciation	Sold after 10 years	At 10 % of cost	10 % of cost/ year	3,500	9.8
Interest on capital	Based on half capital cost	4.5 % annual rate		1,150	3.2
Insurance	2 % of average capital cost			510	1.4
Fuel	18–25 L h^{-1} for 100–180 hp	£0.38 L^{-1}, March 2009		8,170	22.9
Repairs and maintenance	Not directly related to use	Some repairs needed with age	8 % of capital cost	4,080	11.4
Labour	£9–10 h $^{-1}$, 1,000 h per year	Slower work rate	But most time to and from store	9,500	26.6
Total tractor costs				*26,910*	*75.5*
Spreader depreciation	Sold after 6 years	At 10 % of cost	15 % of cost/ year	4,200	11.8
Maintenance	7 % of costs for first 200 h	2 % of costs for each extra 200 h	Hence 13 % for 500 h	3,640	10.2
Interest	As per tractor			630	1.8
Insurance	As per tractor			280	0.8
Total spreader costs				*8,750*	*24.5*
Total costs				**35,660**	

All the assumptions noted were used in the calculation. The assumptions and the outputs from the calculations are discussed in the sub-sections below

---

[1] e.g. http://www.rossfarm.co.uk/pdf/rfm-dec-08.pdf or http://www.farmersguardian.com/improving-operating-efficiency/19979.article

total cost estimate. It is worth considering in turn the assumptions and uncertainties of each of the major factors contributing to the cost estimate: labour; fuel; spreader depreciation; tractor repair and maintenance; spreader repair and maintenance; tractor depreciation. All other factors account for no more than 5 % each of the total estimated cost and will not be discussed further.

6.4.1 Labour

The major point to be considered is not the labour cost *per se*, although if those are under- or over-estimated it will compound any other errors, but the increase in labour requirement arising from using RAE machinery. The point was made by Webb et al. (2010) that while there is a difference in the output of the machinery in the field, when spreading with a tanker-based system, most of the time taken is a result of the travel to and from the field. Even with slower application rates of RAE spreaders compared with surface spreading, the overall increase in work rate is relatively small. Table 6.3 is an expanded version of Table 6.1 and shows the difference in number of tanker loads over an 8 h period between two different spreading systems with variation in the distance between field and farm.

Of course the original estimates of time spent travelling will not represent all farms and for some (or perhaps many in some parts of the UK) the average travel time may be at least double that indicated in Table 6.1. Considering first the impact of less travelling time (unlikely though that may be), if travel time is taken to be just 2 min each way, then the number of tanker loads that can be applied by surface

Table 6.3 Slurry tanker loads per 8 h at two different work rates in the field

	Travel time to field from slurry store	Time to spread slurry in field	Travel time to slurry store from field	Time to fill tanker at slurry store	Efficiency	Tanker loads per 8 h
A	4 min	4 min	4 min	4 min	80 %	24
B	4 min	6 min	4 min	4 min	80 %	22
Difference A–B		+50 %				−9 %
A	2 min	4 min	2 min	4 min	80 %	32
B	2 min	6 min	2 min	4 min	80 %	27
Difference A–B		+50 %				−14 %
A	6 min	4 min	6 min	4 min	80 %	19
B	6 min	6 min	6 min	4 min	80 %	17.4
Difference A–B		+50 %				−9 %
A	10 min	4 min	10 min	4 min	80 %	13.7
B	10 min	6 min	10 min	4 min	80 %	12.8
Difference A–B		+50 %				−7 %

Source: Derived from Frost and Mulholland (2004)

application in 8 h will increase to 32 (at 80 % efficiency), while those for RAE spreading will increase to 27, a difference between the two methods of application of 14 %. Conversely if the farm is larger (a more likely scenario in the UK) and the travel time is increased by 50 %, the number of loads that can be spread within 8 h are 19 and 17 respectively, still a difference of 9 %. Increasing the journey time to 10 min reduces the difference to only 7 %.

In conclusion, while the original assumption of 4 min travel time might seem rather short, increasing the travel time does not much diminish the estimated difference in time between the surface and RAE spreaders.

When fitted to umbilical systems the reduction in work rate compared to splash plates will be more marked, since travel time is eliminated.

6.4.2 Fuel

In April 2010 the price of agricultural diesel oil increased to £0.55/L. The impact of this change on spreading costs was estimated. However, while fuel costs were the second largest cost component, as with labour, much or most of the fuel consumption will be required for travel and hence a c. 50 % increase in price will not give rise to a 50 % increased in costs of RAE spreading.

6.4.3 Spreader and Tractor Depreciation

During discussion of the initial report within the TFRN some concern was expressed at the lack of difference in the estimates of spreading costs for conventional and RAE spreaders. This topic is discussed in more detail below in Sect. 6.4.4. The depreciation costs were based on an average price for a reduced NH_3 emission spreader of £28,000, which remains a reasonable estimate of a typical cost. The range of prices quoted in the UK goes up to c. £40,000 for a tanker-mounted machine. By comparison conventional splash plate spreaders can be bought for c. £12,000. Hence this value is also used in the sensitivity analysis.

Another difference in the assumptions underlying these estimates was that in the Defra Pilot study (Laws et al. 2003) spreaders were depreciated over times related to the soil type for each farm (8–12 years) and not 6 years as in the 2009 review.

In the sensitivity analysis no estimate was made of potential reduced costs, since this does not seem likely. However, the greater power needed to pull some RAE machines could perhaps reduce the life of a tractor from 10 to 8 years, and the resultant impact on depreciation rate has been estimated.

Tractor and spreader repair and maintenance costs are shown in Table 6.4 (based on Defra Pilot Study, Laws et al. 2003).

Table 6.4 How the components of cost estimate, based on machine costing £28,000, might change with reasonable changes to the assumptions

	Change	Current estimate	Potential decrease in estimate	Potential increase in estimate	% of total cost
Tractor depreciation	Sold after 10 years	3,500	NA	4,375	9.8–12.2
Interest on capital	Based on half capital cost	1,150	NA	NA	2.6–4.0
Insurance	2 % of average capital cost	510	NA	NA	1.1–1.8
Fuel	Price increase from £0.38 to £0.55 L^{-1}	8,170	NA	11,825	22.9–28.6
Fuel	Based on 25 L per h, not 21.5	8,170	NA	9,500	
Repairs and maintenance	Not directly related to use	4,080	2,550	NA	8.9–11.4
Labour	Increase or decrease in travel time of 50 % (Table 6.3)	9,500	7,740	12,010	26.6–27.1
Total tractor costs		*26,910*	*23,620*	*33,950*	*75.5–82.6*
Spreader depreciation	Prices of £12,000 and £40,000	4,200	1,800	6,000	6.3–13.5
Maintenance	7 % of costs for first 200 h	3,640	2,275	NA	8.0–10.2
Interest	As per tractor	630	NA	NA	1.4–2.2
Insurance	As per tractor	280	NA	NA	0.6–1.0
Total spreader costs		*8,750*	*4,985*	*10,550*	*17.4–24.5*
Total costs		**35,660**	**28,605**	**44,500**	

The impacts of additional power requirements for injectors were not explicitly discussed in the 2009 review. Nor is any data on this aspect available from the Defra Pilot study (Laws et al. 2003).

From the Defra Pilot Study annual maintenance costs of the new slurry applicators were estimated to range from £185 to £2,640, with a mean of £1,100, compared with £3,640 in the 2009 review. Estimated costs of repairs and maintenance for splash plate spreaders in the Pilot study averaged £310. However, consistent with the view expressed in the 2009 review that most farmers do not keep records of repair and maintenance costs, those for the Pilot study were mostly estimates based on the capital cost of the equipment. Costs for splash plate machines were generally 2 % of capital cost, while those for RAE spreaders were 4–5 % of cost. Repair and maintenance costs for the 2009 review were taken to be 8 % of the capital cost. For the sensitivity analysis a reduced estimate of 5 % of capital cost was also used.

The sensitivity analysis indicates that the majority (75–80 %) of costs of slurry application were associated with the costs of the tractor that hauls the spreader. These costs are dominated by fuel and labour costs.

6.4.4 Fuel

No explicit mention was made of additional fuel consumption arising from pulling the reduced-emission spreaders. Average fuel consumption of 18–25 L/working hour are quoted for 100–180 Horse Power tractors. Based on these consumption rates fuel costs per hour of between £6.84 and £9.50 were calculated as typical. Feedback from the Defra Pilot study indicated that tanker-mounted applicators require a tractor of 100+ H.P. A revised calculation might be made assuming £6.84 per hour for splash plate application and £9.50 for RAE spreaders. Three of the contractors involved in the Defra Pilot farms study indicated they were able to use the same tractors. One bought a 200 hp tractor. This may be a major reason why the 2009 review reported contractors were only charging an extra £0.35 m^{-3} for application by RAE spreaders. Since tractor costs represent the majority of spreading costs if the reduced-NH$_3$ spreaders can be hauled by tractors already used this major cost will remain largely unchanged.

6.4.5 Labour

As indicated above, the baseline scenario assuming 4 min travel time (each way) between store and field, might have over-estimated the difference between splash plate and reduced-NH$_3$ application. Thus the estimate of labour costs in the 2009 review is not likely to have underestimated the difference in costs.

6.5 The Consequences for the Final Cost Numbers (Euro/kg NH$_3$-N Abated)

Table 6.5 includes a detailed comparison with the cost estimates reported by Webb et al. (2010).

The additional costs of applying slurry by RAE spreaders needs to be revised taking into account the increase in fuel price and possible increased fuel consumption per hour. We also present the consequences of a possible increase in labour costs due to the previously assumed travel time between the slurry store and the field being too great.

Of these three changes, only the increased rate of fuel consumption applies to spreading by splash plate. The costs of applying slurry using a splash plate machine were estimated in 2009 to be £1.13 m^{-3}. Although we have presented in Table 6.5 a range of revised cost options we conclude that this cost needs to be increased only by the additional cost of fuel (£0.14) making a revised cost for 2010 of £1.23 m^{-3}.

Although we include in Table 6.6 an estimate of increased labour costs in the event of travel time from the farm to field being underestimated, we consider this to

Table 6.5 Conversion of above costs into cost per m^3 slurry spread

Tractor running costs	Annual cost (£)		Hourly rate assuming 1,000 h (£)		m^3 rate assuming 30 m^3 spread per hour (£)	
	2010	Update	2010	Update	2010	Update
Tractor depreciation (Diminishing balance over 10 years)	3,500	3,500	3.50	4.38	0.12	0.12
Interest on capital	1,150	NA	1.15	NA	0.04	0.04
Insurance	510	NA	0.51	NA	0.02	0.02
Fuel[a]	8,170	11,825	8.17	11.83	0.27	0.40
Repairs and maintenance	4,080	NA	4.08	NA	0.14	0.14
Labour	9,500	[b]12,010	9.5	12.01	0.31	0.31–0.40
Total tractor costs	26,910		26.91		0.90	1.03–1.10
Splash plate spreader running costs						
Splash plate spreader depreciation	1,800	NA	3.60	NA	0.12	0.12
Splash plate spreader repairs (average of all three types)	1,200	NA	2.40	NA	0.08	0.08
Interest on capital	270	NA	0.54	NA	0.02	0.02
Insurance	120	NA	0.24	NA	0.01	0.01
Splash plate spreader costs	3,390	NA	6.78	NA	0.23	0.23
Total package			**33.69**		**1.13**	**1.26–1.33**

Combined running costs per hour for a 150–180 HP £51,000 initial purchase price tractor (1,000 h/year.) plus £12,000 tanker-based splash plate spreader running for 500 h per year
[a]£0.38 in 2009, £0.51 in 2010
[b]This represents the worst case scenario by assuming the estimate of time taken for transport to and from the field made in 2010 was too great. However, as the discussion in Sect. 6.3.1 indicates, it is more likely that in 2010 we underestimated travel time. Hence in the final column two estimates are given to indicate the possible uncertainty

be unlikely and do not include in the revised total. Thus the additional cost of spreading using RAE spreading equipment range from £0.52 in the 2009 review to £0.65 in this update.

6.5.1 Rapid Incorporation of Manures

This section applies to both solid manures and slurries, although this is most relevant to solid manures.

To avoid the challenging logistics identified above, an extra pass of an incorporation machine should be planned for. The use of such an approach would ensure maximum work rates of all other machines and labour units associated with manure

Table 6.6 Conversion of above costs into cost per m^3 slurry spread

Tractor running costs	Annual cost (£)		Hourly rate assuming 1,000 h (£)		m^3 rate assuming 27 m^3 spread per hour (£)	
	2010	Update	2010	Update	2010	Update
Tractor depreciation *(diminishing balance over 10 years)*	3,500	[a]4,375	3.50	4.38	0.13	0.16
Interest on capital	1,150	NA	1.15	NA	0.04	0.04
Insurance	510	NA	0.51	NA	0.02	0.02
Fuel (£0.38 L^{-1})	8,170	11,825	8.17	11.83	0.30	0.44
Fuel (21.5 L/h)	8,170	[b]13,750	8.17	13.75	0.30	0.51
Repairs and maintenance	4,080	NA	4.08	NA	0.15	0.15
Labour	9,500	[c]12,010	9.5	12.01	0.35	0.35–0.44
Total tractor costs	26,910		26.91		1.0	1.23–1.32
RAE spreader running costs			**Hourly rate assuming 500 h (£)**			
RAE spreader depreciation	4,200	NA	8.4	NA	0.31	0.31
RAE spreader repairs (average of all three types)	3,640	NA	7.28	NA	0.27	0.27
Interest on capital	630	NA	1.26	NA	0.05	0.05
Insurance	280	NA	0.56	NA	0.02	0.02
RAE spreader costs	8,750	NA	17.5	NA	0.65	0.65
Total package			**44.41**		**1.65**	**1.88–1.97**

Combined running costs per hour for a 150–180 HP £51,000 initial purchase price tractor (1,000 h/year.) plus £28,000 tanker-based RAE spreader running for 500 h per year

[a]This represents a greater rate of depreciation due to increased power demand to pull RAE spreading machines

[b]This estimate is used in the revised calculation of RAE spreading costs to account for additional power demand

[c]This represents the worst case scenario by assuming the estimate of time taken for transport to and from the field made in 2009 was too great. However, as the discussion in Sect. 6.3.1 indicates, it is more likely that in 2009 we underestimated travel time

application. Tractor costs for such an operation would be similar to those identified for the band spreading machines. Incorporation would be typically via a surface cultivator. The running costs of which are shown in Table 6.7.

It would be very difficult to come up with a robust set of assumptions re opportunity cost savings or loss resulting from such a small change in the interval between spreading and incorporation.

Table 6.7 Annual and hourly based incorporation costs for a £12,000 cultivator working 200 h per year

	Annual running costs (£)	Cost per hour (£)	Cost per ha[a]	Cost per m^3/tonne[b]
Tractor	26,910	26.91	15.4	0.31
Cultivator depreciation (15 %)	1,800	9	5.14	0.10
Cultivator repairs and maintenance (14 % of purchase price)	1,680	8.4	4.8	0.10
Interest on capital	270	1.35	0.77	0.02
Insurance	120	0.6	0.34	0.01
Total cultivator cost	2,790	19.95	11.4	0.23
Total package		**46.86**	**26.77**	**0.54**

Annual work rate of 1,000 hours for tractor and 200 h for cultivator
[a]Assumed work rate of 1.75 ha/hr
[b]Assumed application rate of 50 m^3/50 tonnes/ha

6.5.2   Comparison with the Results of the Defra Pilot Study (WA0710)

While Webb et al. (2010) used 8 % of capital cost to estimate repairs, the Defra Pilot study used 4–5 %; this leads to a big difference.

There was a wide range of costs depending on the volume of slurry to be applied and the type of new systems adopted (Table 6.8). The pre-abatement costs of slurry spreading ranged from £0.26 to £1.59 per m^3 slurry. Following adoption of reduced-emission spreading equipment, costs increased to between £0.64 and £3.75 per m^3, an overall average increase of 128 %.

Table 6.9, taken from the Pilot Farms study, reports the estimated spreading costs before and after adoption of the reduced emission slurry spreaders.

Thus, as can be seen from Table 6.9, on four of the nine farms studies the additional costs were within the estimate of additional costs made in the 2009 review. There was a general trend for the increase in costs from reduced-application slurry application to decrease with the increased volume of slurry to be spread.

This is to be expected as the fixed costs are spread over greater spreading activity. Given the increase in the average size of dairy herds in the UK since the Pilot study was carried out this would be a factor accounting for the smaller estimate of the increase in costs made in the 2009 review. The reduction in interest rates used for the calculation from 8.0 to 4.5 % will also have led to a reduction in estimated costs. However, the reduction in length of the depreciation period used in the 2009 review (6 years) compared with the 8–12 years used in the Pilot study report will have tended to balance these reductions. *The most likely reason for the difference may be the different estimates of repair and maintenance. Given that the*

Table 6.8 Financial costs of slurry management and application procedures before and after adopting ammonia abatement techniques

	Old system £ total	New system £/m³	Change £ total	Total slurry applied (m³) £/m³	£ total	£/m³	
Dairy 1ª/Dairy 2ª	15,525	1.12	10,840	1.18	−4,685	0.06	5,600
Dairy 3	5,430	0.33	13,963	0.85	8,533	0.52	7,600
Dairy 4	2,109	0.58	5,945	1.63	3,836	1.05	4,100
Dairy 5ª	1,100	0.54	3,662	1.79	2,562	1.25	2,000
Dairy 7	2,406	0.78	5,664	1.83	3,258	1.05	5,200
Pig 1ª	5,975	0.41	9,340	0.64	3,365	0.23	15,500
Pig 2ª	7,639	1.59	8,989	3.75	1,351	2.15	4,400
Pig 3	2,182	0.26	5,677	0.68	3,495	0.42	9,200
Pig 4	3,452	1.55	8,142	3.66	4,690	2.11	5,100
Mean		0.80		1.78		0.98	
Mean excluding farms with covered stores		0.70		1.73		1.03	
Contractor 1						0.27	
Contractor 2						0.16	
Contractor 3						6.69	
Contractor 4						0.23	
Contractor mean						1.84	

Annual costs are reported in total for the old and new systems, per m³ slurry generated and per kg⁻¹ ammonia abated

ªCovered slurry store

NA, not available from final report

Table 6.9 RAE spreaders, slurry volumes and additional spreading costs recorded in the Pilot Farms study

	Spreader type	Slurry volume m³	Additional cost £/m³
Dairy 3	Joskin trailing hose	16,000	0.52
Pig 3	Joskin injector	8,300	0.42
Dairy 4	Duport injector	3,500	1.05
Dairy 7	Duport injector	3,100	1.05
Pig 4	Joskin arable/grass injector tanker	2,200	2.11
Pig 1ª	Veenhuis	14,500	0.23
Dairy 2ª	Duport 4.4 m injector	6,500	0.06
Dairy 1ª	Duport 4.4 m injector	5,000	0.06
Pig 2ª	Joskin injector	4,800	2.15
Dairy 5ª	Pichon injector	3,000	1.25
Mean		**6,690**	**0.89**

ªCovered slurry store

Pilot study estimates were, in some cases at least, based on actual farm experience, these ought to be considered more robust. However, the Pilot study is now quite old, RAE spreading equipment is more widely available and hence the costs of maintenance may well have reduced. Moreover, in most cases repair and maintenance costs reported in the Defra Pilot study were only estimated, and hence not inherently more reliable than the estimates made in the 2009 review.

With respect to the estimated costs of slurry application for Contractor 2, the Pilot Farm study report made the point that 'the difficulty with the injection system is that the hours used are not great enough. If the usage could be increased to 500 h per year and the working life remained 8 years then the cost per hour would be comparable [to surface application] at about £17.65 per hour [the cost of surface application]'.

For Contractor 4 using an umbilical grassland injector, the only difference in charges was that the equipment was charged at an extra £50 per day. At the daily spreading rate of 33 m^{-3}, this was equivalent to a charge of £1.50 m^{-3}, with an increase in cost of only c. £0.02 m^{-3} (Table 6.10).

Table 6.10 Reporting maintenance cost estimated as part of Defra project WA0710

	Splash plate		Reduced emission	
Contractor 1	6 year life	Est. £85/year	Est 8 years	Est. £185/year
Contractor 2	6 year life	£1,170	Est 8 years	£2,160
Contractor 3	5 year life	£12,400[a]	Est 10 years	£2,640
Contractor 4	5 year life	£50	Est 3 years	£666
Dairy 1 and 2	Used injector before		Est 12 years	Est £1,300 (4 % of cost)
Dairy 3	Umbilical system		Est 8 years	No estimate
Dairy 4	Contractor		Est 10 years	Est £1,175 (5 % of cost)
Dairy 5	Contractor		Est 10 years	£799 (4.5 % of costs)
Dairy 7	20 year life	£180 (2 % cost)	Est 12 years	Est £1,080 (4 % of cost)
Pig 1	Umbilical system	£120	Umbilical inj	Est £480 (4 % of cost)
Pig 2	10 year life	£600		Est £950
Pig 3	15 year life	£165	Est 12 years	Est £868 (4 % of cost)
Pig 4	15 year life	£110 (2 % cost)	Est 10 years[b]	Est £960 (4 % of cost)
Poultry 1	Spread 40–46 ha/day		Spread 12/14 ha/day	Contractor needed
Beef 1				Contractor needed

[a]Umbilical system, spares and repairs estimated at 40 % of capital cost
[b]Stony soil

6.5.3 Capital Cost of Tanker-Mounted RAE Spreaders

Following circulation of the 2009 review (Webb et al. 2010), there was some surprise that we had taken the capital costs of all three types of RAE machine (TH; TS; Inj) to be the same. However, the information we gathered from vendors indicated there was no clear difference in price among those generic types of machine. The variation was between manufacturers and the detailed specification: bout width etc. The issue was further compounded by offers of discounts to obtain a sale. Insofar as possible we attempted to obtain quotes for machines of similar bout widths, tanker capacity etc. However, this can also be misleading as TH machines are designed to spread over a much greater width than the other machines. This adds to the cost but does provide a greater work rate.

To explore this issue further we report below the information on capital costs we used for the 2009 review, some costs obtained in autumn 2010 and some historic data from the UK Pilot Farms study.

The table below reports cost information used for the 2009 review. It is grouped to provide an easy comparison of machines from the same supplier with similar specification (Table 6.11).

These prices do not indicate that TS machines are any less expensive than injectors.

For comparison, tractors were estimated to cost £51,000 and conventional splash plate spreaders £12,000. Given that the cost, and hence depreciation, of tanker-mounted spreaders was not one of the largest factors in determining the cost of slurry spreading it is perhaps not worth exploring this issue much further.

Table 6.11 Quoted prices for a range of tanker and umbilical mounted band spreaders as at February 2009

Machine	Type	Maker	Price (£)
Trailing shoe	Tanker 5.2 m	Major	35,200
Shallow injection	Tanker 5.2 m	Major	31,900
Trailing shoe	Tanker 6.0 m	Major	36,300
Shallow injection	Tanker 6.4 m	Major	33,000
Trailing shoe	6 m, to fit on 11,000 L tanker	Joskin	28,000
Shallow injection	Double disc 6 m	Samson	28,000
Trailing shoe	10,000 L tanker, 7 m	Schuitemaker	41,000
Trailing shoe	Tanker, 7.5 m	Hi-Spec	33,000
Shallow injection	To mount on tanker, 4 m	Spreadwise	14,000
Trailing shoe, average			34,700
Injector, average			26,725
Trailing shoe	Umbilical, 6 m	Tramspread/Joskin	13,500
Shallow injection	Umbilical, 4 m	Spreadwise	14,500

6.6 Comparison of UK Estimates of the Additional Cost of Spreading with Costs Reported for Other Countries

The estimates for Spain are for the additional costs only, hence x indicates the cost of surface application.

Given the large difference between the cost of RAE spreading reported for the UK and Germany in Table 6.12, we examined the source of the estimates for Germany (Döhler and Eurich-Menden 2004) to determine the cause of the difference from the findings reported here. Crucially Döhler and Eurich-Menden (2004) found the unit costs of slurry application by RAE techniques decreased markedly with the annual volumes of slurry to be spread (Table 6.13).

Note that the volumes examined range from 500 to 3,000 m^3 years^{-1}. For comparison, the annual volumes reported for the UK Pilot farms study ranged

Table 6.12 Comparison of estimates of the additional cost of RAE spreaders produced for the UK with those estimated for other countries

	UK	Denmark	Spain	Germany
Surface	1.4	NA	x	2.3–5.2
Trailing hose	2.0	2.0	x + 1.2[a]	NA
Trailing shoe	2.0	NA	x + 1.4[a]	+1.4–4.1[b]
Open slot injection	2.0	NA	x + 1.0–1.4	+2.6–5.1[b]
Deep injection grass		NA	NA	NA
Deep injection arable, no crop		2.5	NA	NA
Deep injection, arable with crop		3.0	NA	NA
Incorporation		72.0/ha		
Incorporation, slurry			x + 0.6[a]	x + 0.8
Incorporation, FYM	0.65		NA	x + 0.9
Incorporation, poultry manure			x = 0.7–2.5	x + 0.9

Costs are in Euros at an exchange rate of 1.2 €/£. 2009 data
[a]From Pineiro et al. (2006)
[b]See information below on effect of slurry volume

Table 6.13 Results from Döhler and Eurich-Menden (2004) on the effect of annual slurry production on the costs of reduced-NH_3 spreading

	500 m^3 year^{-1}	1,000 m^3 year^{-1}	3,000 m^3 year^{-1}
Splash plate	5.2	3.9	2.3
Trailing hose	6.8	5.5	3.0
Trailing shoe	9.3	7.5	3.9
Open slot injection	11.7	8.9	4.5
Incorporation	6.0	4.7	3.0

Table 6.14 Comparison of estimates of the cost of abating 1 kg of NH$_3$-N using the unit cost estimates made for the UK by Webb et al. (2010) with current GAINS output (Amann et al. 2008)

	Input cost	Abatement efficiency	TAN in manure	Emission surface spread	NH$_3$-N conserved	UK estimate	GAINS estimate for UK	
							High efficiency	Low efficiency
	$€$ m^{-3}	Proportion of emission abated	kg m^{-3}		kg m^{-3}	$€$ kg^{-1} NH$_3$-N conserved		
Trailing hose, dairy slurry	0.80	0.30	1.3	0.68	0.20	3.7	–	9.9
Trailing hose, beef slurry	0.80	0.30	2.0	1.04	0.31	2.4	–	12.1
Trailing hose, pig slurry	0.80	0.30	2.3	1.20	0.36	2.2	–	12.5
Trailing shoe, dairy slurry	0.80	0.60	1.3	0.68	0.41	1.9	4.9	–
Trailing shoe, beef slurry	0.80	0.60	2.0	1.04	0.62	1.2	6.1	–
Trailing shoe, pig slurry	0.80	0.60	2.3	1.20	0.72	1.1	5.9	–
Slot injection, dairy slurry	0.80	0.70	1.3	0.68	0.47	1.6	4.9	–
Slot injection, beef slurry	0.80	0.70	2.0	1.04	0.73	1.1	6.1	–
Slot injection, pig slurry	0.80	0.70	2.3	1.20	0.84	1.0	5.9	–
Incorporation, dairy FYM	0.65	0.90	0.4	0.28	0.25	2.6	10.1	–
Incorporation, dairy slurry	0.65	0.95	1.3	0.68	0.64	1.0	4.9	–
Incorporation, beef FYM	0.65	0.90	0.4	0.28	0.25	2.6	10.0	–
Incorporation, pig FYM	0.65	0.90	1.3	0.91	0.82	0.9	NA	–
Incorporation, pig slurry	0.65	0.95	2.3	1.20	1.14	0.6	5.9	–
Incorporation, layer manure	0.65	0.90	9.0	4.50	4.28	0.2	0.7	–
Incorporation, broiler manure	0.65	0.90	9.8	4.90	4.66	0.15	1.3	–

Farmer-owned machines

Low efficiency is considered an appropriate comparison for application by TH for which the abatement is typically only 30 %. All other abatement options have efficiencies of at least 60 % and hence may be considered high efficiency options

Surface spreading: slurry, 52 % of TAN; FYM, 70 % of TAN; Poultry manure, 50 % of TAN

Table 6.15 Comparison of estimates of the cost of abating 1 kg of NH_3-N using the unit cost estimates made for the UK by Webb et al. (2010) with current GAINS output (Amann et al. 2008)

	Input cost	Abatement efficiency	TAN in manure	Emission surface spread	NH_3-N conserved	UK estimate of abatement cost	GAINS estimate for UK	
							High efficiency	Low efficiency
	$€$ m^{-3}	Proportion of emission abated	kg m^{-3}	52 % of TAN	kg m^{-3}	$€$ per kg NH_3-N conserved		
Trailing hose, dairy slurry	0.50	0.30	1.3	0.68	0.20	3.0	–	9.9
Trailing hose, beef slurry	0.50	0.30	2.0	1.04	0.31	1.9	–	12.1
Trailing hose, pig slurry	0.50	0.30	2.3	1.20	0.36	1.6	–	12.5
Trailing shoe, dairy slurry	0.50	0.60	1.3	0.68	0.41	1.4	4.9	–
Trailing shoe, beef slurry	0.50	0.60	2.0	1.04	0.62	1.0	6.1	–
Trailing shoe, pig slurry	0.50	0.60	2.3	1.20	0.72	0.8	5.9	–
Slot injection, dairy slurry	0.50	0.70	1.3	0.68	0.48	1.3	4.9	–
Slot injection, beef slurry	0.50	0.70	2.0	1.04	0.73	0.8	6.1	–
Slot injection, pig slurry	0.50	0.70	2.3	1.20	0.84	0.7	5.9	–

Contractor-owned machines

Low efficiency is considered an appropriate comparison for application by TH for which the abatement is typically only 30 %. All other abatement options have efficiencies of at least 60 % and hence may be considered high efficiency options

from 2,200 to 16,000 m^3 years^{-1}. The rate of slurry application used in the 2009 review totals 13,500 m^3 years^{-1}, based on an annual workload of 500 h spreading 27 m^3 h^{-1}. Hence it follows that UK estimates of spreading costs, expressed as m^{-3}, would be less than those from other European countries in which farms are generally smaller than in the UK (Webb et al. 2009).

6.6.1 Comparison with GAINS Output

Finally comparison was made with the abatement cost estimates, expressed as $€$ kg^{-1} NH_3-N conserved, made using the GAINs model (Amann et al. 2008) (Tables 6.14 and 6.15).

The estimated contractor costs have been increased pro-rata to the increase estimated for farm costs, i.e. from £0.35 m^{-3} to £0.42.

6.7 Conclusions

Estimates of the additional costs of applying livestock manures by RAE techniques will inevitably vary due to differences among farms and contracting operations with respect to the volumes of manure to be spread, differences in fuel and labour costs and in depreciation and interest rates. Nevertheless, we conclude that the estimates produced in the 2009 review, of £1.65 m^{-3} for slurry and £0.54 t^{-1} for manures spread using farm equipment are broadly reliable and have not been underestimated costs by more than c. 20 %. We also conclude that the costs of RAE techniques for manures applied by contractor will be substantially less than for manures applied by farm machinery.

References

Amann M, Bertok I, Cofala J, Heyes C, Klimont Z, Rafaj P, Schöpp W, Wagner F (2008) National emission ceilings for 2020 based on the 2008 climate & energy package. NEC scenario analysis report nr. 6. International Institute for Applied Systems Analysis (IIASA), Laxenburg, Austria

Anon. (2002) Defra project WA0710, evaluation of ammonia mitigation options. http://randd. defra.gov.uk/Document.aspx?Document=WA0710_558_FRP.doc. Accessed 14 Aug 2014

Anon. (2003) Integrated Pollution Prevention and Control (IPPC) reference document on best available techniques for intensive rearing of poultry and pigs, July 2003. http://eippcb.jrc.es. Accessed 17 Oct 2014

Bittman S, Kowalenko CG, Hunt DE, Schmidt O (1999) Surface-banded and broadcast dairy manure effects on tall fescue yield and nitrogen uptake. Agron J 91(5):826–833. doi:10.2134/ agronj1999.915826x

Döhler H, Eurich-Menden B (2004) Costs of ammonia abatement techniques in Germany on two model farms. UNECE Ammonia Expert Group, Poznan

Frost P, Mulholland M (2004) Alternative slurry spreading systems. Department of Agriculture and Rural Development (Northern Ireland). Technical note

Laws A, Pain BF, Webb J, Forrester A (2003) Added benefits in adopting techniques to abate ammonia emissions on UK farms. In: British Grassland Society 7th research conference, University of Wales, Aberystwyth, UK

Pineiro C, Pérez AI, Illescas P, Montalvo G, Herrero M, Giráldez R, José Sanz M, Bigeriego M (2006) Spanish Ministry of Agriculture, Fisheries and Food project for the IPPC Directive implementation in Spain. Results of 2004-2005 and future work. In: Petersen SO (ed) 12th Ramiran international conference technology for recycling of manure and organic residues in a whole-farm perspective, vol I, DIAS report Plant production no. 122, August 2006, pp 175–178

Webb J, Menzi H, Pain BF, Misselbrook TH, Dämmgen U, Hendriks H, Döhler H (2005) Managing ammonia emissions from livestock production in Europe. Environ Pollut 135:399–406

Webb J, Ryan M, Anthony SG, Brewer A, Laws J, Aller FA, Misselbrook TH (2006) Identifying the most cost-effective means of reducing ammonia emissions from UK agriculture using the NARSES model. Atmos Environ 40:7222–7233

Webb J, Morgan J, Pain B, Eurich-Menden B (2009) Review of the efficiency of methods to reduce emissions of ammonia following the application of manures to land, their costs, potential agronomic benefits and impacts on emissions of nitrous oxide. Report to the task force on

reactive nitrogen. http://www.clrtap-tfrn.org/sites/clrtap-tfrn.org/files/documents/EPMAN%
20Documents/SpreadRev_All_151009.pdf. Accessed 14 Aug 2014
Webb J, Pain B, Bittman S, Morgan J (2010) The impacts of manure application methods on
emissions of ammonia, nitrous oxide and on crop response—a review. Agric Ecosyst Environ
137:39–46

Chapter 7
Co-benefits and Trade-Offs of Between Greenhouse Gas and Air Pollutant Emissions for Measures Reducing Ammonia Emissions and Implications for Costing

Vera Eory, Cairistiona F.E. Topp, Bronno de Haan, and Dominic Moran

Abstract Both ammonia and greenhouse gases have been in the environmental research and policy spotlight in the past decades. Scientific evidence from the natural sciences and from economics have informed policy development and lead to different forms of regulations and policies both on ammonia and greenhouse gas emissions from agriculture, a sector which is an important source these pollutants. Not only agriculture is an important source of these pollutants, but the biophysical and management processes create a situation whereby the emission of these gases are linked, commonly resulting in synergies and trade-offs in mitigation practices. An understanding of these synergies and trade-offs is key in designing efficient integrated policies. This chapter contributes to that effort by providing an overview of the greenhouse gas co-effects of some of the key ammonia mitigation options and presenting an example of integrated cost-effectiveness analysis.

Evidence suggests that some win-win solutions are available where both ammonia and greenhouse gas emissions can be reduced on farms; these include improving nitrogen use efficiency in livestock and crop production, low-emission livestock housing design, slurry acidification and urease inhibitors. Conversely, pollution swapping (trade-off between ammonia and greenhouse gas reduction) is likely to occur with ammonia mitigation in other cases, for example if the amount of starch and sugar in animal feeds is increased, if changes to housing and manure management systems are made, if slurry is separated to a solid and a liquid fraction or if solid manure is aerated during storage. The effects of some measures e.g. low-trajectory manure

V. Eory (✉) • D. Moran
Land Economy, Environment & Society, Scotland's Rural College (SRUC), King's Buildings, West Mains Road, Edinburgh EH9 3JG, UK
e-mail: vera.eory@sruc.ac.uk; dominic.moran@sruc.ac.uk

C.F.E. Topp
Crop & Soil Systems, Scotland's Rural College (SRUC),
King's Buildings, West Mains Road, Edinburgh, EH9 3JG, UK
e-mail: kairsty.topp@sruc.ac.uk

B. de Haan
PBL Netherlands Environmental Assessment Agency, Bilthoven, The Netherlands

© Springer Science+Business Media Dordrecht 2015
S. Reis et al. (eds.), *Costs of Ammonia Abatement and the Climate Co-Benefits*,
DOI 10.1007/978-94-017-9722-1_7

spreading, covering slurry stores and manure heaps, and anaerobic digestion of animal waste are currently uncertain and require further investigation.

Keywords Ammonia emissions • Emission abatement • Agriculture • Greenhouse gases • Cost-benefit analysis

7.1 Introduction

Ammonia (NH_3) pollution is one of many environmental burdens arising from human activities. Both globally and in Europe the main source of NH_3 emission is agriculture, particularly animal husbandry (European Environment Agency 2013, van Vuuren et al. 2011); cattle and swine populations contributed by 54 % to NH_3 emissions in the EU-27 in 2011, while another 20 % of emissions originated from synthetic nitrogen (N) fertiliser use. In the same year, agriculture's share of EU-27 greenhouse gas (GHG) emissions was 10 %, mostly as nitrous oxide (N_2O) and methane (CH_4), not including the carbon dioxide (CO_2) emissions and carbon (C) sequestration effects of land use and land use change (European Environment Agency 2014). The agricultural emissions of CH_4, N_2O and NH_3 are interrelated: they have common sources and their emission rates depend on common factors, such as farm management, weather conditions and soil type.

N is an important element in agricultural production, and was the limiting factor in crop production before inorganic fertilisers became widespread (Smil 1999). N_2O and ammonia are parts of the nitrogen cascade, whereby the captured atmospheric di-nitrogen (N_2) is transformed into various forms of reactive N (N_r) (Galloway et al. 2003). Moreover, they can be transformed into each other in biochemical processes. The agricultural activities responsible for N_2O, NH_3 and CH_4 emissions overlap; animal husbandry emitting NH_3, N_2O and CH_4 and crop production being responsible mainly for NH_3 and N_2O emissions (Fig. 7.1). This complex relationship between biophysical and management processes make synergies and trade-offs inherent in the system.

The potential synergies and trade-offs affect our mitigation efforts and need to be taken into account when optimising abatement activities. Focusing on a single pollutant can lead to under- or overestimating the total benefit of pollution control, and thus to suboptimal mitigation effort (Nemet et al. 2010). Economic efficiency is an important consideration in environmental policy formulation. Regulatory interventions should aim to reduce pollution at least cost, or at least in ways where costs are demonstrably outweighed by benefits; the latter quantified in terms of avoided damages. This criterion involves a comparison of private and what economists terms social costs, which are essentially the wider environmental costs and benefits of pollution control.

Most decisions in livestock systems design, animal feeding, manure management and crop fertilisation are likely to affect more than one of the gases mentioned above. To support policy decisions, integrated assessment of the mitigation of NH_3 and GHG is needed. This chapter reviews current knowledge on the positive and negative co-effects of NH_3 abatement measures in agriculture, focusing on the GHG N_2O and CH_4.

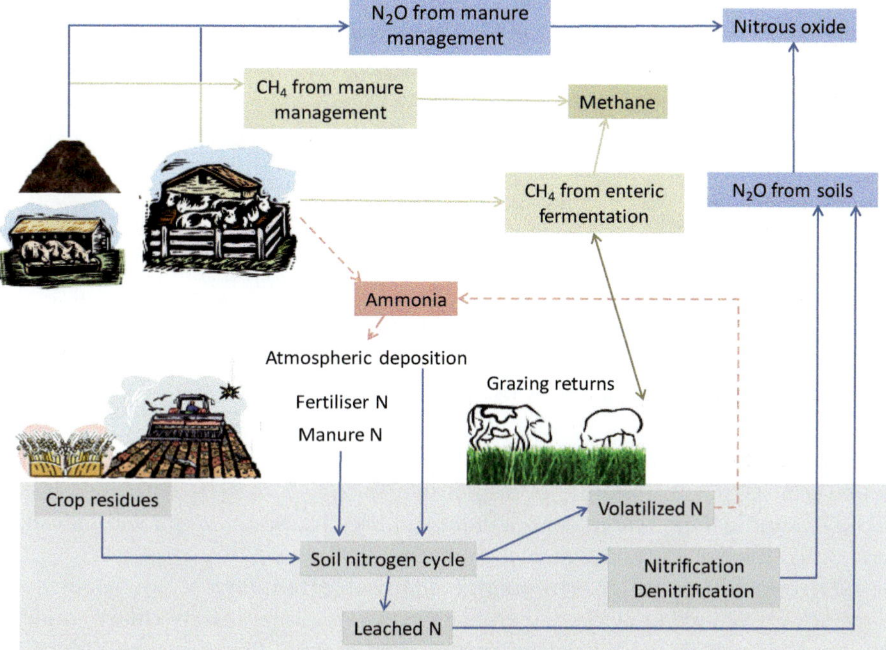

Fig. 7.1 NH$_3$ and GHG emissions from farming activities (Figure courtesy of T. Misselbrook, Rothamsted Research)

The next section provides background on agricultural emissions of NH$_3$ and GHGs, Sects. 7.4, 7.5, 7.6 and 7.7 discuss the likely co-effects of NH$_3$ mitigation measures in various areas of farm management, an example of integrated cost-effectiveness analysis is presented in Sect. 7.8, and conclusions drawn in the last section.

This review focuses on the farm gate pollutants arising related to practices on temperate farms in Europe. Nevertheless, some important implications on emissions beyond the farm gate, for example GHG emissions from fertiliser production, and changes in the soil C stock are mentioned, and the experimental evidence reviewed goes beyond Europe.

7.2   Ammonia and Greenhouse Gas Emissions in Agriculture

7.2.1   Ammonia

NH$_3$ contributes to acidification and eutrophication in marine and terrestrial ecosystems, and it also has detrimental effects on human health (Smart et al. 2011). A

Table 7.1 European N_2O, CH_4, CO_2 and NH_3 emissions and the contribution of main agricultural activities to the agricultural emissions (in EU-27, in year 2011)

Emissions	N_2O (Mt CO_2e)	CH_4 (Mt CO_2e)	CO_2 (Mt CO_2)	NH_3 (kt NH_3)
Total emissions	337	397	3,747	3,635
Agricultural emissions	275	197	0	3,394
Contribution to agricultural emissions				
Enteric fermentation	0 %	74 %	n/a	0 %
Manure management	11 %	24 %	n/a	74 %
Rice cultivation	0 %	1 %	n/a	0 %
Agricultural soils	89 %	0 %	n/a	25 %
Field burning	0 %	0 %	n/a	0 %

Source: European Environment Agency (2013) and European Environment Agency (2014)

small part of the NH_3 released into the environment is converted into N_2O, which is a powerful GHG. Agriculture is responsible for 94 % of NH_3 emissions in the EU-27 countries, the remainder coming from road transport, waste and industrial processes (European Environment Agency 2013) (Table 7.1).

NH_3 originates both from livestock and arable farming. N not retained by livestock is excreted in faeces and urine excreta; the former mainly contain organic N compounds, while the N in urine is mainly non-protein N (mostly urea) (Monteny and Erisman 1998). Birds excrete uric acid which is readily hydrolysed to urea (Webb 2001). The urea can be quickly hydrolysed into NH_3 by the enzyme urease, which can be found in the faeces, on fouled surfaces and in soil. On the other hand, the protein-N of faeces first has to go through the slow process of mineralisation to become part of the total ammoniacal N (TAN) pool (i.e. NH_3 and ammonium (NH_4^+)), therefore NH_3 volatilisation is much lower from faeces (Bussink and Oenema 1998). All in all, the N content of the excreta is partially lost as NH_3 from the livestock houses and manure stores and from the fields either after being deposited during grazing or having been applied to soils as a fertilizer. As for cropping activities, inorganic N fertilisers are also sources of NH_3 emissions, but a great difference exist according to the type of fertilizer and the application method (Hutchings et al. 2001; Misselbrook et al. 2000).

Various physical and biological factors have an effect on what proportion of the N in livestock excreta and in inorganic fertilizers is being lost as NH_3. NH_3 emissions are positively correlated with pH, temperature and air velocity and also increase with higher urease concentration (Bouwmeester and Vlek 1981; Carmona et al. 1990; Sommer et al. 1991). At the same time the NH_3 can be converted into other N compounds by processes like microbial immobilisation, assimilation by plants, and nitrification (Rennenberg et al. 2009), reducing the TAN content and thus NH_3 emissions.

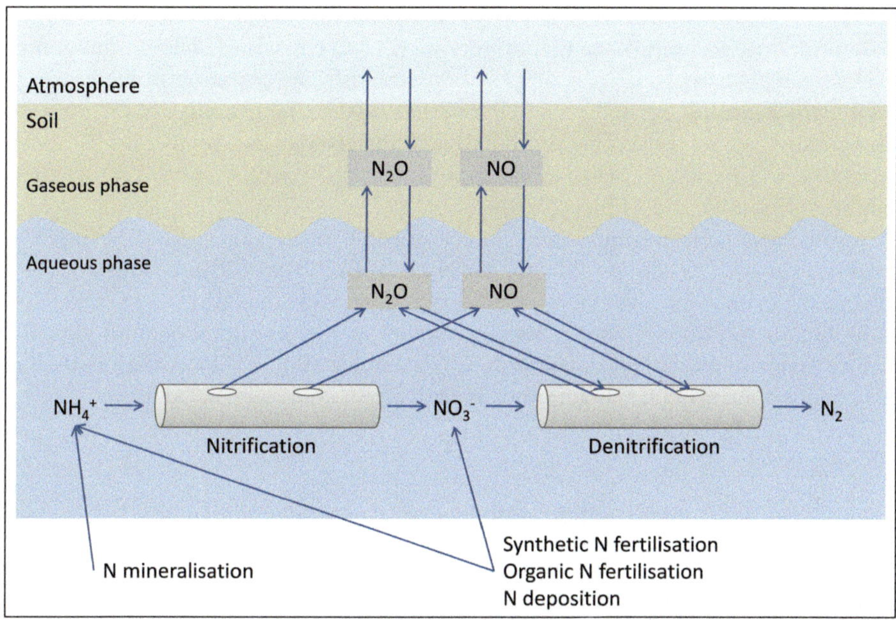

Fig. 7.2 'Hole-in-the-pipe' model of the regulation of trace-gas production and consumption by nitrification and denitrification. NH_4^+: ammonium, NO_3^-: nitrate, N_2: di-nitrogen, N_2O: nitrous oxide, NO: nitric oxide (Bouwman 1998)

7.2.2 Nitrous Oxide

N_2O is a potent greenhouse and, at the same time, the most important ozone depleting substance (Ravishankara et al. 2009). Primary human-related sources of this gas are agriculture, and, to a lesser extent, combustion and industrial processes, the former contributing 50 % of the total N_2O emissions in Europe (European Environment Agency 2014), and 75 % of the global total (EPA 2012). Most of the agricultural emissions are produced in soils, with a lesser amount generated during manure management (Table 7.1): the N added to soils (e.g. inorganic and organic fertilisation, crop residues, atmospheric deposition, livestock excreta on pastures) and excreted by livestock in animal houses are the main sources of N_2O. Additionally, soluble N compounds leached into water bodies and gaseous NH_3 emissions can also be converted into N_2O.

The two main processes of N_2O generation are nitrification and denitrification (Fig. 7.2). In nitrification, in aerobic conditions NH_4^+ is transformed into nitrite and then into nitrate (NO_3^-), and, particularly in low oxygen concentration, N_2O is emitted (Bremner and Blackmer 1978). Subsequently, denitrifying bacteria convert NO_3^- into N_2 gas in anoxic conditions. However, if the concentration of molecular oxygen increases, the formation of N_2O rather than N_2 is promoted through incomplete denitrification (Firestone et al. 1980). As the nitrifying and denitrifying bacteria require different oxygenation level, aerobic and anaerobic pockets being in close proximity to each other favour very high N_2O emissions. The production of N_2O depends on the NH_4^+ and other N compounds' concentration in the environment (which are all related to soil properties and manure composition) and on temperature: warm conditions promote bacterial growth, but temperatures above approx. 50 °C inhibit it, because nitrifying and denitrifying bacteria are not thermophilic (Sommer and Moller 2000).

Agricultural soils are important sources of N_2O; on average 1 % of N added to the soils escapes to the air directly as N_2O (IPCC 2006). These emissions are enhanced during wet and warm conditions. Livestock operations generate N_2O emissions mainly through solid manure storage and in livestock bedding, but the surface layers of slurry can also emit N_2O (Chadwick et al. 2011). Additionally, the NH_3 emitted by agricultural activities is an indirect source of N_2O.

7.2.3 Methane

Globally and in the EU-27 approximately half of anthropogenic CH_4 emissions originate from agriculture, dominated by enteric fermentation; while the other half mainly arises from gas drilling, coal mining and landfill (European Environment Agency 2014). In 2010 global agricultural CH_4 emissions were dominated by enteric fermentation (62 %), followed by rice cultivation (17 %) and livestock waste (7 %) (EPA 2012). The pattern in Europe is similar, with the notable difference of emissions from rice cultivation being marginal (Table 7.1).

CH_4 is produced by anaerobic respiration of methanogen microorganisms. This process happens when the breakdown of organic material takes place in the lack of oxygen and no other electron acceptors are present but small organic compounds and CO_2. The release of CH_4 intensifies with higher temperatures (Khan et al. 1997), even above 50 °C, as many methanogens are thermophile microorganisms (Sommer and Moller 2000). Eventually the emitted CH_4 is oxidised back to CO_2 in the atmosphere.

Environments favouring methanogenesis occur in the digestives system of animals, in manure stores and in anoxic soils, like wetlands and rice paddies. In animals methanogenesis happens during bacterial fermentation of feedstuff in the rumen of cattle, sheep and other ruminants, and also occurs, to a lesser extent, in the large intestine of all livestock. Manure management is also responsible for CH_4 emissions, where these emissions originate from the anaerobic decomposition of livestock bedding and manure, especially in liquid manure stores. The manure composition (especially the proportion of volatile solids) and the length of the anaerobic storage period are important factors in determining the CH_4 emissions. While the CH_4 emissions of ruminants are mainly produced in the rumen, those from pigs and poultry are mostly manure-born (Table 7.2).

Table 7.2 Contribution of the livestock species to CH_4 emissions from enteric fermentation and manure management (EU-28, 2011)

	Enteric fermentation (%)	Manure management (%)
Cattle	82	49
Sheep	12	1
Pig	3	44
Poultry	0	4

Source: European Environment Agency (2014)

7.2.4 Carbon Dioxide

Although considering all anthropogenic GHG emissions CO_2 contributes the most to global warming, its importance in agriculture is tertiary to N_2O and CH_4, its main sources being land use, land use change and fossil fuel combustion. Agricultural land use activities, particularly changes in the land use, e.g. from cropland to grassland or vice versa, result in a positive or negative change in the soil C stocks. The former process removes CO_2 from the atmosphere (C sequestration), the latter releases CO_2, for example cropland and grassland related land use and land use change added 78 Mt CO_2e emissions to the EU-27 inventory in 2011 (European Environment Agency 2014). In the same year fossil fuel combustion (agriculture together with forestry and fisheries) contributed with a further 75 Mt CO_2e to the total emissions (European Environment Agency 2014).

7.3 Dietary Options

Animal nutrition has considerable effects on NH_3 and GHG emissions, both directly and indirectly. Optimal feed composition and additional factors (e.g. water and feed availability, temperature in the stalls) facilitates higher energy and protein use efficiency and improves animal health (Roche 2006; VandeHaar and St-Pierre 2006), and thus reduces waste directly at the animal level. Good feeding practice can boost the physical efficiency at the farm level as well, reducing waste indirectly. For example, dietary factors play an important role in both the age of first breeding and in the fertility of dairy cattle (VandeHaar and St-Pierre 2006), impacting on the length of unproductive periods and on the need for replacement heifers in the herd. This section focuses on dietary measures targeting N intake and briefly presents two additional options relevant for piggeries. These options impact on the whole N cascade, and thus they effect direct and indirect N_2O emissions, and in some cases they also effect enteric and manure CH_4 emissions.

Removal of the excess N from the feed is a widely proposed feeding measure to control NH_3 emissions through the reduction of N excreta. Though significant progress have been made in some European regions in this respect (Dalgaard et al. 2012; Groot et al. 2006), farmers still often feed livestock with excess protein in order to avoid the risk of reduced production due to inadequate N intake (Aarnink and Verstegen 2007; Aberystwyth University 2010). One survey in the USA showed that farmers, on average, fed 6.6 % more N than was recommended by the National Research Council, resulting in an increase of 16 % and 2.7 % in urinary N and faecal N excretion, respectively (Jonker et al. 2002). Such excess in N inputs can be avoided without productivity loss (Hristov et al. 2011; Rotz 2004).

A range of practical solutions have been suggested to achieve the aforementioned reduction in N intake, which translates into a reduction in the protein content of the diet with a parallel increase of non-protein substrates (often carbohydrates). Monogastric animals are fed with compound feeds, where the crude protein (CP) content reduction can be achieved by replacing part of the high N content feed components with components rich in energy or fibre. In the case of ruminants there is scope to alter the ratio of forage versus concentrate feeds (the former usually richer in proteins), the CP content of the concentrates by altering their composition (e.g. more components rich in starch) and the CP content of the forage (for example by providing starch-rich maize silage, changing the grass varieties or reducing the grass fertilisation rate). Where the low-protein diet is limited in essential amino acids (AAs) then supplementing these to balance AA composition might be needed to maintain production levels (Aarnink and Verstegen 2007; VandeHaar and St-Pierre 2006). For ruminants the AAs must be in rumen-undegradable form to go under enzymatic digestion and absorbed by the animal itself rather than its microorganisms (Broderick et al. 2008).

A reduced N intake translates into reduced N excretion; there is a linear relationship between dietary CP content and N excretion in dairy and beef cattle (Hristov et al. 2011; Waldrip et al. 2013). The drop in the N excretion is mostly due to a decrease in the urinary N, while the faecal N remains relatively constant (Bussink and Oenema 1998). As NH_3 volatilisation is much higher from the urine than from the faeces, the saving in NH_3 emissions can be proportionally higher than the savings in the N excretion (Rotz 2004). Hristov et al. (2011) provides a summary of experiments reporting 28–50 % reduction in NH_3 emissions from cattle manure storage and parallel reduction in NH_3 emissions after soil application of the manure when the CP content of the diet was altered from high level (between 15.4 and 17.5 %) to low level (between 12.5 and 14.8 %). The relationship between N intake and excretion and NH_3 volatilisation is similar for pigs and poultry to that of cattle. Reducing the CP content of the diet while administering essential AAs can reduce N excretion in pigs and poultry and hence leads to a reduction in NH_3 emissions (Rotz 2004). An additional effect of the reduced protein intake in pigs is that the manure becomes more acidic, further decreasing NH_3 volatilisation (Canh et al. 1998a).

The GHG effects of the reduced N intake are multiple. The reduced NH_3 emissions imply lower indirect N_2O emissions from NH_3 volatilisation, and the reduced N excretion is expected to translate into reduced direct N_2O emissions from manure storage and application, for example the IPCC Tier 2 calculations assume a linear relationship between direct manure storage N_2O emissions and CP intake (IPCC 2006). However, experimental evidence is not conclusive in this respect (see below) (Philippe and Nicks 2015). Furthermore, the changes in the feed composition can alter the CH_4 emissions both from manure storage and from enteric fermentation. Regarding the latter, if in the low protein feed the energy replaced comes from fibre or sugars, more enteric CH_4 is likely to be produced, whereby if it comes from starch or fat, methane emissions can be reduced (Dijkstra et al. 2011).

Looking at GHG effects of reduced CP content in the diet, Philippe et al. (2006) found a net increase of 19 % in GHG emissions from the buildings of pigs kept on deep litter. The two-phase diet CP content was 18.1 % and 17.5 %, respectively, for growers and finishers in the high protein group and was 15.5 % and 14.0 %, respectively, for the two growth stages in the low protein group, (the latter diets were supplemented with AAs). While NH_3 emissions from the low protein group were significantly lower than from the high protein group, N_2O emissions doubled for the former group, and this was only partially offset by the reduced CH_4 emissions. Conflicting results exist on the consequence of low protein diet on GHG emissions from manure storage: Külling et al. (2001) reported increased CH_4 and reduced N_2O emissions with zero net GHG effect for dairy manure, Velthof et al. (2005) found reduced CH_4 emissions from pig manure, whereas there was no statistical difference for either GHGs in a third experiment (Lee et al. 2012) on dairy manure. Kreuzer and Hindrichsen (2006) imply that the C:N ratio of the manure is a more important factor in the CH_4 emissions from manure storage than the N content, with a low C:N ration resulting in higher CH_4 emissions. The complex effect of reduced CP content on GHG emissions from manure application directly affected by further factors such as volatile fatty acid content of the manure (Sommer et al. 2004), the type of manure management, soil characteristics and weather conditions. The direction of change in N_2O emissions in a soil incubation study varied with soil type after application of pig manure (Velthof et al. 2005), while no statistical difference was observed between GHG emissions from manure application following feeding dairy animals with high and low protein diets (Lee et al. 2012); Misselbrook et al. (1998) found no change in N_2O emissions, although CH_4 emissions were reduced with lower CP content (however, CH_4 emissions from manure application are marginal).

The CP content of the diet is often reduced with a correspondent increase in the starch content. This has a positive side-effect on GHG emissions, more starch also leads to lower enteric CH_4 emissions in ruminants (Aberystwyth University 2010; Mc Geough et al. 2010a, b; Moe and Tyrrell 1979), since CH_4 production originates mainly from the by-products of structural polysaccharide (e.g. cellulose) fermentation (Ellis et al. 2008). It should also be noted that too much starch is detrimental to the animal health as it causes rumen acidosis (Owens et al. 1998), and feeding high levels of concentrates diminishes the main environmental benefit of cattle: converting structural polysaccharide (not only grass, but fibrous by-products, like almond hulls, citrus pulp) into high-quality protein for human use (Oltjen and Beckett 1996; VandeHaar and St-Pierre 2006). Additionally, the net GHG saving achievable with this method is questionable, as the soil C content of land under arable cultivation (i.e. silage maize) is lower than that of grasslands, and such a change in land use results in CO_2 emissions from soil (Beauchemin et al. 2010; Vellinga and Hoving 2011).

If the CP content is partially replaced by dietary fats in ruminant diets, enteric CH_4 emissions are reduced, partly due to a suppression of some of the rumen microflora and to a lower extent due to unsaturated fatty acids acting as hydrogen sinks in the rumen (Johnson and Johnson 1995; Martin et al. 2010). The savings in enteric CH_4 emissions is proportional to the amount of fat in the diet (Beauchemin et al. 2008; Eugene et al. 2008; Grainger and Beauchemin 2011) and can be increased up to 5–6 % without adverse nutritional effects. According to Hristov et al. (2013) and Martin et al. (2010), the question of persistence of the mitigation effect has not been adequately addressed yet: some studies do report long-term effects, but data are inconsistent. In addition, two mechanisms might (partially) offset the savings in enteric CH_4 emissions: potential increases in manure storage CH_4 emissions (Kulling et al. 2002) and in emissions related to the production of feedstuff, especially if they induce a land use change deteriorating soil C stocks, for example via an increase in palm oil plantations.

As Peyraud and Astigarraga (1998) summarise in a review, with decreasing amount of N fertiliser applied, the protein content of the grass substantially decreases and the water soluble carbohydrate content increases. Livestock feeding on such grass excrete markedly reduced urinal N, thus related NH_3 emissions are lower. With this method, fertiliser related NH_3 and N_2O emission savings are also achieved through reduced fertiliser use per land area, though this benefit can be outweighed by the lower grass yield and thus the potential reduction in soil C stocks, if maintaining livestock production leads to a conversion or woodlands or wetlands into grazing land. Similarly, using high sugar content grass varieties can also improve N efficiency of cattle by increasing the capture of N into microbial protein, and thereby increasing milk protein outputs and at the same time reducing urinary N excretion (Moorby et al. 2006). However, lowering the N fertilisation of the grass affects ruminants' enteric CH_4 emissions, though research in this respect is so far inconclusive (Dijkstra et al. 2011). Similarly, contrasting results exist on the enteric CH_4 effects of the high sugar content grasses; a modelling exercise by Ellis et al. (2011), presented variable results on CH_4 emissions, depending on the concurrent changes in the diet and the measurement unit, i.e. whether results were expressed as percentage of gross energy intake or grams per kg of milk.

As discussed earlier, farmers often perceive a high risk of a reduction in productivity in response to lower protein intake. Falling production results in both financial losses to the farmers and a possible increase in pollutant load per production unit (Weiske 2005). An increasing reliability of feed recommendation systems should help to provide the confidence to farmers in better diet formulation (Cuttle et al. 2004). In addition, stricter quality control of feed materials could also help to balance nutrients (Nahm 2002). However, as St-Pierre and Thraen showed (1999), there might be a discrepancy between the maximum physical efficiency and the maximum economic efficiency, causing overfeeding of the animals, though this discrepancy varies not only with livestock and crop varieties but also with the changes in the relative price level of N inputs and products.

Beyond lowering the N intake of the animals, two more dietary options aiming to reduce NH_3 emissions from pig farms are discussed here. First, providing a higher non-starch polysaccharide content diet (e.g. sugar beet pulp) with a constant CP concentration decreases both the urinary-N/faecal-N ratio and the faeces pH (Canh et al. 1997), potentially lowering NH_3 volatilisation. But with more fermentable polysaccharides, volatile fatty acid concentration increases, and thus increases CH_4 emissions from slurry (Velthof et al. 2005). A further economic and environmental disadvantage of this option is that using higher digestibility raw materials decreases the use of low-cost by-products (Edwards et al. 2002). Second, the urine pH can be lowered by replacing calcium carbonate with calcium sulphate in the pig diet, reducing NH_3 volatilisation (Canh et al. 1998b). At the same time the lower ileal pH might reduce CH_4 emissions from liquid manure stores (Kim et al. 2004). Indeed, Velthof et al. (2005) found that this technique reduced both CH_4 and NH_3 emissions from anaerobic storage, but N_2O emissions from soil incubation were variable, depending on the soil type.

7.4 Livestock Housing

During periods spent in houses animals excrete the urine and faces onto hard surfaces, onto the bedding or into slurry pits, where a substantial part of the TAN content volatilises – housing emissions constitute for ¼ of agricultural NH_3 emissions in the UK (Misselbrook et al. 2012). A range of housing characteristics impacts on the gaseous emissions, including the manure handling system (liquid or solid), the housing design (e.g. ventilation type and airflows, floor surface, inside or outside manure pit), management decisions (e.g. frequency of manure removal, manure additives) and climatic conditions. Numerous management and technical options exists both in slurry-based and litter-based systems to reduce NH_3 emissions, and in many of them affect GHG emissions as well. Below five options are summarised.

In litter-based cattle and pig systems increasing the amount of straw has a positive effect on both NH_3 and GHGs. Adding extra straw bedding (25–50 % above typical practice), targeting especially the wetter and more fouled areas is effective on in-house and manure storage related NH_3 emissions (Gilhespy et al. 2009; IGER 2005). More straw reduces air-flow and consequently volatilisation while at the same time the higher C:N ratio enhances immobilisation of NH_4-N, reducing NH_3 emissions considerably (Dewes 1996). The additional straw efficiently reduces N_2O emissions (Sommer and Moller 2000; Yamulki 2006) and might reduce or increase CH_4 emissions, as the better aeration inhibits anaerobic methanogens, but the additional carbohydrates provide extra substrate for methanogens (Philippe and Nicks 2015).

Converting fully-slatted floors in pig houses to partly slatted floors can reduce the fouled surface area by encouraging the pigs to dung over the slatted area, and reduces air exchange between the pit and the house – thus reducing NH_3 emissions

(Aarnink et al. 1997). Concerning both CH_4 and N_2O emissions contradictory results have been published, with no consensus yet on the effects (Philippe and Nicks 2015).

In the straw-flow systems in pig houses (Bruce 1990) the straw is added at the top of a sloped lying area and it travels down the slope towards an excretion area, where it mixes with the dung. The manure then leaves the excretion area either into an underneath pit or onto a scraped passage, depending on the actual design. According to the amount of added straw the produced manure is either slurry or solid manure. Amon et al. (2007) experienced reduced in-house NH_3, CH_4 and N_2O emissions from this system compared to fully slatted floor systems, while Philippe et al. (2007) found only the CH_4 emissions to be lower from straw flow system compared to houses with fully slatted floors, with N_2O emissions being at the same level and NH_3 emission being 2.5 times higher.

Since NH_3 emissions are positively correlated with temperature, frequent removal of manure from pig and cattle in-house storage to outdoor storage reduces NH_3 emissions, given the temperature is lower outside than inside (Hartung and Phillips 1994). As CH_4 emissions increase with temperature, they can also be reduced with this option by 10–19 % from both liquid and solid systems, and though N_2O emissions might increase, they stay negligible compared to CH_4 (Philippe and Nicks 2015). Pit flushing and scraping in piggeries can concurrently reduce CH_4 and N_2O emissions, the more frequent the flushing or scraping, the higher GHG savings to achieve. Regarding poultry, keeping the poultry manure dry or drying it on manure belts saves NH_3 emissions, but a study comparing laying hen housing systems with aerated deep pit manure storage and with forced drying manure belt removal found higher CH_4 emissions from the removal system (Fabbri et al. 2007).

Finally, installing air scrubbers or biofilters to remove NH_3 from animal houses is a very efficient end-of-pipe technology used for mechanically ventilated houses (Melse et al. 2009). Air scrubbers work on a chemical basis: NH_3 is captured as NH_4^+ salt in an acidic solution (mostly sulphuric acid). Biofilters use microorganisms to convert NH_3 into NO_3^-. In both cases, the discharge water can be used as fertiliser. Slightly increased CO_2 emissions arise from the increased energy usage if mechanical ventilation is only installed for the air filtering purposes. Furthermore, in biofilters there is a risk of increased N_2O emissions due to the nitrification process in the filter (Melse and van der Werf 2005); one study found that 20 % of the biofilter's N content was released as N_2O (Maia et al. 2012).

7.5 Manure Storage

Animal excreta are a crucial source of gaseous emissions from agriculture, either as deposited on pastures during grazing, or collected, stored, and subsequently applied as fertiliser. The emission profile of various manure handling systems are markedly different: liquid systems are generally an important source of NH_3 and CH_4

emissions, while solid systems emitting more N_2O and less NH_3 and CH_4. In both cases several factors play important roles in the emissions, like the initial composition of manure (e.g. N content, TAN content, C:N ratio, water content, dry matter content, volatile solid content), the manure store characteristics (e.g. covered or not), the management decisions (e.g. length of in-house storage and outside storage, aeration level, additives) and the environmental variables (e.g. temperature, rainfall) all affect the gaseous emissions and leaching from manure (Monteny et al. 2006; Sommer et al. 2004). Solutions favouring lower NH_3 emissions in both types of systems are very likely to have synergies or trade-offs with GHGs.

7.5.1 Liquid Manure

In conventional liquid manure storage, where the slurry is not aerated, the anaerobic environment does allow denitrification happening only at a very low rate, close to the surface, producing only small amount of N_2O and holding back subsequent denitrification which would also be a source of N_2O (Sommer et al. 2000; Zhang et al. 2005). On the other hand, the anaerobic environment is ideal for methanogen microorganisms, making slurry stores an important source of CH_4 emissions. The GHG effects of covering slurry stores, separating the slurry into solid and liquid fractions and slurry acidification is presented here.

Covering slurry stores substantially reduces NH_3 emissions (VanderZaag et al. 2008 923 /id). As a result, the TAN content of the slurry increases, and it will be susceptible to elevated emission levels after having been spread on the soil, unless low NH_3-emission spreading techniques (see Sect. 7.7) are implemented. The effects of covering slurry stores on GHGs are less explored than the consequences on NH_3, and the results are highly variable and inconclusive (VanderZaag et al. 2008 923 /id), as presented below via selected examples from the wide range of covering options.

Floating covers can be made of organic (e.g. straw, vegetable oil), inorganic (expanded clay) or synthetic materials. If manure properties allow and the slurry is not agitated, natural crust can develop on the surface, especially on cattle slurry (Chadwick et al. 2011). The crust development can be artificially enhanced by covering the surface with straw or Leca (expanded clay) pebbles. Though greatly reducing NH_3 emissions, crust and straw cover provides suitable conditions to nitrifying bacteria and thus provoke a dramatic increase in N_2O emissions, especially in dry weather (Berg et al. 2006; Sommer et al. 2000). At the same time these surface layers can be colonised by methanotroph bacteria, oxidising part of the methane to CO_2 (Petersen and Ambus 2006; Petersen et al. 2005): Sommer et al. (2000) observed a similar 28 % reduction with straw, leca and crust cover and VanderZaag (VanderZaag et al. 2009 924 /id /d) also noted 24–28 % savings in CH_4 emissions with straw cover. However, a reduction in the CH_4 emissions is not always observed (Berg et al. 2006; Hudson et al. 2006; Petersen et al. 2013). On the other hand, permeable synthetic cover though reduces N_2O emissions, the overall

GHG emissions are not affected substantially due to no significant effect on CH_4 emissions (VanderZaag et al. 2010 925 /id).

Rigid covers (e.g. wooden or concrete lids or tent structures) may also be promising for CH_4 emission reductions: Clemens et al. (2006) reported 14–16 % savings in CH_4 emissions from crusted cattle slurry if covered with wooden lid, Amon et al. (2006) found that CH_4 emissions were 18 % lower from lid-covered than from straw-covered cattle slurry, though CH_4 emissions might increase as well (Silsoe Research Institute 2000). The effect of solid covers on N_2O emissions is more variable, some research showing benefits others disadvantages (Amon et al. 2006; Clemens et al. 2006; Petersen et al. 2009; Silsoe Research Institute 2000).

Finally, impermeable floating or rigid covers can be equipped with gas pipes and pumping system to collect the gas produced. In such systems most of the CH_4 is captured and converted to CO_2 either by direct flaring, reducing the global warming potential substantially, or by purification and use in electricity or heat generation, providing further GHG benefits by replacing non-renewable energy sources (Petersen and Miller 2006).

The mechanical or chemical separation of slurry produces a solid and a liquid fraction with markedly different gaseous emission patterns; the solid fraction akin to untreated solid manure while the liquid fraction is similar to slurry. So far the results show contrasting effects on NH_3 and GHG emissions: the former is often higher from the separated slurry than from the unseparated slurry (Amon et al. 2006; Dinuccio et al. 2008, 2011; Fangueiro et al. 2008), while the overall CH_4 and N_2O emission is reduced by separation by as much as 26–37 % (Amon et al. 2006; Fangueiro et al. 2008), though Dinuccio et al. (2008) observed higher GHG emissions as well. The GHG is attributable to a drop in CH_4 emissions, usually counter-balancing the – sometimes considerably – increased N_2O emissions. Regarding emissions from the application of separated and not separated slurries, Amon et al.(2006) found higher overall GHG emissions from the separated slurry, but as field application GHG emissions were only 1.3 % of the total GHG emissions, the increased only slightly reduced the net GHG benefits.

Slurry acidification can reduce NH_3 emissions from housing, storage and application by 10–60 % (Kai et al. 2008; Monteny and Erisman 1998), though Berg et al. (2006) reported increased NH_3 emissions from acidified slurry covered by perlite or Leca. They also found 43–76 % less CH_4 emissions from the acidified tanks, and an earlier research showed that pH below 5.0 substantially reduced CH_4 emissions, while $pH < 4.5$ almost completely mitigated them (Berg and Hornig 1997). A recent paper investigated the effects of acidification on NH_3 and CH_4 emissions, and found 93–98 % and 67–87 % reductions, respectively (Petersen et al. 2012). When applied on land, acidification delays nitrification and N_2O formation, and total emissions of N_2O might also be reduced if the slurry is also separated (Fangueiro et al. 2010).

7.5.2 Solid Manure

Traditionally, solid manure was usually composted, i.e. the manure was aerated by turning the heap several times during the storage period. As composting progresses, the physical and biological circumstances and microbial communities change substantially, leading to a temporal pattern in the gaseous substances generated. The compost heap also has a significant spatial heterogeneity, supplemented with a prominent temperature and oxygen gradient from the surface to the centre. Furthermore, climatic conditions modify the surface layer of the heap, altering mainly the N_2O emissions (Petersen et al. 1998).

The first phase of composting is characterised by high microbial activity, quick decomposition of easily degradable substances, high CO_2 emissions, intensive heat production, depletion of acidic components, with very low CH_4 emissions and decreasing N_2O emissions (Hellmann et al. 1997). The second phase (thermophilic phase) is a high temperature phase, with quickly declining CO_2 emissions but high CH_4 emissions from the centre of the heap. N_2O emissions are restricted to the surface in this phase first because the nitrifying and denitrifying bacteria are not thermophilic (Sommer and Moller 2000), and secondly because of the anaerobic environment in the centre of the heap. Close to the surface anaerobic and aerobic pockets close to each other allow for the NH_4^+ to be nitrified and the NO_3^- to be denitrified (Hansen et al. 2006; Hellmann et al. 1997). In the third phase (curing phase) the CH_4 emissions are decreasing and the N_2O emissions are increasing due to the lower temperature, whereas the CO_2 emissions remain low (Hellmann et al. 1997).

The physical and biological circumstances of solid manure storage can be altered by different practices, like waterproof cover, anaerobic storage (airtight cover), compression at the beginning of storage, cut and mix before storage, aeration, or adding extra straw to the manure. Consequences of these options on NH_3 and GHG emissions are discussed below.

By compaction or airtight covering an increased proportion of the solid manure heap becomes anaerobic, reducing NH_3 emissions by 19–98 % (Amon et al. 1997, 2001; Chadwick 2005; Kirchmann and Witter 1989), though with variable effects on GHGs. Chadwick (2005) reported about inconclusive GHG effects of simultaneous compaction and covering of the manure heap. N_2O emission changes ranged from -71 % to 19-fold increase, the effect on CH_4 emissions were from -78 to $+139$ %. Sommer's (2001) results for porous covering and compacting the heaps showed that these options increased both GHG emissions, porous covering resulting in moderate increase, while compacting leading to 1.5 and 5.5-fold increase in N_2O and CH_4 emissions, respectively. Amon et al. (2001) compared anaerobically stacked heap with one composted, and found higher GHG emissions from the stacked heap (+7 % in winter, and +347 % in summer). On the other hand, results from Hansen et al. (2006) suggest that the gaseous emissions from the separated solids of anaerobically digested pig slurry can simultaneously be reduced: airtight covering decreased the emissions of NH_3, N_2O, CH_4 and CO_2 by 12 %,

99 %, 88 % and 93 %, respectively. The authors explained the reduction in CH_4 emissions despite the anaerobic conditions by the lower temperature of the heap, which is not favourable to methanogens.

Adding extra straw to the farmyard manure reduces the density, increases the C: N ratio, increases the porosity of the manure and enhances the airflow within the heap (affecting both the oxygen supply and the removal of volatile compounds). The high C:N ratio enhances the immobilisation of NH_4^+-N and thereby reduces NH_3 volatilisation (Dewes 1996; Kirchmann and Witter 1989). The lower level of available NH_3 restricts nitrification, while the aeration further hinders denitrification, reducing N_2O production by 42–99 %; additionally the higher oxygen level impedes methanogenic activity, abating 45–99 % of CH_4 emissions (Sommer and Moller 2000; Yamulki 2006). According to Fukumoto et al. (2003), stocking the manure into smaller pile sizes also increases the oxygenation rate of the heap, and leads to a decrease in both GHG and NH_3 emissions (67 %, 77 % and 64 % of N_2O, CH_4 and NH_3 emissions were eliminated, respectively).

Forced aeration also reduces the number and volume of anaerobic sites in the centre of the heap (Fukumoto et al. 2003), and controls GHG emissions (reductions up to 90 %), although the NH_3 emission levels increase linearly with the air flow rate (Osada et al. 2000). Similarly, the effect on NH_3 emissions of turning the manure heaps more frequently is not favourable, with 44–100 % increase in the emissions (El Kader et al. 2007; Parkinson et al. 2004; Szanto et al. 2007) and substantial increase in the leached NH_4-N as well (Parkinson et al. 2004), though Hassouna et al. (2008) found no significant effect on NH_3 emissions. At the same time, the results regarding N_2O are inconclusive (Chadwick et al. 2011). CH_4 emissions are reported to be hugely decreased by frequent turning by one author (Szanto et al. 2007), and another research found no significant change in them (Hassouna et al. 2008).

7.6 Soil Fertilisation

With the widespread use of inorganic fertilisers agricultural soils become an important source of N_2O emissions, which is produced during the nitrification and denitrification of the N_r added to the soils. Additionally, urea and NH_4^+-based synthetic fertilisers, along with livestock manures spread on land and excreta deposited on pastures are significant sources of NH_3 emissions. On the other hand, as most agricultural soils in Europe have predominantly oxidised environment, the CH_4 emissions observable after fertilisation originate from the methane generated during the storage of liquid organic fertilisers (Sommer et al. 2009).

Synthetic fertiliser and manure spreading techniques which minimise the contact of manure with air are efficient ways to abate NH_3 emissions. Trailing hose spreaders apply slurry in narrow bands on top of the surface, while trailing shoe applicators have shoe-like attachments to deposit the slurry below the crop canopy. Injection techniques make shallow or deep cuts in the soil where slurry or other liquid fertiliser is placed. Finally, fertiliser can be incorporated into the soils by ploughing. These low-trajectory spreading techniques decrease NH_3-N-losses, leaving more N available for subsequent processes, including nitrification and denitrification, and often producing wetter environment in the soils right around the fertiliser – increasing the likelihood of enhanced N_2O emissions. Nevertheless, this is not always the case – as summarised in reviews by Webb et al. (2010) and Chadwick et al. (2011). Webb et al. (2010) draw attention to the savings in indirect N_2O emission achieved by the NH_3 abatement, which are likely to be higher than the increase in the direct N_2O emissions. As Chadwick et al. (2011) note, soil and climatic conditions favourable for denitrification (i.e. warm and/or wet weather, heavy soil structure and/or high moisture content) might result in increased N_2O emissions from slurry injection compared to broadcasting, but other conditions offer the opportunity of reducing N_2O and NH_3 emissions simultaneously. Thorman et al. (2008) have a contrasting opinion, suggesting that conditions beneficial for denitrification might provide win-win situation for solid manure incorporation, while in other conditions incorporation is likely to increase N_2O emissions.

Using urease inhibitors along with urea fertiliser is another efficient way of reducing NH_3 emissions from soils (Zaman et al. 2009). Though a meta-analysis of studies published in 2008 found no significant effect of urease inhibitors on N_2O emissions (Akiyama et al. 2010), in recent years many studies were published about their beneficial effects on N_2O emissions (Dawar et al. 2011; Halvorson et al. 2010; Halvorson et al. 2011; Sanz-Cobena et al. 2012; Vistoso et al. 2012).

Finally, changing the type of inorganic fertilisers can bring benefits for ammonia savings: urea has the highest potential for generating NH_3 emissions, followed by NH_4^+-based fertilisers, while NO_3^--based fertilisers generate the lowest NH_3 emissions (Bussink and Oenema 1998; Misselbrook et al. 2000). However, N_2O emissions can be variable, as summarised by Snyder et al. (2009) and Harrison and Webb (2001), and Stehfest and Bouwman (2006) found no pronounced differences between most fertiliser types in terms of N_2O emissions after the sample had been balanced for other factors, like rate of application, crop type, climate, soil pH.

7.7   Integrated Assessment: A Case Study

The above described synergies and trade-offs are only some examples of the complex biophysical processes in agriculture. Capturing these co-effects is important in economic assessment of technical options and policy instruments. This section presents a case study about the economic evaluation of the trade-off and synergies between multiple environmental goals, namely GHGs, NH_3, NO_3^--leaching, phosphorus (P) and sediment pollution (see as well Eory et al. 2013). By including the monetary values of NH_3, NO_3^-, P and sediment into the cost-effectiveness analysis of the GHG mitigation options the co-effects can be explicitly taken into account in decision making.

The cost-effectiveness is assessed by marginal abatement cost curves. MACCs show the cost of reducing pollution by one additional unit as a function of the cumulative pollution reduction happening against a business as usual scenario (Moran et al. 2011b). When compared to the marginal benefits from GHG mitigation, the economic optimum of pollution reduction is defined as the intercept of these two curves (Pearce and Turner 1989). The marginal benefits of mitigation can be approximated by a C price, for example by the shadow price of carbon (SPC) (Price et al. 2007).

7.7.1 Methodology

The analysis develops the GHG MACC elaborated in Moran et al. (2011a). The 2022 maximum technical potential, optimistic scenario for England and Wales is used as an illustrative basis for adjustment in this paper, assuming full uptake of measures by farmers and using the more optimistic estimates for costs and GHG saving potential. NO_3^- leaching, NH_3 emissions, P and sediment pollution co-effects are included in the cost-effectiveness metric (Eqs. 7.1 and 7.2).

$$Social\,cos\,t_i = \sum_{j=1}^{k} Change\,in\,pollution\,load_{i,j} * Damage\,cos\,t_j \tag{7.1}$$

$$CE(ext)_i = \frac{Private\,cos\,t_i + Social\,cos\,t_i}{GHG\,saved_i} \tag{7.2}$$

Where:

Social cost$_i$: the total social cost of MM$_i$ (£/year)

Change in pollutant load$_{i,j}$: change in pollution load of pollutant$_j$ caused by MM$_i$ (t pollutant/year)

Damage cost$_j$: unitary damage cost of pollutant$_j$ (£/t pollutant)

CE(ext)$_i$: CE with co-effects of MM$_i$ (£/CO_2e)

Private cost$_i$: financial cost of MM$_i$ (£/year)

GHG saved$_i$: GHG saved by MM$_i$ (CO_2e /year)

i: refers to MM$_i$

j: refers to pollutant$_j$

The quantity of associated co-effects were derived from Anthony et al. (2008). The monetary value of the pollutants were approximated by their damage costs (estimates of the damage an extra unit of a pollutant causes to society), using five sets of unitary damage costs (Table 7.3). The C price benchmark used is the SPC: 34.3 £ tCO_2e^{-1} in 2022 (Price et al. 2007).

Table 7.3 Unit damage costs

Damage value set	NO$_3$-N [£/t]	P [£/t]	Sediment [£/t]	NH$_3$-N [£/t]	MACC
None	–	–	–	–	GHG-MACC
A	217[a]	9,634[a]	25[a]	1,804[b]	MP-MACC-A
B	672[c]	45,144[c]	108[c]	1,804[b]	MP-MACC-B
C	4,287[d]	9,634[a]	25[a]	17,699[e]	MP-MACC-C
D	4,287[d]	45,144[c]	108[c]	17,699[e]	MP-MACC-D
E	20,577[d]	45,144[c]	108[c]	52,055[e]	MP-MACC-E

[a]Values derived by Anthony et al. (2008) after Spencer et al. (2008)
[b]Defra damage cost Defra (2008) as used in Anthony et al. (2008)
[c]Values derived by Anthony et al. (2008), based on Spencer et al. (2008) and Baker et al. (2007)
[d]Values based on Brink et al. (2011)
[e]Values based on Holland et al. (2005)

7.7.2   Results and Discussion

While the annual private (financial) cost of the measures fall in a range of £ −811 million to £ 1,650 million (negative values denoting saving), the annual value of external impacts is smaller: varies from £ −16 million to £ 0 and from £ −512 million to £ 0 calculated with damage cost set A and set E, respectively (the data available on external impacts imply that no pollution swapping occurs with the mitigation options analysed here). Changing between damage cost sets from A to E increases the value of social benefits, with bigger increases happening when NH$_3$ and NO$_3{}^-$ damage costs are increased in comparison to when the damage costs of P and sediment are increased. Consequently, the cost-effectiveness of the measures improves (Fig. 7.3).

In 2022, in England and Wales, the economically feasible GHG abatement, assuming full uptake of measures by the farmers, is 11.9 Mt CO$_2$e y^{-1}. This is 36 % of agricultural GHG emissions, which are expected to be 32.6 Mt CO$_2$e in that year (Defra 2011). Adding the co-effects valued with the damage cost sets A and B has a small effect on the MACC (Fig. 7.4), changing the cumulative GHG abatement of measures with less than 0.15 Mt CO$_2$e y^{-1}. For both damage value sets A and B, the annual non-GHG abatement potential up to the economically efficient GHG mitigation are 38 kt NO$_3{}^-$-N, 0.7 kt P, 198 kt sediment and 14 kt NH$_3$-N (14 %, 18 %, 11 % and 9 % of annual load from agriculture, respectively, based on annual loads estimated by Anthony et al. (2008)).

Applying higher damage values (damage value sets C and D), again leads to a slight change in the MACC, increasing the GHG abatement potential by 0.07 Mt CO$_2$e and the NH$_3$ abatement potential by 0.4 kt NH$_3$-N. With even higher damage values for NO$_3{}^-$ and NH$_3$ in damage value set E "Using biological fixation to provide N inputs" becomes economically efficient, increasing the cumulative annual GHG savings by 1.8 Mt, and the NH$_3$ savings by 3.4 kt (Fig. 7.5).

The omission of external impacts has been highlighted as a drawback of GHG MACC analysis in policy making. The evidence presented here shows how the

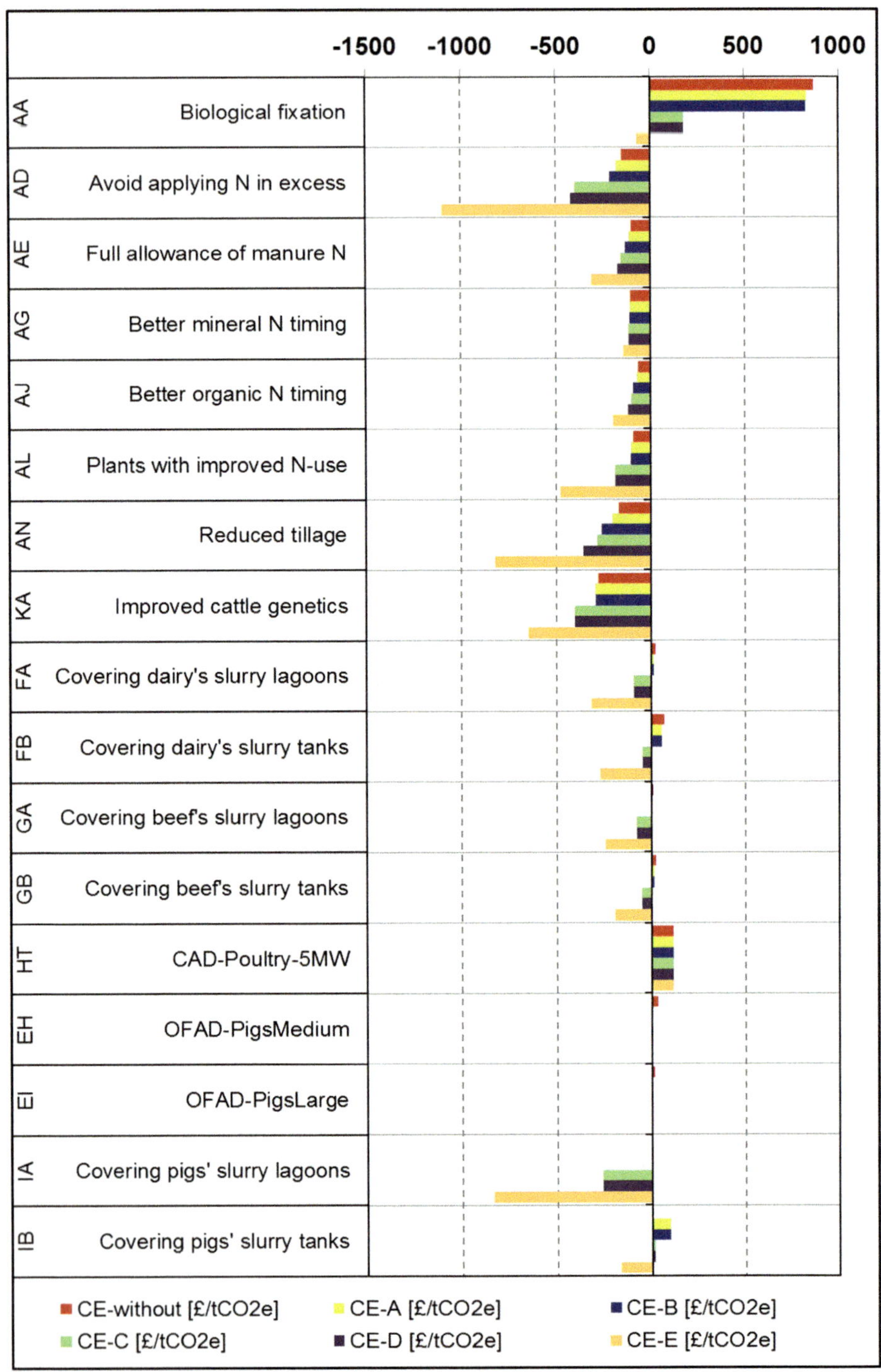

Fig. 7.3 Cost-effectiveness of measures on the MACCs with data on co-effects. CE-without, CE-A, CE-B, CE-C, CE-D and CE-E represents cost-effectiveness values as calculated in GHG-MACC, MP-MACC-A, MP-MACC-B, MP-MACC-C, MP-MACC-D and MP-MACC-E, respectively. OFAD: on-farm anaerobic digestion, CAD: centralised anaerobic digestion

inclusion of external impacts can alter the economic efficiency of environmental measures, though in this case these impacts are small, especially with the more conservative damage value estimates (sets A–D). Very high damage costs (set E) would justify the implementation of almost all the GHG measures which have positive co-effects.

The low impact can be explained partly by the relative monetary values of the GHG and non-GHG pollution loads in England and Wales, the former being substantially higher than the latter, using damage value sets A and B. Increasing the damage values makes a difference, but this couldn't be fully realised in this analysis due to the lack of data on external impacts for many measures, which is the second reason of making little impact on the MACC.

On the other hand, there are some measures currently with no data on co-effects which might have negative effects on one or more of the other four pollutants. For example, "Use composts, straw-based manures in preference to slurry" might increase NH_3 emissions from housing and storage (Chadwick et al. 2011; Jungbluth et al. 2001). Some other important caveats are the gaps in the monetary valuation of the co-effects. The issues of displaced production and full life-cycle costing are further critiques of existing MACCs, which we have not been addressed here.

Notwithstanding the data gaps, the multiple pollutant MACC can offer specific policy messages for agencies trying to interpret MACC information. The first is to focus any further analysis on options that are slightly above the threshold on the GHG MACCs, as they most probably have co-effects which could make their implementation worthwhile. The second message is to explore thoroughly any possible negative external impacts of those GHG measures that are cost-effective on the GHG MACCs and become cost-effective on the MP MACCs. In these cases it may be useful to consider effects beyond those analysed here, like biodiversity, soil quality, human health and social effects (e.g. food security, resilience of rural communities).

7.8 Conclusions

Inter-dependencies between agricultural NH_3 and GHG emissions mean that almost all mitigation options have effects on more than one gaseous emission. Identifying the synergies and the trade-offs is crucial in supporting an integrated policy approach.

Higher production efficiency in livestock and crop production (using less input per unit of production) can reduce most of the environmental burdens arising from agricultural production, including NH_3 and GHG pollution per unit of crop and/or livestock output. Examples include improved cattle fertility, stricter pest and disease monitoring, and optimised grazing management. Ongoing genetic selection in crops and animals can advance resource use efficiency in a variety of goals, including such as N, energy and water use efficiency, disease resistance, fertility

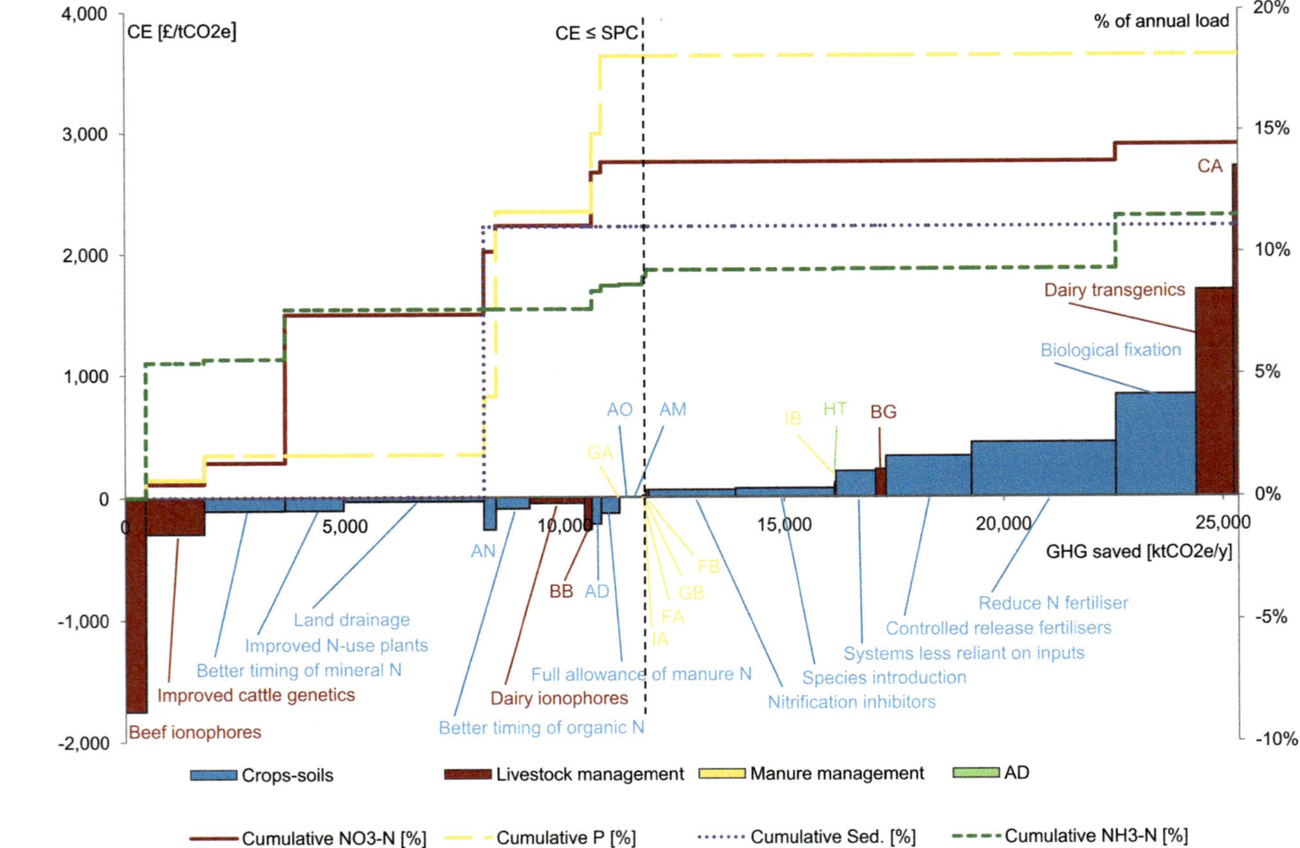

Fig. 7.4 MP-MACC-B. The bars represent the MP-MACC while the lines represent the cumulative savings in the annual load of the four pollutants Sed.: sediment. See Fig. 7.3 for the names of the measures with abatement potential less than 400 kt CO_2e y^{-1}

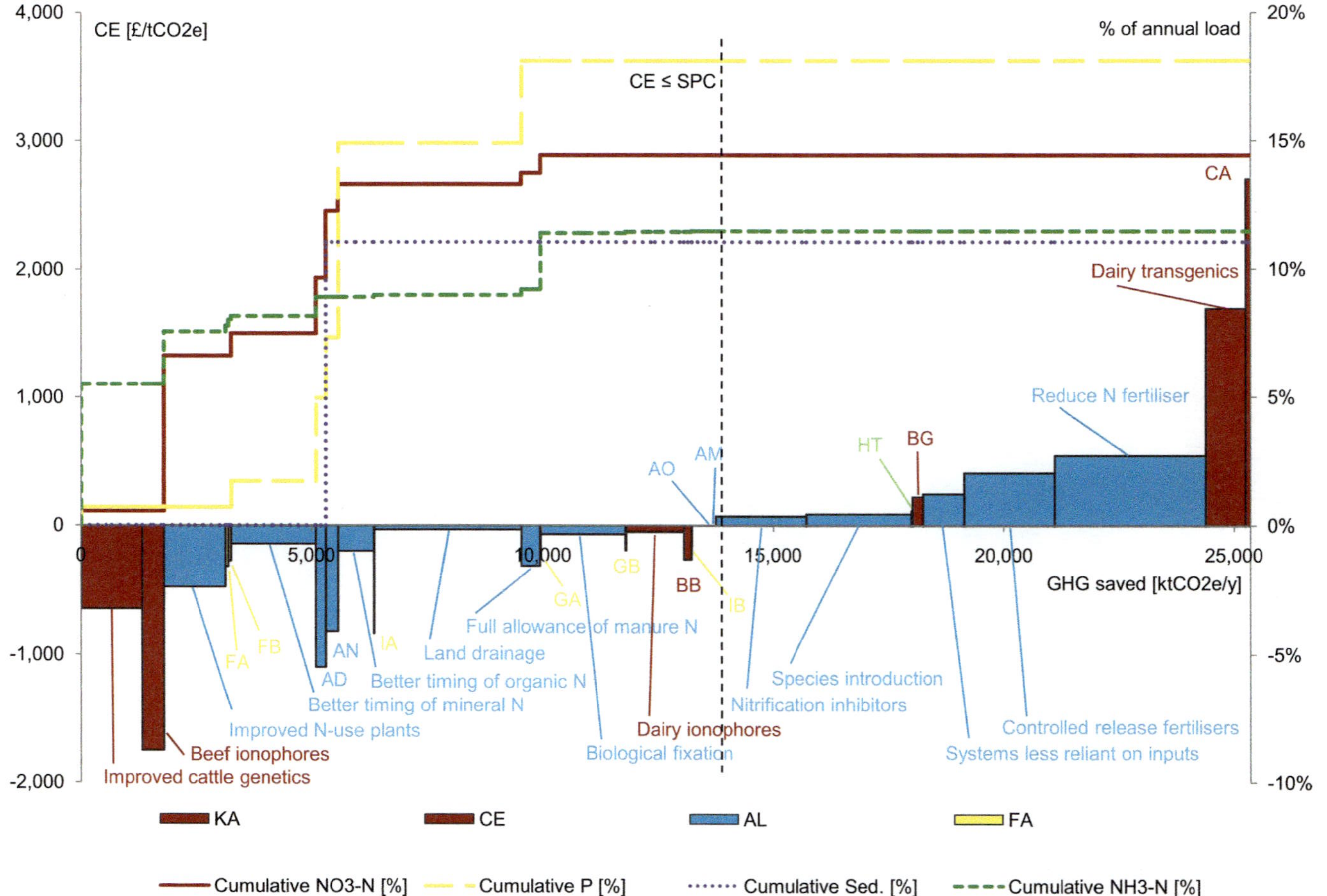

Fig. 7.5 MP-MACC-E. The bars represent the MP-MACC while the lines represent the cumulative savings in the annual load of the four pollutants. Sed.: sediment. See Fig. 7.3 for the names of the measures with abatement potential less than 400 kt CO$_2$e y^{-1}

and longevity. To best adapt new varieties and breeds to their environment, climate change in an important consideration when employing genetic selection.

Animal feeding techniques targeting N input are usually win-win solutions for NH_3 and N_2O, and in some cases also for CH_4, though some feeding techniques, e.g. higher starch or sugar content, can provoke land use change, with negative implications on biodiversity, food security and lifecycle GHG emissions from land converted to croplands.

Livestock system changes (changing the proportion of the time spent outdoors or changing between solid and liquid manure systems) can result in pollution swapping: the reduced NH_3 and CH_4 emissions are accompanied by increased N_2O emissions or vice versa. The majority of NH_3 emissions from agriculture originate from bedding and manure storage, while N_2O emissions from the same source contribute only 6 % to total agricultural GHG in Europe, the evident trade-offs should be considered. Furthermore for both solid and liquid manure systems various efficient NH_3 and GHG mitigation options are available, some providing savings in both pollutants. But when comparing the different housing options it is important to consider that most of the feeding, housing, and manure storage mitigation options are unavailable for the time what animals spend outdoors.

Changed housing design and in in-house manure management practices can offer both NH_3 and GHG benefits. Win-win options for both liquid and solid manure storage exist, including slurry acidification and airtight covering of solid manure heaps. Nevertheless, many other mitigation measures targeting manure storage show variable results for the different gases, making the outcomes uncertain.

To reduce NH_3 emissions from soil fertilisation, low trajectory manure spreading and urease inhibitors are important options. The former has uncertain effects on N_2O emissions, the effects depending on local conditions. The latter might prove to be an option to reduce both GHG and NH_3 emissions.

The most promising win-win measures, delivering improvements in both NH_3 and GHG emissions, are improving production efficiency and N-use efficiency. Low-emission housing design and management (with attention to all types of gaseous emissions) is also likely to deliver multiple benefits and is becoming more important due to the increasing concentration of livestock production and an emerging need for climate-adapted housing. Slurry acidification, urease inhibitors and the choice of inorganic fertilisers are also potential win-win options.

There is a risk of pollution swapping when the amounts of starch and sugar in animal feeds are increased, when changing indoor/outdoor housing and liquid/solid manure management systems, from separating slurry and from increasing the aeration of solid manure. When choosing between these alternatives and current practices, the negative and positive effects of the different pollutants have to be weighted and compared.

Some options require further investigation; for example low-trajectory manure spreading could be a win-win solution in some circumstances. The effectiveness of covering slurry stores and manure heaps is highly dependent on the type of material and method used, and could offer opportunities for a concurrent reduction in GHGs and NH_3. Anaerobic digestion of animal waste has important positive consequences beyond the farm gate, and might be a win-win measure if efficient NH_3 mitigation measures are applied in the storage and spreading of the digestate and if the substrates do not contribute to reducing carbon stocks via land use change.

In many cases the mitigation options have effects on the whole farm, potentially impacting on yield, product quality or gaseous emissions from other parts of the system. Whole-farm biophysical and economic models can help understanding these interdependencies. Beyond the farm gate changes are also possible. For example reduced grass fertilisation rates imply lower synthetic fertiliser production and therefore reduced CO_2 and N_2O emissions from industrial processes. Optimising the diet can also lead to off-farm emission changes from fertiliser related emissions of feed crops or in the soil C stock if the land use pattern changes. Beyond gaseous emissions, the financial costs, other – usually locally and regionally important – environmental effects (e.g. biodiversity, water pollution, soil degradation) and social consequences have to be considered.

References

Aarnink AJA, Verstegen MWA (2007) Nutrition, key factor to reduce environmental load from pig production. Livest Sci 109(1–3):194–203

Aarnink AJA, Swierstra D, van den Berg AJ, Speelman L (1997) Effect of type of slatted floor and degree of fouling of solid floor on ammonia emission rates from fattening piggeries. J Agric Eng Res 66(2):93–102

Aberystwyth University (2010) Ruminant nutrition regimes to reduce methane and nitrogen emissions. Defra AC0209. Available via: http://randd.defra.gov.uk/Document.aspx?Document=AC0209_10114_FRP.pdf

Akiyama H, Yan X, Yagi K (2010) Evaluation of effectiveness of enhanced-efficiency fertilizers as mitigation options for N_2O and NO emissions from agricultural soils: meta-analysis. Glob Chang Biol 16(6):1837–1846

Amon B, Boxberger J, Amon T, Zaussinger A, Pollinger A (1997) Emission data of NH_3, CH_4 and N_2O from fattening bulls, milking cows and during different ways of storing solid manure. In: Voermans JAM, Monteny GJ (eds) Proceedings of the international symposium on ammonia and odour control from animal production facilities, Rosmalen, 1997

Amon B, Amon T, Boxberger J, Alt C (2001) Emissions of NH_3, N_2O and CH_4 from dairy cows housed in a farmyard manure tying stall (housing, manure storage, manure spreading). Nutr Cycl Agroecosyst 60(1):103–113

Amon B, Kryvoruchko V, Amon T, Zechmeister-Boltenstern S (2006) Methane, nitrous oxide and ammonia emissions during storage and after application of dairy cattle slurry and influence of slurry treatment. Agric Ecosyst Environ 112(2–3):153–162

Amon B, Kryvoruchko V, Frohlich M, Amon T, Pollinger A, Mosenbacher I, Hausleitner A (2007) Ammonia and greenhouse gas emissions from a straw flow system for fattening pigs: housing and manure storage. Livest Sci 112(3):199–207

Anthony S, Duethman D, Gooday R, Harris D, Newell-Price P, Chadwick D, Misselbrook T (2008) Quantitative assessment of scenarios for managing trade-off between economics, environment and media. Defra WQ0106 (Module 6). Available via: http://randd.defra.gov.uk/Default.aspx?Menu=Menu&Module=More&Location=None&ProjectID=14421&FromSearch=Y&Publisher=1&SearchText=WQ0106&SortString=ProjectCode&SortOrder=Asc&Paging=10

Baker B, Metcalfe P, Butler S, Gueron Y, Sheldon R, East J (2007) The benefits of water framework directive programmes of measures in England and Wales. Defra CRP 4b/c. Available via: http://www.wfdcrp.co.uk/pdf%5CCRPSG%204bcd%20Final.pdf

Beauchemin K, Kreuzer M, O'Mara F, McAllister T (2008) Nutritional management for enteric methane abatement: a review. Aust J Exp Agric 48(1–2):21–27

Beauchemin KA, Henry Janzen H, Little SM, McAllister TA, McGinn SM (2010) Life cycle assessment of greenhouse gas emissions from beef production in western Canada: a case study. Agric Syst 103(6):371–379

Berg W, Hornig G (1997) Emission reduction by acidification of slurry investigations and assessment. In: Voermans JAM, Monteny GJ (eds) Proceedings of the international symposium on ammonia and odour control from animal production facilities, Rosmalen, 1997

Berg W, Brunsch R, Pazsiczki I (2006) Greenhouse gas emissions from covered slurry compared with uncovered during storage. Agric Ecosyst Environ 112(2–3):129–134

Bouwman AF (1998) Nitrogen oxides and tropical agriculture. Nature 392(6679):866–867

Bouwmeester RJB, Vlek PLG (1981) Rate control of ammonia volatilization from rice paddies. Atmos Environ 15(2):131–140

Bremner JM, Blackmer AM (1978) Nitrous oxide – emission from soils during nitrification of fertilizer nitrogen. Science 199(4326):295–296

Brink C, van Grinsven H, Jacobsen BH, Rabl A, Gren IM, Holland M, Klimont Z, Hicks WK, Brouwer R, Dickens R, Willems J, Termansen M, Velthof GL, Alkemade R, van Oorschot M, Webb J (2011) Costs and benefits of nitrogen in the environment. In: Sutton MA, Howard CM, Erisman JW, Billen I, Bleeker A, Grennfelt P, van Grinsven H, Grizzetti B (eds) The European nitrogen assessment. First Cambridge University Press, Cambridge, pp 513–540

Broderick GA, Stevenson MJ, Patton RA, Lobos NE, Olmos Colmenero JJ (2008) Effect of supplementing rumen-protected methionine on production and nitrogen excretion in lactating dairy cows. J Dairy Sci 91(3):1092–1102

Bruce JM (1990) Straw-flow – a high welfare system for pigs. Farm Building Prog 102:9–13

Bussink DW, Oenema O (1998) Ammonia volatilization from dairy farming systems in temperate areas: a review. Nutr Cycl Agroecosyst 51(1):19–33

Canh TT, Verstegen MWA, Aarnink AJA, Schrama JW (1997) Influence of dietary factors on nitrogen partitioning and composition of urine and feces of fattening pigs. J Anim Sci 75 (3):700–706

Canh TT, Aarnink AJA, Schutte JB, Sutton A, Langhout DJ, Verstegen MWA (1998a) Dietary protein affects nitrogen excretion and ammonia emission from slurry of growing-finishing pigs. Livest Prod Sci 56(3):181–191

Canh TT, Aarnink AJA, Verstegen MWA, Schrama JW (1998b) Influence of dietary factors on the pH and ammonia emission of slurry from growing-finishing pigs. J Anim Sci 76(4):1123–1130

Carmona G, Christianson CB, Byrnes BH (1990) Temperature and low concentration effects of the urease inhibitor N-(n-butyl) thiophosphoric triamide (nBTPT) on ammonia volatilization from urea. Soil Biol Biochem 22(7):933–937

Chadwick DR (2005) Emissions of ammonia, nitrous oxide and methane from cattle manure heaps: effect of compaction and covering. Atmos Environ 39(4):787–799

Chadwick D, Sommer S, Thorman R, Fangueiro D, Cardenas L, Amon B, Misselbrook T (2011) Manure management: implications for greenhouse gas emissions. Anim Feed Sci Technol 166–67:514–531

Clemens J, Trimborn M, Weiland P, Amon B (2006) Mitigation of greenhouse gas emissions by anaerobic digestion of cattle slurry. Agric Ecosyst Environ 112(2–3):171–177

Cuttle S, Shepherd MA, Lord EI, Hillman J (2004) Literature review of the effectiveness of measures to reduce nitrate leaching from agricultural land. Defra NT2511. Available via: http://randd.defra.gov.uk/Default.aspx?Menu=Menu&Module=More&Location=None& ProjectID=11431&FromSearch=Y&Publisher=1&SearchText=NT2511& SortString=ProjectCode&SortOrder=Asc&Paging=10#Description

Dalgaard T, Bienkowski JF, Bleeker A, Dragosits U, Drouet JL, Durand P, Frumau A, Hutchings NJ, Kedziora A, Magliulo V, Olesen JE, Theobald MR, Maury O, Akkal N, Cellier P (2012)

Farm nitrogen balances in six European landscapes as an indicator for nitrogen losses and basis for improved management. Biogeosciences 9(12):5303–5321

Dawar K, Zaman M, Rowarth J, Blennerhassett J, Turnbull M (2011) Urease inhibitor reduces N losses and improves plant-bioavailability of urea applied in fine particle and granular forms under field conditions. Agric Ecosyst Environ 144(1):41–50

Defra (2008) Damage cost estimates. Available via: http://archive.defra.gov.uk/environment/quality/air/airquality/panels/igcb/guidance/damagecosts.htm. Accessed 22 June 2011

Defra (2011) Greenhouse gas emission projections for UK agriculture to 2030. PB 13622. Available via: http://www.defra.gov.uk/publications/2011/08/02/pb13622-ghg-emission-projections/

Dewes T (1996) Effect of pH, temperature, amount of litter and storage density on ammonia emissions from stable manure. J Agric Sci 127:501–509

Dijkstra J, Oenema O, Bannink A (2011) Dietary strategies to reducing N excretion from cattle: implications for methane emissions. Curr Opin Environ Sustain 3(5):414–422

Dinuccio E, Berg W, Balsari P (2008) Gaseous emissions from the storage of untreated slurries and the fractions obtained after mechanical separation. Atmos Environ 42(10):2448–2459

Dinuccio E, Berg W, Balsari P (2011) Effects of mechanical separation on GHG and ammonia emissions from cattle slurry under winter conditions. Anim Feed Sci Technol 166–167:532–538

Edwards S, Edge H, Cain P, Guy J, Hazzledine M, Gill P (2002) A review and workshop to agree advice on reduced nutrient pig diets – Final Report. Defra WA0321. Available via: http://randd.defra.gov.uk/Default.aspx?Menu=Menu&Module=More&Location=None&ProjectID=10332&FromSearch=Y&Publisher=1&SearchText=wa0321&SortString=ProjectCode&SortOrder=Asc&Paging=10#Description

El Kader NA, Robin P, Paillat JM, Leterme P (2007) Turning, compacting and the addition of water as factors affecting gaseous emissions in farm manure composting. Bioresour Technol 98(14):2619–2628

Ellis J, Dijkstra J, Kebreab E, Bannink A, Odongo N, McBride B, France J (2008) Aspects of rumen microbiology central to mechanistic modelling of methane production in cattle. J Agric Sci 146:213–233

Ellis JL, Dijkstra J, Bannink A, Parsons AJ, Rasmussen S, Edwards GR, Kebreab E, France J (2011) The effect of high-sugar grass on predicted nitrogen excretion and milk yield simulated using a dynamic model. J Dairy Sci 94(6):3105–3118

Eory V, Topp CFE, Moran D (2013) Multiple-pollutant cost-effectiveness of greenhouse gas mitigation measures in the UK agriculture. Environ Sci Pol 27:55–67

EPA (2012) Global anthropogenic non-CO_2 greenhouse gas emissions: 1990–2030. EPA 430-R-12-006. Available via: http://www.epa.gov/climatechange/EPAactivities/economics/nonco2projections.html

Eugene M, Masse D, Chiquette J, Benchaar C (2008) Meta-analysis on the effects of lipid supplementation on methane production in lactating dairy cows. Can J Anim Sci 88(2):331–337

European Environment Agency (2013) European Union emission inventory report 1990–2011 under the UNECE Convention on Long-range Transboundary Air Pollution (LRTAP). EEA Technical Report No 10/2013. Available via: http://www.eea.europa.eu//publications/eu-emission-inventory-report-lrtap

European Environment Agency (2014) Annual European Union greenhouse gas inventory 1990–2012 and inventory report 2014. EEA Technical report No 9/2014. Available via: http://www.eea.europa.eu/publications/european-union-greenhouse-gas-inventory-2014, http://www.eea.europa.eu/data-and-maps/data/data-viewers/greenhouse-gases-viewer

Fabbri C, Valli L, Guarino M, Costa A, Mazzotta V (2007) Ammonia, methane, nitrous oxide and particulate matter emissions from two different buildings for laying hens. Biosyst Eng 97(4):441–455

Fangueiro D, Coutinho J, Chadwick D, Moreira N, Trindade H (2008) Effect of cattle slurry separation on greenhouse gas and ammonia emissions during storage. J Environ Qual 37(6):2322–2331

Fangueiro D, Ribeiro H, Coutinho J, Cardenas L, Trindade H, Cunha-Queda C, Vasconcelos E, Cabral F (2010) Nitrogen mineralization and CO_2 and N_2O emissions in a sandy soil amended with original or acidified pig slurries or with the relative fractions. Biol Fertil Soils 46 (4):383–391

Firestone MK, Firestone RB, Tiedje JM (1980) Nitrous oxide from soil denitrification: factors controlling its biological production. Science 208(4445):749–751

Fukumoto Y, Osada T, Hanajima D, Haga K (2003) Patterns and quantities of NH_3, N_2O and CH_4 emissions during swine manure composting without forced aeration – effect of compost pile scale. Bioresour Technol 89(2):109–114

Galloway JN, Aber JD, Erisman JW, Seitzinger SP, Howarth RW, Cowling EB, Cosby BJ (2003) The nitrogen cascade. Bioscience 53(4):341–356

Gilhespy SL, Webb J, Chadwick DR, Misselbrook TH, Kay R, Camp V, Retter AL, Bason A (2009) Will additional straw bedding in buildings housing cattle and pigs reduce ammonia emissions? Biosyst Eng 102(2):180–189

Grainger C, Beauchemin KA (2011) Can enteric methane emissions from ruminants be lowered without lowering their production? Anim Feed Sci Technol 166–167:308–320

Groot JCJ, Rossing WAH, Lantinga EA (2006) Evolution of farm management, nitrogen efficiency and economic performance on Dutch dairy farms reducing external inputs. Livest Sci 100(2–3):99–110

Halvorson AD, Del Grosso SJ, Alluvione F (2010) Nitrogen source effects on nitrous oxide emissions from irrigated no-till corn. J Environ Qual 39(5):1554–1562

Halvorson AD, Del Grosso SJ, Jantalia CP (2011) Nitrogen source effects on nitrous oxide emissions from strip-till corn. J Environ Qual 40(6):1775–1786

Hansen MN, Henriksen K, Sommer SG (2006) Observations of production and emission of greenhouse gases and ammonia during storage of solids separated from pig slurry: effects of covering. Atmos Environ 40(22):4172–4181

Harrison R, Webb J (2001) A review of the effect of N fertilizer type on gaseous emissions. Adv Agron 73:65–108

Hartung J, Phillips VR (1994) Control of gaseous emissions from livestock buildings and manure stores. J Agric Eng Res 57(3):173–189

Hassouna M, Espagnol S, Robin P, Paillat JM, Levasseur P, Li Y (2008) Monitoring NH_3, N_2O, CO_2 and CH_4 emissions during pig solid manure storage – effect of turning. Compost Sci Util 16(4):267–274

Hellmann B, Zelles L, Palojarvi A, Bai QY (1997) Emission of climate-relevant trace gases and succession of microbial communities during open-window composting. Appl Environ Microbiol 63(3):1011–1018

Holland M, Pye S, Watkiss P, Droste-Franke B, Bickel P (2005) Damages per tonne emission of $PM_{2.5}$, NH_3, SO_2, NO_x and VOCs from each EU25 Member State (excluding Cyprus) and surrounding seas. Clean Air for Europe (CAFE) Programme. Available via: http://www.cafe-cba.org/assets/marginal_damage_03-05.pdf

Hristov A, Hanigan M, Cole A, Todd R, McAllister T, Ndegwa P, Rotz A (2011) Review: ammonia emissions from dairy farms and beef feedlots. Can J Anim Sci 91(1):1–35

Hristov A, Oh J, Lee C, Meinen R, Montes F, Ott T, Firkins J, Rotz A, Dell C, Adesogan A, Yang WZ, Tricarico J, Kebreab E, Waghorn GC, Dijkstra J, Oosting S (2013) Mitigation of greenhouse gas emissions in livestock production – a review of technical options for non-CO2 emissions, FAO Animal Production and Health Paper 177. FAO, Rome

Hudson N, Duperouzel D, Melvin S (2006) Assessment of permeable covers for odour reduction in piggery effluent ponds. 1. Laboratory-scale trials. Bioresour Technol 97(16):2002–2014

Hutchings NJ, Sommer SG, Andersen JM, Asman WAH (2001) A detailed ammonia emission inventory for Denmark. Atmos Environ 35(11):1959–1968

IGER (2005) Evaluation of targeted or additional straw use as a means of reducing NH_3 emissions from buildings housing cattle and pigs. Defra AM0103. Available via: http://randd.defra.gov.uk/Default.aspx?Menu=Menu&Module=More&Location=None&ProjectID=9637&

FromSearch=Y&Publisher=1&SearchText=am0103&SortString=ProjectCode&
SortOrder=Asc&Paging=10#Description

IPCC (2006) 2006 IPCC guidelines for national greenhouse gas inventories, Prepared by the National Greenhouse Gas Inventories Programme, Volume 4: Agriculture, forestry and other land use. Available via: http://www.ipcc-nggip.iges.or.jp/public/2006gl/vol4.html

Johnson KA, Johnson DE (1995) Methane emissions from cattle. J Anim Sci 73(8):2483–2492

Jonker JS, Kohn RA, High J (2002) Dairy herd management practices that impact nitrogen utilization efficiency. J Dairy Sci 85(5):1218–1226

Jungbluth T, Hartung E, Brose G (2001) Greenhouse gas emissions from animal houses and manure stores. Nutr Cycl Agroecosyst 60(1–3):133–145

Kai P, Pedersen P, Jensen JE, Hansen MN, Sommer SG (2008) A whole-farm assessment of the efficacy of slurry acidification in reducing ammonia emissions. Eur J Agron 28(2):148–154

Khan RZ, Muller C, Sommer SG (1997) Micrometeorological mass balance technique for measuring CH_4 emission from stored cattle slurry. Biol Fertil Soils 24(4):442–444

Kim IB, Ferket PR, Powers WJ, Stein HH, van Kempen TATG (2004) Effects of different dietary acidifier sources of calcium and phosphorus on ammonia, methane and odorant emission from growing-finishing pigs. Asian-Australas J Anim Sci 17(8):1131–1138

Kirchmann H, Witter E (1989) Ammonia volatilization during aerobic and anaerobic manure decomposition. Plant Soil 115(1):35–41

Kreuzer M, Hindrichsen IK (2006) Methane mitigation in ruminants by dietary means: the role of their methane emission from manure. Int Congr Ser 1293:199–208

Kulling DR, Menzi H, Krober TF, Neftel A, Sutter F, Licher P, Kreuzer M (2001) Emissions of ammonia, nitrous oxide and methane from different types of dairy manure during storage as affected by dietary protein content. J Agric Sci 137(2):235–250

Kulling DR, Dohme F, Menzi H, Sutter F, Lischer P, Kreuzer M (2002) Methane emissions of differently fed dairy cows and corresponding methane and nitrogen emissions from their manure during storage. Environ Monit Assess 79(2):129–150

Lee C, Hristov A, Dell C, Feyereisen G, Kaye J, Beegle D (2012) Effect of dietary protein concentration on ammonia and greenhouse gas emitting potential of dairy manure. J Dairy Sci 95(4):1930–1941

Maia GDN, Day V, Gates RS, Taraba JL (2012) Ammonia biofiltration and nitrous oxide generation during the start-up of gas-phase compost biofilters. Atmos Environ 46:659–664

Martin C, Morgavi D, Doreau M (2010) Methane mitigation in ruminants: from microbe to the farm scale. Animal 4(3):351–365

Mc Geough E, O'Kiely P, Foley P, Hart K, Boland T, Kenny D (2010a) Methane emissions, feed intake, and performance of finishing beef cattle offered maize silages harvested at 4 different stages of maturity. J Anim Sci 88(4):1479–1491

Mc Geough E, O'Kiely P, Hart K, Moloney A, Boland T, Kenny D (2010b) Methane emissions, feed intake, performance, digestibility, and rumen fermentation of finishing beef cattle offered whole-crop wheat silages differing in grain content. J Anim Sci 88(8):2703–2716

Melse RW, van der Werf AW (2005) Biofiltration for mitigation of methane emission from animal husbandry. Environ Sci Technol 39(14):5460–5468

Melse RW, Ogink NWM, Rulkens WH (2009) Air treatment techniques for abatement of emissions from intensive livestock production. Open Agric J 3:6–12

Misselbrook TH, Chadwick DR, Pain BF, Headon D (1998) Dietary manipulation as a means of decreasing N losses and methane emissions and improving herbage N uptake following application of pig slurry to grassland. J Agric Sci 130(2):183–191

Misselbrook TH, Van der Weerden TJ, Pain BF, Jarvis SC, Chambers BJ, Smith KA, Phillips VR, Demmers TGM (2000) Ammonia emission factors for UK agriculture. Atmos Environ 34(6):871–880

Misselbrook T, Chadwick D, Ghilespy SL, Chambers B, Smith KA, Williams J, Dragosits U (2012) Inventory of ammonia emissions from UK agriculture – 2011. Defra AC0112. Available via: http://naei.defra.gov.uk/reports/reports?report_id=724

Moe PW, Tyrrell HF (1979) Methane production in dairy-cows. J Dairy Sci 62(10):1583–1586

Monteny GJ, Erisman JW (1998) Ammonia emission from dairy cow buildings: a review of measurement techniques, influencing factors and possibilities for reduction. Neth J Agric Sci 46(3–4):225–247

Monteny GJ, Bannink A, Chadwick D (2006) Greenhouse gas abatement strategies for animal husbandry. Agric Ecosyst Environ 112(2–3):163–170

Moorby JM, Evans RT, Scollan ND, Macraet JC, Theodorou MK (2006) Increased concentration of water-soluble carbohydrate in perennial ryegrass (Lolium perenne L.). Evaluation in dairy cows in early lactation. Grass Forage Sci 61(1):52–59

Moran D, MacLeod M, Wall E, Eory V, McVittie A, Barnes A, Rees R, Topp CFE, Moxey A (2011a) Marginal abatement cost curves for UK agricultural greenhouse gas emissions. J Agric Econ 62(1):93–118

Moran D, MacLeod M, Wall E, Eory V, McVittie A, Barnes A, Rees RM, Topp CFE, Pajot G, Matthews R, Smith P, Moxey A (2011b) Developing carbon budgets for UK agriculture, land-use, land-use change and forestry out to 2022. Clim Chang 105(3–4):529–553

Nahm KH (2002) Efficient feed nutrient utilization to reduce pollutants in poultry and swine manure. Crit Rev Environ Sci Technol 32(1):1–16

Nemet G, Holloway T, Meier P (2010) Implications of incorporating air-quality co-benefits into climate change policymaking. Environ Res Lett 5(1):9

Oltjen JW, Beckett JL (1996) Role of ruminant livestock in sustainable agricultural systems. J Anim Sci 74(6):1406–1409

Osada T, Kuroda K, Yonaga M (2000) Determination of nitrous oxide, methane, and ammonia emissions from a swine waste composting process. J Mater Cycles Waste Manag 2(1):51–56

Owens FN, Secrist DS, Hill WJ, Gill DR (1998) Acidosis in cattle: a review. J Anim Sci 76 (1):275–286

Parkinson R, Gibbs P, Burchett S, Misselbrook T (2004) Effect of turning regime and seasonal weather conditions on nitrogen and phosphorus losses during aerobic composting of cattle manure. Bioresour Technol 91(2):171–178

Pearce DW, Turner RK (1989) Economics of natural resources and the environment. Johns Hopkins University Press, Baltimore

Petersen S, Ambus P (2006) Methane oxidation in pig and cattle slurry storages, and effects of surface crust moisture and methane availability. Nutr Cycl Agroecosyst 74(1):1–11

Petersen SO, Miller DN (2006) Greenhouse gas mitigation by covers on livestock slurry tanks and lagoons? J Sci Food Agric 86(10):1407–1411

Petersen SO, Lind AM, Sommer SG (1998) Nitrogen and organic matter losses during storage of cattle and pig manure. J Agric Sci 130:69–79

Petersen SO, Amon B, Gattinger A (2005) Methane oxidation in slurry storage surface crusts. J Environ Qual 34(2):455–461

Petersen SO, Skov M, Droscher P, Adamsen APS (2009) Pilot scale facility to determine gaseous emissions from livestock slurry during storage. J Environ Qual 38(4):1560–1568

Petersen SO, Andersen AJ, Eriksen J (2012) Effects of cattle slurry acidification on ammonia and methane evolution during storage. J Environ Qual 41(1):88–94

Petersen SO, Dorno N, Lindholst S, Feilberg A, Eriksen JO (2013) Emissions of CH_4, N_2O, NH_3 and odorants from pig slurry during winter and summer storage. Nutr Cycl Agroecosyst 95 (1):103–113

Peyraud JL, Astigarraga L (1998) Review of the effect of nitrogen fertilization on the chemical composition, intake, digestion and nutritive value of fresh herbage: consequences on animal nutrition and N balance. Anim Feed Sci Technol 72(3–4):235–259

Philippe FX, Nicks B (2015) Review on greenhouse gas emissions from pig houses: production of carbon dioxide, methane and nitrous oxide by animals and manure. Agric Ecosyst Environ 199:10–25

Philippe FX, Laitat M, Canart B, Farnir F, Massart L, Vandenheede M, Nicks B (2006) Effects of a reduction of diet crude protein content on gaseous emissions from deep-litter pens for fattening pigs. Anim Res 55(5):397–407

Philippe F, Laitat M, Canart B, Vandenheede M, Nicks B (2007) Comparison of ammonia and greenhouse gas emissions during the fattening of pigs, kept either on fully slatted floor or on deep litter. Livest Sci 111(1–2):144–152

Price R, Thornton S, Nelson S (2007) The social cost of carbon and the shadow price of carbon: what they are, and how to use them in economic appraisal in the UK. Available via: http://www.decc.gov.uk/en/content/cms/what_we_do/lc_uk/valuation/shadow_price/shadow_price.aspx

Ravishankara AR, Daniel JS, Portmann RW (2009) Nitrous oxide (N_2O): the dominant ozone-depleting substance emitted in the 21st century. Science 326(5949):123–125

Rennenberg H, Dannenmann M, Gessler A, Kreuzwieser J, Simon J, Papen H (2009) Nitrogen balance in forest soils: nutritional limitation of plants under climate change stresses. Plant Biol 11:4–23

Roche JF (2006) The effect of nutritional management of the dairy cow on reproductive efficiency. Anim Reprod Sci 96(3–4):282–296

Rotz CA (2004) Management to reduce nitrogen losses in animal production. J Anim Sci 82 (13 suppl):E119–E137

Sanz-Cobena A, Sanchez-Martin L, Garcia-Torres L, Vallejo A (2012) Gaseous emissions of N_2O and NO and NO_3^- leaching from urea applied with urease and nitrification inhibitors to a maize (Zea mays) crop. Agric Ecosyst Environ 149:64–73

Silsoe Research Institute (2000) The effects of covering slurry stores on emissions of ammonia, methane and nitrous oxide. Defra WA0625. Available via: http://randd.defra.gov.uk/Default.aspx?Menu=Menu&Module=More&Location=None&Completed=0&ProjectID=6022#RelatedDocuments

Smart JC, Hicks K, Morrissey T, Heinemeyer A, Sutton MA, Ashmore M (2011) Applying the ecosystem service concept to air quality management in the UK: a case study for ammonia. Environmetrics 22(5):649–661

Smil V (1999) Detonator of the population explosion. Nature 400(6743):415

Snyder CS, Bruulsema TW, Jensen TL, Fixen PE (2009) Review of greenhouse gas emissions from crop production systems and fertilizer management effects. Agric Ecosyst Environ 133 (3–4):247–266

Sommer SG (2001) Effect of composting on nutrient loss and nitrogen availability of cattle deep litter. Eur J Agron 14(2):123–133

Sommer SG, Moller HB (2000) Emission of greenhouse gases during composting of deep litter from pig production – effect of straw content. J Agric Sci 134:327–335

Sommer SG, Olesen JE, Christensen BT (1991) Effects of temperature, wind speed and air humidity on ammonia volatilization from surface applied cattle slurry. J Agric Sci 117 (01):91–100

Sommer SG, Petersen SO, Sogaard HT (2000) Greenhouse gas emission from stored livestock slurry. J Environ Qual 29(3):744–751

Sommer SG, Petersen SO, Moller HB (2004) Algorithms for calculating methane and nitrous oxide emissions from manure management. Nutr Cycl Agroecosyst 69(2):143–154

Sommer SG, Olesen JE, Petersen SO, Weisbjergz MR, Valli L, Rodhe L, Beline F (2009) Region-specific assessment of greenhouse gas mitigation with different manure management strategies in four agroecological zones. Glob Chang Biol 15(12):2825–2837

Spencer I, Bann C, Moran D, McVittie A, Lawrence K, Caldwell V (2008) Environmental accounts for Agriculture. Defra SFS0601. Available via: http://www.dardni.gov.uk/environmental-accounts.pdf

Stehfest E, Bouwman L (2006) N_2O and NO emission from agricultural fields and soils under natural vegetation: summarizing available measurement data and modeling of global annual emissions. Nutr Cycl Agroecosyst 74(3):207–228

St-Pierre NR, Thraen CS (1999) Animal grouping strategies, sources of variation, and economic factors affecting nutrient balance on dairy farms. J Anim Sci 77:72–83

Szanto G, Hamelers H, Rulkens W, Veeken A (2007) NH_3, N_2O and CH_4 emissions during passively aerated composting of straw-rich pig manure. Bioresour Technol 98(14):2659–2670

Thorman R, Webb J, Yamulki S, Chadwick DR, Bennett G, McMillan S, Kingston H, Donovan N, Misselbrook T (2008) The effect of solid manure incorporation on nitrous oxide emissions. In: Proceedings of the 13th international conference of the FAO RAMIRAN: potential for simple technology solutions in organic manure management. Ambrozia NT Ltd., Albena, Bulgaria, 2008

van Vuuren DP, Bouwman LF, Smith SJ, Dentener F (2011) Global projections for anthropogenic reactive nitrogen emissions to the atmosphere: an assessment of scenarios in the scientific literature. Curr Opin Environ Sustain 3(5):359–369

VandeHaar MJ, St-Pierre N (2006) Major advances in nutrition: relevance to the sustainability of the dairy industry. J Dairy Sci 89(4):1280–1291

VanderZaag AC, Gordon RJ, Glass VM, Jamieson RC (2008) Floating covers to reduce gas emissions from liquid manure storages: a review. Appl Eng Agric 24(5):657–671

VanderZaag AC, Gordon RJ, Jamieson RC, Burton DL, Stratton GW (2009) Gas emissions from straw covered liquid dairy manure during summer storage and autumn agitation. Trans Asabe 52(2):599–608

VanderZaag AC, Gordon RJ, Jamieson RC, Burton DL, Stratton GW (2010) Permeable synthetic covers for controlling emissions from liquid dairy manure. Appl Eng Agric 26(2):287–297

Vellinga TV, Hoving I (2011) Maize silage for dairy cows: mitigation of methane emissions can be offset by land use change. Nutr Cycl Agroecosyst 89(3):413–426

Velthof GL, Nelemans JA, Oenema O, Kuikman PJ (2005) Gaseous nitrogen and carbon losses from pig manure derived from different diets. J Environ Qual 34(2):698–706

Vistoso E, Alfaro M, Saggar S, Salazar F (2012) Effect of nitrogen inhibitors on nitrous oxide emissions and pasture growth after an autumn application in volcanic soil. Chilean J Agric Res 72(1):133–139

Waldrip HM, Todd RW, Cole NA (2013) Prediction of nitrogen excretion by beef cattle: a meta-analysis. J Anim Sci 91(9):4290–4302

Webb J (2001) Estimating the potential for ammonia emissions from livestock excreta and manures. Environ Pollut 111(3):395–406

Webb J, Pain B, Bittman S, Morgan J (2010) The impacts of manure application methods on emissions of ammonia, nitrous oxide and on crop response – a review. Agric Ecosyst Environ 137(1–2):39–46

Weiske A (2005) Survey of technical and management-based mitigation measures in agriculture. MEACAP WP3 D7a. Available via: http://www.ieep.eu/publications/pdfs/meacap/WP3/WP3D7a_mitigation.pdf

Yamulki S (2006) Effect of straw addition on nitrous oxide and methane emissions from stored farmyard manures. Agric Ecosyst Environ 112(2–3):140–145

Zaman M, Saggar S, Blennerhassett JD, Singh J (2009) Effect of urease and nitrification inhibitors on N transformation, gaseous emissions of ammonia and nitrous oxide, pasture yield and N uptake in grazed pasture system. Soil Biol Biochem 41(6):1270–1280

Zhang G, Strom JS, Li B, Rom HB, Morsing S, Dahl P, Wang C (2005) Emission of ammonia and other contaminant gases from naturally ventilated dairy cattle buildings. Biosyst Eng 92(3):355–364

Chapter 8
Country Case Studies

Mark A. Sutton, Clare Howard, Stefan Reis, Diego Abalos,
Annelies Bracher, Aleksandr Bryukhanov, Rocio Danica Condor-Golec,
Natalia Kozlova, Stanley T.J. Lalor, Harald Menzi, Dmitry Maximov,
Tom Misselbrook, Martin Raaflaub, Alberto Sanz-Cobena, Edith von
Atzigen-Sollberger, Peter Spring, Antonio Vallejo, and Brian Wade

Abstract In this chapter, we present a series of country case studies, addressing
specific challenges of reducing ammonia emissions and managing nitrogen on farm
and field scale. Section 8.1 introduces nitrogen management activities in an intensively farmed region of Italy, while Sect. 8.2 addresses aspects of animal feed in
Swiss pig farming. The following Sect. (8.3) illustrates N management in cattle and
poultry operations in Switzerland. The assessment of ammonia abatement cost in

M.A. Sutton (✉)
NERC Centre for Ecology & Hydrology (CEH), Bush Estate, Penicuik EH26 0QB, UK
e-mail: ms@ceh.ac.uk

C. Howard
NERC Centre for Ecology & Hydrology (CEH), Bush Estate, Penicuik EH26 0QB, UK

School of GeoSciences, University of Edinburgh, Grant Institute, Kings Buildings Campus,
West Mains Road, Edinburgh EH9 3JW, UK
e-mail: cbritt@ceh.ac.uk

S. Reis
NERC Centre for Ecology & Hydrology (CEH), Bush Estate, Penicuik EH26 0QB, UK

University of Exeter Medical School, Knowledge Spa, Truro TR1 3HD, UK
e-mail: srei@ceh.ac.uk

D. Abalos • A. Sanz-Cobena
ETSI Agrónomos, Technical University of Madrid, Ciudad Universitaria, 28040 Madrid, Spain
e-mail: diego.abalos@upm.es; a.sanz@upm.es

A. Bracher
School of Agricultural, Forestry and Food Sciences HAFL, Bern University of Applied
Sciences (BFH), Länggasse 85, CH-3052 Zollikofen, Switzerland

Institute for Livestock Sciences, Agroscope, Route de la Tioleyre 4, Case postale 64
CH-1725 Posieux, Switzerland
e-mail: annelies.bracher@agroscope.admin.ch

A. Bryukhanov • N. Kozlova • D. Maximov
North-West Research Institute of Agricultural Engineering and Electrification (SZNIIMESH),
Filtrovskoe shosse, 3, St-Petersburg-Pavlovsk 196625, Russia
e-mail: sznii@yandex.ru; natalia.kozlova@sznii.ru; maximov@sznii.ru

© Springer Science+Business Media Dordrecht 2015

S. Reis et al. (eds.), *Costs of Ammonia Abatement and the Climate Co-Benefits*,
DOI 10.1007/978-94-017-9722-1_8

dairy farming in the Russian Federation is covered in Sect. 8.4, with Sect. 8.5 discussing the costs of adoption of low ammonia emission slurry application methods on grassland in Ireland. A further case study on slurry application addresses the costs incurred by the trailing hose technique and by slurry dilution with water under Swiss frame conditions (Sect. 8.6). Section 8.7 highlights the estimated cost of abating volatilized ammonia from urea by urease inhibitors in the EU, and finally Sect. 8.8 discusses potential N_2O reduction associated with the use of urease inhibitors in Spain (Authors of this section: Stefan Reis[1,2], Mark A. Sutton[1], Clare Howard[1,3] *(1) NERC Centre for Ecology & Hydrology, Bush Estate, Penicuik, EH26 0QB, UK; (2) Knowledge Spa, University of Exeter Medical School, Truro, TR1 3HD, UK; (3) School of Geosciences, University of Edinburgh, Institute of Geography, Drummond Street, Edinburgh, EH8 9XP, UK).*

R.D. Condor-Golec
Istituto Superiore per la Protezione e la Ricerca Ambientale, ISPRA, Via Vitaliano Brancati 48, 00144 Rome, Italy

Food and Agriculture Organization of the United Nations (FAO), Viale delle Terme di Caracalla, 00153 Rome, Italy
e-mail: rocio.condor@isprambiente.it; rocio.condor@fao.org

S.T.J. Lalor
Grassland AGRO, Dock Rd., Limerick, Ireland

Formerly Teagasc, Johnstown Research Centre, Wexford, Ireland
e-mail: slalor@grassland.ie

H. Menzi
School for Agricultural, Forestry and Food Sciences (HAFL), Bern University of Applied Sciences (BFH), Länggasse 85, CH-3052 Zollikofen, Switzerland

Agroscope, Tioleyre 4, 1725 Posieux, Switzerland
e-mail: harald.menzi@agroscope.admin.ch

T. Misselbrook
Rothamsted Research, North Wyke, Okehampton, Devon EX20 2SB, UK
e-mail: tom.misselbrook@rothamsted.ac.uk

M. Raaflaub • E. von Atzigen-Sollberger • P. Spring
School of Agricultural, Forestry and Food Sciences HAFL, Bern University of Applied Sciences (BFH), Länggasse 85, CH-3052 Zollikofen, Switzerland
e-mail: martin@raaflaub.net; peter.spring@bfh.ch

A. Vallejo
College of Agriculture, Technical University of Madrid, Ciudad Universitaria. Avenida Complutense s/n 28040, Madrid, Spain
e-mail: antonio.vallejo@upm.es

B. Wade
Agrotain International, AGROTAIN International LLC, 1 Angelica Street, St. Louis, MO 63147, USA
e-mail: bwade@agrotain.com

Keywords Ammonia emissions • Emission abatement • Agriculture • Greenhouse gases • Nitrogen management

8.1 Research, Demonstrative Farm and Dissemination Activities Related to N Management in Italy

Rocio Danica Condor-Golec (✉)
Food and Agriculture Organization of the United Nations (FAO),
Viale delle Terme di Caracalla, 00153 Rome, Italy

Istituto Superiore per la Protezione e la Ricerca Ambientale, ISPRA
Via Vitaliano Brancati 48, 00144 Rome, Italy

The largest and most intensive agricultural area in Italy is the Po River catchment with the following characteristics: high crop yields due to climatic factors, double cropping system adopted by livestock farms, flooded rice fields, high livestock density and animal production that keep animals in stables all the year (Bassanino et al. 2011; Bechini and Castoldi 2009). In 2012, 63 % of cattle and 83 % of swine production were located in Piedmont, Lombardy, Emilia-Romagna, and Veneto Regions (Northern Italy), while 37 % of Protected Designation of Origin/Protected Geographical Indication products are produced in these regions (ISTAT 2013, ISTAT several years). Research experience, which includes data collection (farm-gate/soil-surface nutrient balance) through direct/detailed interviews with farmers, demonstrative farms and dissemination activities are presented. Nutrient balances have demonstrated to be useful agri-environmental indicators that could encourage farmers to go for economic and environmental targets, through the adoption of best practices on crop and livestock management, without compromising the farm production.

8.1.1 Farm-Gate N Balance

Farm-gate N balances are used to study agriculture farming systems in demonstrative farms. The LIFE project OptiMa-N "Optimisation of nitrogen management for groundwater quality improvement and conservation" was a farm level initiative, which includes monitoring activities and dissemination to farmers/agricultural technicians between October 2004 and September 2007 (CRPA 2011). The goal was to test the impact of N fertilisation on water pollution, and to promote the conduction of farming practices with reduced environmental impact through the awareness of farmers. The project involved Emilia-Romagna Region (Parma, Reggio Emilia and Modena provinces). Deliverables include, for instance, SIM.

Table 8.1 Farm-gate N balance of a Parmigiano Reggiano milk farm from Parma Province, Emilia-Romagna Region (Italy)

Farm-gate balance	2005 ($kg\ N\ ha^{-1}\ year^{-1}$)	2005 (%)	2006 ($kg\ N\ ha^{-1}\ year^{-1}$)	2006 (%)
INPUTS				
Feed	61	20	62	23
Fodder	4	1	11	4
Straw	5	2	3	1
Mineral fertiliser	123	41	85	31
Organic fertiliser (horse bedding)	24	8	33	12
Seeds and seedlings	1	0	1	0
Atmospheric deposition	16	5	16	6
N fixation by legumes	65	22	65	24
Totals	*300*	*100*	*276*	*100*
OUTPUTS				
Plant product (tomato and grain)	29	48	41	53
Milk sold	28	48	29	38
Animals (cows at the end and calves)	2	4	3	4
Cattle manure	–		4	5
Totals	*59*	*100*	*77*	*100*
Farm-gate N balance (N surplus)	**240**		**199**	
NUE × 100 (%)	**20**		**28**	

Source: Modified from Mantovi (2007)

BA-N[1] a free software developed to allow farmers/technicians to optimise the N application rate for crops. In a specific task the calculation of the N balance at different levels (farm, barn, single plot) was conducted in a "demonstrative farm" (Mantovi 2007). N surplus was reduced from 240 kg N ha^{-1} $year^{-1}$ to 199 kg N ha^{-1} $year^{-1}$ (see Table 8.1), lower than the representative N surplus for South European intensive systems and similar to that of less intensive systems of France and Scotland (Raison and Pflimlin 2006). NUE (%) was improved from 20 to 28 % through the reduction in the use of mineral fertiliser and the improvement of organic fertiliser utilisation. Mantovi (2007) concluded that a realistic N management intervention to increase environmental sustainability without negative economic impacts is given by actions on the feed and the mineral and organic fertilisers. Results obtained with LIFE OptiMa-N have promoted the presentation of **LIFE + AQUA** "Achieving good water QUality status in intensive Animal production areas" conducted by the CRPA and involving Emilia-Romagna, Piedmont,

---

[1] SIM.BA-N: Simplified Balance –N.

Lombardy, Veneto and Friuli Venezia Giulia regions (CRPA 2014). The aim is to contribute to the reduction of water pollution from nutrients at river basin scale by optimising the utilisation of N/P in livestock farms through increasing the awareness of the business sector (3 year project, started October 2010). Activities include demonstrative farms, where changes will be done in feeding ration to achieve a reduction in N excretion (dairy cattle, beef cattle and pigs); improvement of fertilisation efficiency by the use of innovative techniques for land spreading, and dissemination actions involving stakeholders (farmers, advisors, policy-makers).

Comparative farm-gate N balance on diverse agriculture production systems are available from research studies from Northern Italy. Simon et al. (2000) have evaluated 11 types of farm production systems including stockless farming, cropless intensive rearing, forage crop-livestock integrated farming systems, mixed farming with livestock in North-west Italy and France. Authors obtained a very large variability in N surplus and other indicators (NUE, N conversion, N losses) suggesting for many situations the possibility to reduce environmental impact without drastic changes in the production system. For Piedmont Region, Bassanino et al. (2007) demonstrated that the N surplus varies according to animal categories: higher for pig breeding (486 kg N ha^{-1}), intermediate for beef breeding (257 kg N ha^{-1}) and dairy cow (318 kg N ha^{-1}), and lower for suckling cow (100 kg N ha^{-1}). For Lombardy Region, Fumagalli et al. (2011) presented farmgate N balances for four typical dairy and three cereal farming systems (see Table 8.2). 'Chemical fertilisers' inputs were common for dairy and cereal farms ranging from 20 to 50 %. 'Animal feeding' inputs range from 20 to 70 % in dairy farms, and 'manure' inputs range from 30 to 50 % in cereal farms. For dairy farms N balances range from 113 kg N ha^{-1} to 316 kg N ha^{-1}, while for cereal farms N balances range from 33 kg N ha^{-1} to 339 kg N ha^{-1}. Fumagalli et al. (2011) concluded that high levels of N surplus imply possibility for N management improvements.

8.1.2 Soil-Surface Nutrient Balance

Soil-surface balances combined with other agro-environmental indicators supplied an integrated analysis of farming systems. An environmental/economic accounting for diverse cropping systems in animal and arable farms (maize, rice, permanent meadows, winter wheat, winter barley, Italian ryegrass, triticale, and soybean) was performed in the South Milan Agricultural Park (Lombardy Region) during a 2-year study. Economic, P/N nutrient management, energy management, pesticides and soil management indicators were selected (Bechini and Castoldi 2009). Large variability results were obtained with an integrated set of indicators, including soil-surface N/P balances, which reveal that for maize, rice and permanent meadows there is a concrete potential for management improving (see also Bechini and Castoldi 2006). Similar results with nutrient balances show variability within farms (cattle, dairy, pig and non-livestock) calculated for individual fields in a 4-year study in Piedmont Region (Sacco et al. 2003). This last study highlights the advantages of a regional agronomic information system for estimating nutrient balances such as a scenario analysis and the connection of GIS and other information systems. In fact, it was found that the optimum situation (scenario) is where manure becomes the only source of nutrients for all the crops in the area (efficiency: 77 % N, 61 % K) even the values for P efficiency (49 %) were low. Sacco et al. (2003) concluded that a sustainable livestock production could be achieved with a reduction in P feeding to animals and a very efficient exchange of manure between farms. The soil-surface N balance also provides information on good practices. A 2-year study described a soil-surface balance in the Po Valley located in the Province of Ferrara, Emilia-Romagna Region (Ventura et al. 2008). The authors reported that total inputs and outputs were of similar magnitude (overall balance close to zero), indicating that crop management, in particular, N fertilization techniques, reached a sustainable level.

From a practical point of view, research studies are supporting decision making at regional level in Italy. Lombardy Region financed the "Sustainable nitrogen management at farm level project"[2] aiming to survey and optimise the management of representative farming/cropping systems, providing recommendations for improved N management.

Economic, agronomic, and agro-ecological indicators were used for comparing current management practices and alternative practices leading to sustainable practices (Regione Lombardia 2008). For dairy farms, soil-surface N balances range from 60 kg N ha^{-1} to 163 kg N ha^{-1} due to high amounts of chemical and organic fertilisers applied mainly on maize, and for cereal farms values ranges from 27 kg N ha^{-1} to 339 kg N ha^{-1} also because of the fertilisation (Fumagalli et al. 2011). Fumagalli (2009) has estimated total management costs of production for the current practise and theoretical alternatives (rational use of fertiliser or crop allocation/rotation) for improving N management (see Table 8.2). Total costs (TC) of production ranges from 867 € ha^{-1} to 2,094 € ha^{-1} for dairy farms, and from 796 € ha^{-1} to 1,502 € ha^{-1} for cereal farms (current management).

The rational use of fertiliser, generally, decreased the N surplus with the exception for DAI-INT (data not shown). It was also verified a small reduction of TC for CER-DIG and DAI-EXT + alf determining an increase of the gross margin between 100 and 250 € ha^{-1} (decrease N dose applied). The rotation alternative determine slightly increase of N surplus in all farms, and the additional operations required by cropping systems increased TC with one exception. Regione Lombardia (2008) highlighted that alternative practices has shown that in almost all cases there is possibility for improving farming sustainability, without compromising their economic productivity or increasing the demand for labour and energy.

Another example, has been developed in Piedmont Region, that has identified five different agro-environments (sub-regional level) and once spatially defined, a set of farm-scale characteristics (farm types, stocking rates, land use and

---

[2] GAZOSA Gestione dell'azoto sostenibile a scale aziendale.

Table 8.2 Farm-gate/soil-surface N balances of agricultural systems in the Lombardy Region *(in kg ha^{-1})*

Current agricultural management	DAI-EXT +alf	DAI-DRY	DAI-INT	DAI-EXT	CER-SS	CER-DIG	CER-IND
Dairy cow ha^{-1}	1.2	1	3	1.3			
Usable agricultural land (ha)	114	50	120	169	0	0	0
INPUTS							
Manure	68	0	0	0	153	266	0
Chemical fertilisers	82	90	102	44	81	280	62
Animal feeding	70	157	362	188	0	0	0
Litter	2	11	0	0	0	0	0
Atmospheric deposition	30	30	30	30	30	30	30
Biological N fixation	116	25	12	4	1	0	61
Totals	***368***	***313***	***506***	***265***	***265***	***577***	***153***
OUTPUTS							
Milk	62	51	175	75	0	0	0
Animal sold	6	6	15	10	0	0	0
Cash crops	0	8	0	67	134	238	120
Manure	0	4	0	0	0	0	0
Totals	68	69	190	152	134	238	120
Farm-gate N balance, kg N ha^{-1} (farm area)	300	244	316	113	130	339	33
NUE × 100 (%)	18	22	38	57	51	41	78
Soil-surface N balance[a], in kg N ha^{-1}	156(39)	66 (20)	163 (44)	60 (20)	110 (27)	339 (14)	27(8)
Cost[b] (€ ha^{-1})							
Current management	2,094	1,354	1,435	867	796	1,502	860
Alternative 1	2,007	1,419	1,470			1,245	
Alternative 2	2,178[c]	1,687[c]	1,328[c]	1,080		2,091[c]	
Alternative 3					1,025[d]		783
Alternative 4							1,129

Note: Elaborations were based on representative cropping systems and presented as a weighted average at farm level. Source: elaboration from Fumagalli et al. (2011) and Fumagalli (2009)

Alternative 1: rational use of fertilisers through nutrient management plans, *Alternative 2:* crop allocation/crop rotation, *Alternative 3:* crop allocation/crop rotation (agro-energetic cultivations rape/sunflower/maize), *Alternative 4:* crop allocation/crop rotation (base on alternative 3 substitution of sugar beet by onion)

DAI-EXT + alf dairy extensive with alfalfa, *DAI-DRY* dairy non irrigated, *DAI-INT* dairy intensive, *DAI-EXT* dairy extensive, *CER-SS* cereal with use of sewage sludge, *CER-DIG* cereal with use of digested manure, *CER-IND* cereal and industrial crops

[a]Standard deviation

[b]Costs include: raw material, labour and the working site (variable + fixed costs)

[c]Includes also alternative 1

[d]Includes cash crops

management), and nutrient balances[3] (N, P, K) were calculated (Bassanino et al. 2011). The authors concluded that these indicators allowed to assess the relative importance of different inputs in surplus determination, enhancing the potential for improved management techniques.

8.1.3 Nutrient Balances on Nitrate Vulnerable Zones

The Fertilisation Plan (*Piano di Utilizzazione Agronomica*, PUA) is used as main technical tool for the Nitrate Vulnerable Zone (NVZ) Action Programs from agricultural sources. The PUA has been designed to achieve a balance between N to the soil and the predictable crop needs (farm level N balance). The compilation could be done directly by farmers, however, there are cases in which professional organization services are contracted. Regions have organised the presentation of PUA in different ways. For example, Piedmont Region has implemented the compilation of the PUA[4] through an information system that assist farmers including information available from the Agricultural Registry. The PUA is compiled exclusively in an electronic format, through a free application (interactive site between public administration and citizens for processing administrative compliance). Additionally, there is a call-center to solve questions or problems. But wherever farmers need support from the Agricultural Assistance Centers a cost should be included[5]. Costs depend on the type of animal breeding, size, and cultivated surface (for crop farms) varying from 400 to 2,000 € per farm in Lombardy Region[6]. Further efforts from this region include the development of a software instrument to analyse the current situation and effects of scenarios. At local level, farmers will be supported on N management (feeding, manure treatment, cropping management), and at regional level the effects of policies in the framework of Nitrate Action Programs will be evaluated (Acutis et al. 2009). For an average size farm in Emilia-Romagna Region (around 100 ha) the PUA could cost around 2,000 euro per farm.[7]

Other Italian regions are working on sustainable cropping systems for the NVZ through the ***Progetto PRIN ZVN***[8] initiative (Campania, Friuli Venezia Giulia, Marche and Sardegna Regions). The project considers an experimental phase

---

[3] Indicators defined in the '*Indicator Reporting on the Integration of Environmental Concerns into Agriculture Policy-IRENA*' project from the European Environment Agency, EEA.

[4] http://www.regione.piemonte.it/agri/dirett_nitrati/pua.htm (accessed 08/02/2011).

[5] Personal communication Dr. Paolo Cumino/Dr. ssa Monica Bassanino, Piedmont Region – DG Agriculture, 27/01/2011.

[6] Personal communication Dr. ssa. Marisa Meda, Lombardy Region – DG Agriculture, 01/02/2011.

[7] Personal communication with Dr. Paolo Mantovi, CRPA, 31/01/2011.

[8] http://www.uniss.it/php/zvn.php (accessed 28/01/2011).

involving private farms aimed at specific test of the relationship between farming systems and nitrate leaching (soil-surface and farm-gate N balances), and a modelling phase aimed at verifying the changes expected from various options of cropping systems in terms of crop productivity, dynamics of organic matter and nitrate leaching. The project provides a network of researchers, stakeholders and policy makers which will support the design of new effective normative frameworks for effective agro-environmental schemes and integrate expert's knowledge at different levels (Roggero et al. 2009).

8.2 N-Efficiency and Ammonia Emission Reduction Potential Through Adaptation of Pig Diets and Associated Costs on Swiss Farms

Edith von Atzigen-Sollberger and Peter Spring (✉)
School of Agricultural, Forestry and Food Sciences (HAFL),
Bern University of Applied Sciences (BFH), Länggasse 85,
CH-3052 Zollikofen, Switzerland

Annelies Bracher
School of Agricultural, Forestry and Food Sciences (HAFL),
Bern University of Applied Sciences (BFH), Länggasse 85,
CH-3052 Zollikofen, Switzerland

Institute for Livestock Sciences, Agroscope, Route de la Tioleyre 4,
Case postale 64, CH-1725 Posieux, Switzerland

8.2.1 Introduction

Livestock waste – solid or liquid manure – is an integral part of a farm's nutrient cycle. However, these excretions may have harmful effects on the environment, especially when stocking density is high. Swiss agriculture is supposed to implement measures for reducing ammonia (NH_3) emissions as part of Switzerland's fulfillment of the Gothenburg Protocol. The formation of ammonia is closely related to the protein nutrition and metabolism. Adaptations in feeding strategies change the protein supply and directly influence the nitrogen turnover, but these strategies imply extra costs. A survey of the current feeding practices in Swiss pig production was made to evaluate the potential to reduce nitrogen supply and excretion, including ammonia emissions, and the respective extra cost were calculated.

8.2.2 Material and Methods

A survey of the current Swiss pig feeding practices was used to estimate the potential in reducing ammonia emissions. It is based on three complementary data sources. They include:

- **The larger Swiss feed mills**: The seven most important compound pig feed producers, comprising 80 % of the Swiss pig feed market, provided data about diet specifications and sales volumes.
- **The official feed control agency**: Data from the control of declared nutrient contents in compound feed was used to check for potentially systematic over-formulations of protein.
- **Nutrient balance data** from 1,665 **pig producing farms** situated in the Canton of Lucerne: This database was used to get information about farm specific diet compositions and N-efficiencies.

All data used refer to the year 2008. In regions with high livestock densities, as found in the Canton of Lucerne, standardized nutrient balances with the Excel-based program IMPEX are used to control the on farm nutrient flow. These single farm calculations serve as a proof for the application of feedstuff with reduced content of nitrogen and phosphorus (NPr-feed). A balance between nutrient supply and requirements is a precondition for a farm to qualify for subsidies. Usually, standard values are used to calculate a farm's nutrient outputs. When NPr-feed is applied, nutrient outputs that diverge from the standard values can be asserted. On one hand, such nutrient balances include the N-input to the farm via purchased compound feed, by-product feedstuffs and purchased livestock, on the other hand they contain the N-output via sold animals. Livestock imports or exports are accounted by 24.6 g N kg^{-1} live weight for smaller pigs ($<$60 kg live weight) and 22.2 g N for heavier ones ($>$60 kg live weight). First, the data of 1,665 farms was collected in a database and used for a set of analyses. The composition of the diets was analyzed and specific conclusions were drawn after grouping the data according to farm types, livestock categories, feed types and feeding strategies. Second, N-efficiency (N-output/N-input) was examined at farm level. Third, nitrogen excretion per standard fattening pig was calculated. For integrated pig farms (birth to slaughter), it was not always possible to get a clear-cut identification of the livestock category a specific feed was used for. Therefore, only specialized grower-finisher farms (n = 899) were used for the calculations on nitrogen excretion and -efficiency. A standard fattening pig was assumed to be fattened from 26 to 108 kg during its grower-finisher phase, and – when leaving the farm – to correspond to a nitrogen export of 1.758 kg N ($N_{endweight} - N_{startweight}$). The N-excretion per standardized fattening pig can be calculated from the difference between the N-input and this N-export.

A potential in reducing emissions can only be realized if it is economically sustainable. To estimate the effects of changes in prices of compound feed, the optimization was done per least cost formulation software Opti-Schwein (designed

Table 8.3 Average crude protein (CP), lysine and energy concentration of standard and NPr diets (compound pig feed)

Compound feed type	Crude protein [g CP/kg]	Lysine [g Lys/kg]	Digestible energy [MJ DEP/kg]
Grower-finisher feed, standard	172.95	9.97	13.57
Grower-finisher feed, NPr	158.04	10.12	13.72
Lactation feed, standard	178.85	10.08	13.68
Lactation feed, NPr	164.81	10.04	13.73
Gestation feed, standard	144.97	6.54	12.05
Gestation feed, NPr	139.12	6.67	12.26
Combined sow feed, standard	171.28	9.32	12.87
Combined sow feed, NPr	157.68	9.26	12.89

by HAFL Zollikofen). These estimations took account of the costs for each raw material in a compound feed. The raw material prices were taken from a marked survey that took place in May 2009. Optimized formulas were validated in collaboration with several feed mills which gave evidence that Opti-Schwein provided results corresponding well to commercial formulation practices.

8.2.3   Results and Discussion

8.2.3.1   Feeding Practices and N-Efficiency

The inquiry among feed mills showed that about 70 % of the compound pig feed currently used in Switzerland is classed as NPr-feed. However, this proportion varies largely between regions. The majority of pig producers use a fattening feed type suitable for the entire grower-finisher phase (1-phase feed). For sows, there are three feed types commonly used: lactation feed, gestation feed and combined sow feed. The protein and energy content of these feed types is shown in Table 8.3, in each case for an NPr- and a standard feed.

The potential of differentiating into several feeding phases for fattening pigs or for pregnant sows is not yet fully exploited. If a combined sow feed is used, gestating sows are heavily oversupplied with protein despite some feeding of roughage. Even when an NPr-feed is used, there is a potential for further cutting down the protein supply in the diet for gestating sows as well as for fattening pigs during the finisher phase. Compound feed which is formulated for the finishing phase often does not differ enough in its protein content from a 1-phase feed.

The feed mills vary heavily regarding the proportion of complementary feed within their compound feed production (5–46 %). However, the classification as complete or complementary feed is not always possible. The major part of complementary feed is formulated as a complement to whey. The data from the official

Table 8.4 Average contents of the entire diet and N-efficiency for different farm types with and without by product feedstuffs

Farm type (number of farms)	DEP (MJ/kg)	CP (g/kg)	P (g/kg)	g CP/MJ DEP	N-efficiency (%)
OPP CF (69)	13.51	164.8	4.87	12.20	32.7
OPP BP (373)	13.06	159.1	4.79	12.19	29.3
MPP gestation CF (10)	12.43	141.4	4.37	11.38	14.4
MPP gestation with BP (34)	12.45	143.3	4.66	11.51	15.9
MPP farrowing CF (35)	13.78	169.9	5.07	12.33	35.7
MPP farrowing with BP (39)	13.29	163.7	4.99	12.32	34.7
MPP nursery CF (15)	13.78	163.7	4.99	11.88	46.9
MPP nursery + growing/finishing CF (7)	13.63	162.4	4.47	11.01	36.3
MPP nursery + growing/finishing with BP (3)	13.30	161.8	4.55	12.16	35.9
OPP + growing/finishing CF (20)	13.58	160.9	4.50	11.85	31.9
OPP + growing/finishing BP (138)	13.38	161.2	4.65	12.05	30.5
growing/finishing CF (626)	13.75	159.7	4.16	11.62	31.6
growing/finishing with BP (261)	13.74	158.3	4.39	11.52	32.6
OFFM growing/finishing with BP (11)	13.90	172.3	4.42	12.40	30.4
OFFM OPP + growing/finishing with BP (9)	13.68	165.3	4.57	12.08	30.2

DEP digestible energy pig, *MPP* multisite piglet production, *OPP* one site piglet production, *CF* complete feed, *BP* by-product, *OFFM* on-farm feed manufacturer

feed controls gave no evidence of systematic crude protein over-formulations in NPr-feed with respect to the declared compositions.

The evaluation of the nutrient balances showed that complete feed accounts for 73 % of all feedstuffs used on the sampled farms. The amount and type of by-products used as feedstuffs differ between farm types. Whey is the by-product most often used on grower-finisher farms; on average 10 % of the diet (dry matter) consist of whey. Piglet producing farms often complete their diet with roughage, such as top quality hay and silage of maize or grass. Table 8.4 shows the average contents of the whole diet per farm type weighted by the quantity of feedstuff used. For each farm type, a distinction is made with respect to their use of complete feed (CP) or by-product feedstuffs (BP). Most farms in the sample specialize either in pig fattening or in piglet production. In Swiss piglet production, gestating sows, farrowing sows and weaned piglets (nursery) are commonly kept on the same farm (one site piglet production, OPP). However, more and more producers specialize on one of these steps of production, which leads to a multisite piglet production (MPP). A special case is given by the producers who mix their pig feed on farm (on-farm feed manufacturers).

When interpreting these data, it must be taken into account that the sample was taken from a region with high livestock densities. Therefore, most of the farms take

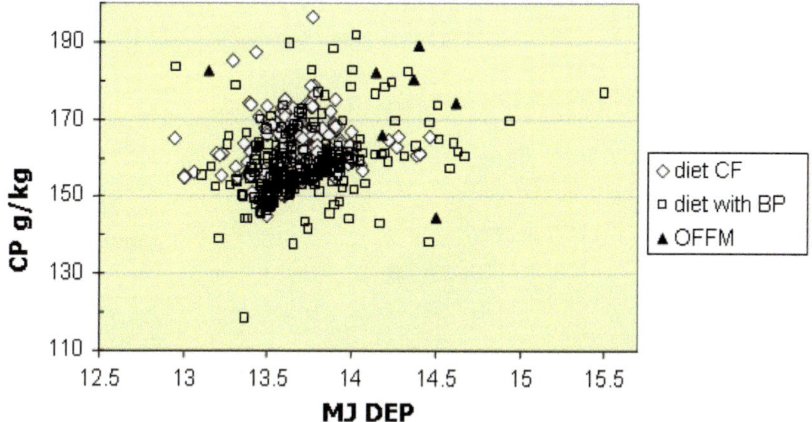

Fig. 8.1 Entire diet energy and crude protein content of all grower-finisher pig farms (n = 899) with or without by-products (*CF* complete feed, *BP* by-product, *OFFM* on-farm feed manufacturer)

part in the NPr-program and use feedstuffs with reduced content of nitrogen and phosphorus. Their data does not correspond to a Swiss standard. The average CP content of the diet is below 170 g/kg among all farm type groups except the on-farm feed manufacturers. The most extreme differences in N-efficiency can be seen between producers taking part in multisite piglet production: Farms specialized on nursery piglets achieve an N-efficiency of 47 % (some individual farms up to 56 %), whereas farms specialized on gestating sows have an N-efficiency of only 15 %. On those farms, the recommendation of 10 g CP/MJ DEP is on average exceeded – which indicates an oversupply of protein. The use of NPr-feed also implies low phosphorus contents. Phytase is commonly applied on nearly all farms.

For the analysis of the range of variation of dietary compositions and possible causes of the variation regarding the N-efficiency, the specialized grower-finisher farms were used. Figure 8.1 shows the crude protein content of the entire diet plotted against its energy content. The mean of each of these two contents does not differ between farms using by-product feedstuffs or complete feed; however the variation within the group of farms using by-product feedstuffs is considerably larger. There is only a small number of on-farm feed manufacturers, but as they are above average size, and as their protein content exceeds the level of the other grower-finisher farms by 10 g on average, they should not be ignored when analyzing emissions.

On the whole, there are just a few farms on which the CP content exceeds 180 g kg^{-1} of the entire diet. Those farms either do not use NPr-feed, or they could optimize their strategy regarding complementary feeding. The extreme value for the energy content (15.5 MJ DEP) refers to a farm using a high proportion of catering by product for its diet, which was still permitted at the time the study was conducted.

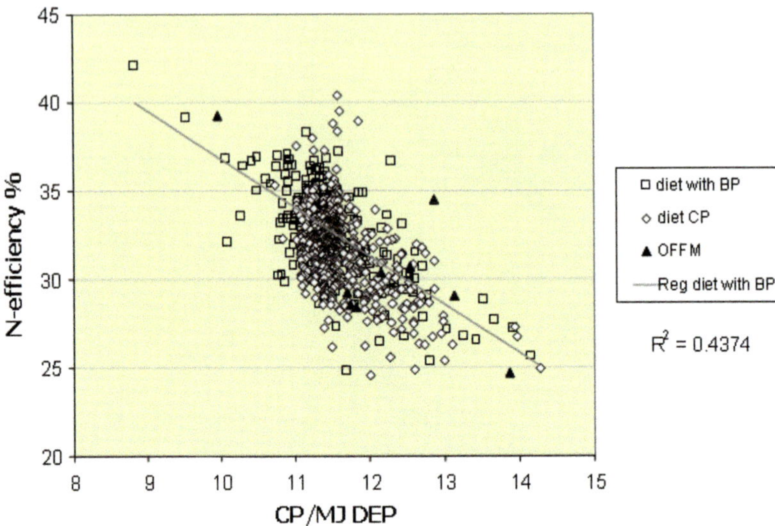

Fig. 8.2 N-efficiency on grower-finisher pig farms (n = 899), depending on feeding regime and the CP content per MJ DEP (*CF* complete feed, *BP* by-product, *OFFM* on-farm feed manufacturer)

The data show that N-efficiency (N-output/N-input) depends on the diet's protein content per energy unit (Fig. 8.2); however this relationship is not very strong. Most of the farms are clumped within a small CP range, and additional causes of variation not linked with protein supply should be considered. More analyses were done by additionally splitting the sample with respect to the type of by-product feedstuffs or feeding strategy (1-phase feed for the whole grower-finisher period or distinguish multiple phases). However, no clear separation effect was found. The expected effects of diets and strategies are likely to be interfered by a large impact of farm characteristics. The variance within feeding strategy groups is larger than between strategies, whereas there is evidence for a slight advantage of grower-finisher farms using whey as the only by-product feedstuff. These farms show an average N-efficiency of 33 % as compared to 31–32 % in other subgroups. More parameters should be tested to identify the causes of variation. Among the potential causes are the level of performance, genetic characteristics, farm sanitation, the health status, technical equipment, and management factors. A large range of variation for N-efficiency (from 24 to 40 %) for the same feeding strategy indicates a high potential for farm specific optimizations.

As mentioned, the calculated N-excretion corresponds to the quantity of nitrogen excreted by a standardized fattening pig during the whole grower-finisher period. In a region with a high proportion of NPr-feed users, the actual N-excretion per fattening pig varies from 2.4 kg N to 5.4 kg N (see Fig. 8.3). Farms using complete feed show an average N-excretion of 3.83 kg N ± 0.38, farms using by-product feedstuffs one of 3.64 kg N ± 0.44, and on-farm feed manufacturers one of 4.06 kg N ± 0.70. As expected, N-excretion grows with higher CP contents of the entire diet; however this

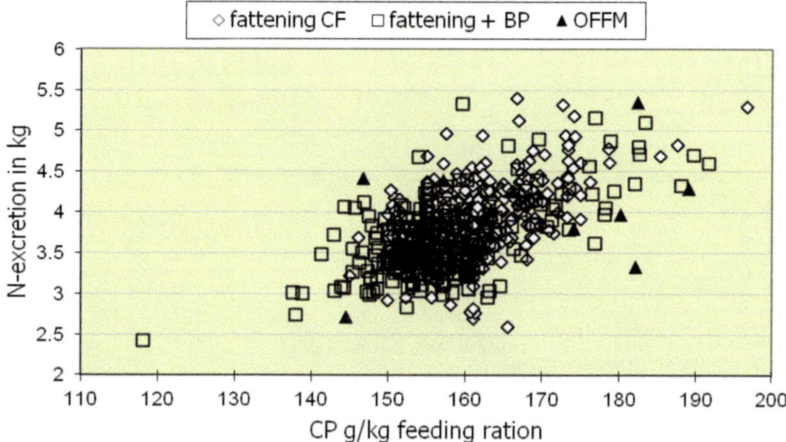

Fig. 8.3 N-excretion during the production of a standard fattening pig (26 – 108 kg live weight) depending on the CP content of the entire diet on grower-finisher farms (n = 899; *CF* complete feed, *BP* by-product, *OFFM* on-farm feed manufacturer)

relationship is not very strong. N-excretion considerably depends on additional factors, such as adaptations of the feeding strategy to the animal's need depending on the phase of fattening or on its health status, as well as on management factors which can have important effects on protein deposition and with it on N-excretion. Further research would be needed to quantify these additional effects.

8.2.3.2 Calculation of Costs Incurred by Low-Protein Feeding Strategies

Ingredient cost was estimated in a series of least cost formulations setting minimal restrictions for either seven or ten amino acids. Additionally, estimations are done with scenarios of 25 % higher cost for protein components and synthetic amino acids. In the case of a pig with 60 kg live weight, each of these estimations reaches minimal costs at a CP content of 160 g kg^{-1} (see Fig. 8.4). The estimation labeled 7AA represents a formulation respecting the contents of the amino acids lysine, methionine (+cystine), threonine, tryptophan, leucine and valine. These restrictions allow a reduction in crude protein content to 130 g kg^{-1} without affecting raw material cost. However, such low crude protein contents lead to a shortage in isoleucine. Therefore, in a second estimation, minimal standards are taken into account for 10 amino acids (10AA). In that case, crude protein contents below 150 g coincide with higher cost for the raw material used in the formulation. To evaluate the consequence of these cost changes, prices of all protein-rich components, including potato protein (which is rich in isoleucine) and of synthetic amino acids are increased by 25 %. These calculations show that higher prices have only a small impact on the price differences between

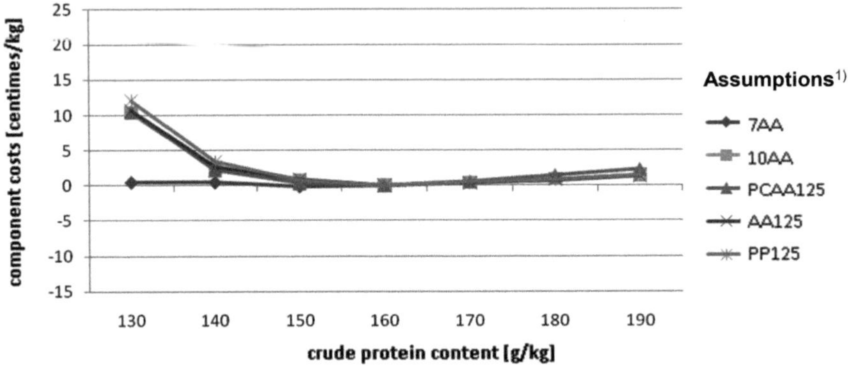

[1] 7AA= minimal requirements for 7 amino acids; 10AA = minimal requirements for 10 amino acids; PCAA125 = prices for protein components and synthetic amino acids 25% higher; AA125 = prices for synthetic amino acids 25% higher; PP125 = prices for potato protein 25% higher

Fig. 8.4 Costs for the raw material used in a formulation depending on varied CP contents; optimized feed for a fattening pig of 60 kg life weight (initial value CP = 160)

diets. When the same optimization is done for pigs with various life weights, it can be seen that the raw material cost increase occurs at lower protein contents for animals at a late stage of fattening than for animals at an early stage. During the growing phase, a crude protein content of 150 g kg^{-1} or less can not be formulated in a way to meet the animal's requirements in isoleucine. It should also be taken into consideration that any further genetical improvement in protein accretion will require diets with higher amino acid and CP concentrations.

Generally, a reduction in crude protein content does not involve drastic changes regarding the diet composition. The proportion of cereals used in the formulation increases, whereas the proportion of protein-rich components, such as soybean, decreases. However, when formulating for 10 amino acids, potato protein use increases with the protein reduction, to provide sufficient isoleucine. As the supply for potato protein is very unsteady, this will pose an additional risk regarding costs. Furthermore, synthetic amino acids are more frequently used when CP-contents are reduced.

Table 8.5 shows the development of feed component costs for reduced crude protein contents. A 1-phase feeding system is compared to 2-phase and 3-phase systems. If the current crude protein content is reduced from an average of 160 g kg^{-1} (entire diet) to 150 g kg^{-1}, the cost change per kg compound feed ranges from −1.80 to +1 centimes. Additional reductions, to a minimum crude protein content of about 140 g kg^{-1} (content level at which the supply of non-essential amino acids may become limiting) lead to increased costs by 1 to 3.4 centimes per kg. This indicates that a reduction of that dimension has a considerable negative economic impact. These calculations refer to the raw material costs only. However, cost changes for processing, transport, storage and feeding might be important, depending on the farm size. Before CP contents for pig feed is reduced at a large scale, it should be tested whether lower CP contents affect the fattening

Table 8.5 Cost differences for the raw material used in the formulation of 1, 2 and 3 phase feeding in centimes/kg for different scenarios regarding CP reductions for fattening pigs

Weight class in kg	Share of total feed (%)	Current average	Recommended CP content[a]			Minimum CP content[b]		
		g CP	*g CP*	*Price change*	*Difference scenarios*[c]	*g CP*	*Price change*	*Difference scenarios*[c]
	100	160	153	0	1	143	1	3
2-phase feeding								
40	43	160	162	0	−0.5	152	2	4
80	57	160	143	−0.5	−1	133	1	1.5
3-phase feeding								
30	27	160	166	−2	−8	156	5	6
60	33	160	153	0	1	143	1	3
90	40	160	138	0	0	128	1	2

[a]Recommendations ALP (Stoll et al. 2004)
[b]Minimal CP content where the supply of nonessential amino acids may become limiting
[c]Difference of price between scenarios to show effect of costlier protein components

performance. All calculations in this article are valid only under the assumption that the formulation of compound feeds with respect to their amino acid contents agrees with the recommended values published by ALP (Stoll et al. 2004). Should the current formulation be higher due to expectations of a higher muscle growth potential, the cost increase would be higher.

Similar raw material cost scenarios were calculated for gestation feed. It turned out, that reductions in crude protein content to 11 g/MJ DEP are without substantial effects on raw material costs. At 10.5 g/MJ DEP, it is still possible to provide all amino acids at moderate additional feed costs (<1 centime/kg feed). However, further reduction to 10 g/MJ DEP lead to considerable cost increases as it becomes challenging to cover all essential amino acids.

Finally, it is important to point out that the whole set of ten essential amino acids must be taken into account when crude protein contents are reduced. For gestating sows, a reduction to 10.5 g CP/MJ DEP seems maintainable from an economic point of view. For fattening pigs, a considerable CP reduction in a multi-phase feeding system to 140 g kg^{-1} result in increased costs by 1.00 to 3.50 centimes per kg feed. However, research is needed to check for potential effects of such a reduction on fattening and slaughtering performances.

8.2.4   Conclusions

- In Swiss regions with high livestock densities, compound feed reduced in nitrogen or phosphorus is frequently used. There is an additional potential for reducing N-inputs by differentiating the diet with respect to the phases of fattening and by better adapting the diet of gestating sows to their need.

- As an analysis of the nutrient balances from 1,665 pig producing farms shows, farm specific factors have a large influence on N-efficiency and N-excretion, which goes beyond the impact of feeding. Presumably, there is a considerable potential to increase N-efficiency by adapting technical equipment and management or by taking into account the pigs' health status.
- Reducing the CP level of fattening feeds to 160 g kg^{-1} (ca. 12 g CP/MJ DEP) at current feedstuff prices induces no extra costs. For further reductions to 150 g kg^{-1} (ca. 11 g CP/MJ DEP) the price increases are moderate for 1-phase-feeding, in multi-phase feeding systems, even cost reductions are incurred.
- Reductions of the CP-level below 150 g kg^{-1} or 11 g/MJ DEP in fattening rations leads to considerable price increases, particularly if the amino acids requirements of 10 essential amino acids according to the recommended minimal values of ALP are respected.
- Multi-phase feeding allows CP-reductions at lower costs or no costs at all regarding feedstuffs: However, the necessary investment in feeding installations should not be overlooked.

8.3 Costs Incurred by the Reduction of the Soiled Surface in Cattle Stables and Manure Belt Drying of Layer Hen Manure Under Swiss Frame Conditions

Martin Raaflaub (✉)
School for Agricultural, Forestry and Food Sciences (HAFL),
Bern University of Applied Sciences (BFH), Länggasse 85,
CH-3052 Zollikofen, Switzerland

Harald Menzi
School for Agricultural, Forestry and Food Sciences (HAFL),
Bern University of Applied Sciences (BFH), Länggasse 85,
CH-3052 Zollikofen, Switzerland

Agroscope, Tioleyre 4, 1725 Posieux, Switzerland

8.3.1 Introduction

Frame conditions regarding economic aspects and farm structure differ considerably between Switzerland and other European countries, even between Switzerland and its neighbours. Namely, fixed housing costs are considerably higher due to the smaller scale of farm operations, higher construction prices and more restrictive animal welfare regulations. Also purchasing prices for equipment are higher, possibly due to lesser competition outside the EU common market. Salaries are higher than in neighboring countries while interest rates are lower. Thus, even though the methodology of abatement cost calculation is consistent with the methodology presented by Montalvo et al. in this document, the resulting abatement costs can differ considerably.

In this contribution, specific cost data are presented for the measures "reduction of the soiled surface in cattle housing" and "belt drying of laying hen manure", as these are major measures propagated in Switzerland for emission mitigation in animal houses.

8.3.2 Methodology, Data Sources, Assumptions

The cost calculation methodology is to a great extent comparable to the method described by Montalvo et al. earlier in this document. The main divergences concern

- **Interest rate**: 1.60 % (interest paid for Swiss Federal Bonds, contract period 10 years, average of the year 2011)
- **Amortization period**: 20 years for buildings, 10 years for equipment
- **Labour cost**: CHF 28/h

The costs of the abatement measures are calculated as the difference to a reference method representing current good agricultural practice.

8.3.3 Reduction of the Soiled Surface in Cattle Stables: Description and Results

The installation of resting boxes in loose housing systems for cattle is considered current practice in stable design. Structuring the feeding area with separation brackets and ground level difference between feeding place and walking area (see Fig. 8.5) is not yet current practice and has a considerable potential to abate emissions by reducing the soiled area.

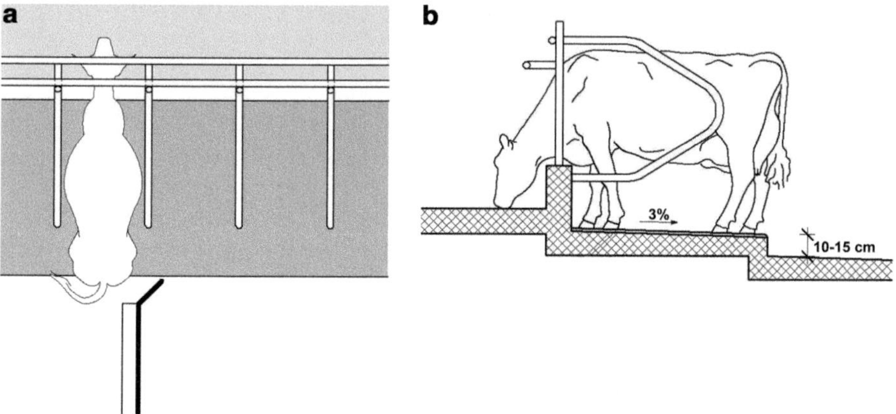

Fig. 8.5 Drawings of structured feeding areas in loose housing systems for cattle: outline (*left*) and profile (*right*) (Source: BAFU und BLW 2012)

Table 8.6 Additional construction costs incurred by the structured feeding area in cattle loose housing systems

Additional stable surface	77 m^2
Additional roof surface	83 m^2
Additional excavation	149 m^3
Excavation, civil engineering and masonry	16,100 CHF
Assembly in lumber, roof, plumbing, lightning protection	14,800 CHF
Stable installation	2,750 CHF
Rounding	350 CHF
Total costs	**34,000 CHF**
Total costs per housed animal	1,000 CHF
Annuity costs per housed animal	58.80 CHF

Source: Bucheli M, Krieger Stallbau AG, Ruswil: plans and cost estimates for cattle stables, pers. comm. 2011

Adjacent to the feeding rack separation brackets are added which reach out towards the aisle. Thus, the area near the feeding rack is no transiting area for the animals anymore, which reduces the soiled area. As Swiss animal welfare regulations require a minimum aisle width (measured from the hind foot of the feeding animals), the building volume of the stable increases. Additionally, a floor level threshold must be provided between the aisle floor level and the feeding area in order to avoid that slurry is being pushed back to the feeding area by the slurry pusher. This adds to the excavation costs.

The extra costs incurred by structuring the feeding area for the construction of a new stable for 34 head of cattle (25 dairy cows and 9 replacement heifers) are presented in Table 8.6.

It is assumed that no additional labour requirement (i.e. for cleaning) is incurred.

Reducing the soiled surface by structuring the feeding area is a relatively expensive measure, particularly with regard to the rather modest emission reduction (estimated at 10 % by Menzi et al. 1997). Only about half of the costs are caused by the measures that aim directly at reducing the soiled surface (separation brackets and ground level threshold), the other half being caused indirectly by the increased space requirement.

8.3.4 Drying of Laying Hen Manure on the Manure Belt – Description and Results

Ammonia emissions of laying hen manure can be significantly reduced by drying it within 48 h after excretion to a dry matter content of 45–65 %. This is done by ventilating the manure belt with air from inside or outside the stable. Using air from outside results in a higher dry matter content, but requires a more expensive ventilation installation (so called air blenders) and generates significantly higher electricity costs, while simple radial ventilators are sufficient when solely air from inside the stables is being used.

Table 8.7 Yearly costs of drying laying hen manure on the manure belt (all costs in CHF)

Farm size ventilation system	6,000 laying hens air mixer ventilator 2.2 kW	6,000 laying hens radial ventilator 2.2 kW	10,000 laying hens air mixer ventilator 4 kW	10,000 laying hens radial ventilator 3 kW
Ventilators, control	8,300	2,500	9,000	5,100
Tubing	2,800	1,300	3,600	1,900
Transport, assembly	5,000	3,000	6,000	4,000
Investment total	16,100	6,800	18,600	11,000
Investment costs/year	1,360	530	1,550	840
Electricity costs/year	300–1,300	300–1,300	600–2,400	500–1,800
Maintenance/ cleaning	100	100	Fr. 100	100
Costs/year	1,700–2,700	900–1,900	2,200–4,000	1,400–2,700
Costs/100 laying hens	29–46	15–32	22–40	14–27

Source: B. Penkhues/Big Dutchman GmbH, own calculations
Fire insurance 0.1 %, electricity uptake 77 % of nominal power output, electricity costs: 24 h average rate 0.09 CHF/kWh, night rate 0.07 CHF/kWh, maintenance/cleaning 4 h/year

The first belt trying system was introduced in Switzerland in 2007. For new laying hen operations it can be considered as good agricultural practice today.

The investment costs can be divided into the following components:

- costs for ventilators and controls
- costs for tubing, chimneys assembling items etc
- costs for assembling the installation.

The operating expenses are dominated by electricity costs. The question whether the ventilator only operates during 8 night hours at the low electricity night rate or during 24 h at the average 24 h electricity rate is of great importance. For the calculation made in Table 8.7, the lower electricity cost value applies to 8 h operating time, the higher value to 24 h operating time.

The drying installation does not require any maintenance other than a periodic cleaning of the radiators. The frequency of cleaning depends mainly on the dust charge and the quality of the aspirated external air.

Drying hen manure on the manure belt is a relatively inexpensive abatement measure, as the costs involved are relatively modest in relation to the overall production costs. However, the costs vary considerably between the drying strategies, mainly related to differences in electricity consumption and investment costs.

Empirical evidence suggests that the costs correlate with the drying efficiency strived for:

- The more expensive air mixers allow the use of external air with considerably lower humidity, thus increasing the drying rate and the final dry matter content.
- Running the ventilators during 24 h instead of limiting their operation to the night hours leads to higher electricity costs by increasing both the amount of electricity used and the average price, but increases the final dry mater content of the manure.

8.4 Assessment of Ammonia Abatement Cost in Dairy Farming of the North-West of the Russian Federation: Case Study

Natalia Kozlova (✉), Dmitry Maximov, and Aleksandr Bryukhanov
North-West Research Institute of Agricultural Engineering
and Electrification (SZNIIMESH), Filtrovskoe shosse, 3,
St-Petersburg-Pavlovsk 196625, Russian Federation

8.4.1 Cattle Housing in RF PEMA

The Gothenburg Protocol (GP) specifies emission ceilings only for the Russian Federation the Designated Pollutant Emissions Management Area for (RF PEMA), which includes seven constituent entities: Murmansk Oblast, Republic of Karelia, Leningrad Oblast, St-Petersburg (which is a separate federal subject), Pskov Oblast, Novgorod Oblast and Kaliningrad Oblast. Their area totals 14 % of the European part of Russia under EMEP.

RF PEMA features efficient agriculture, dairy farming in particular. Statistical data of 2008 shows average per cow milk yield of 4,970 kg in RF PEMA against 3,892 kg in the Russian Federation. In Leningrad Oblast dairy cows account for 38 % of the RF PEMA total cattle stock, in 2008 the average milk yield here was 6,663 kg.

In RF PEMA dairy farm capacity ranges from 100 to 800 head, average farm capacity is from 400 to 600 head. RF PEMA has also farms with the capacity up to 2,000 head. In the area under consideration the operating, newly built or reconstructed farms apply practically all elements of housing systems and equipment used in Europe in various combinations. There are also turnkey farms in RF PEMA participating leading European companies. The general tendency in new farm construction is loose, all year round zero grazing housing of animals.

Around 80 % of farms in RF PEMA use tied housing systems. Loose housing systems are introduced on reconstructed or newly built farms.

8.4.2 Assessment of Ammonia Abatement Costings in GAINS for RF PEMA

The objective of the assessment of ammonia abatement costings in GAINS was to get the initial data to estimate the cost of the Russian Federation's accession to GP. Some results of the estimation are shown in Tables 8.8 and 8.9.

To answer the question whether ammonia abatement cost as currently included in the GAINS model is suitable for Russian conditions the following data were used:

- estimated data of ammonia abatement cost for the Russian Federation as cited in the "Summary of Ammonia Abatement Costings in GAINS (expressed as euro/kg NH_3 as N abated)", prepared by IISA as the background information for the Expert Workshop on the Costs of Ammonia Mitigation and the Climate co-benefits;
- required emission abatement levels for various ammonia emission sources, as stated in the latest version of the Draft Revised Technical Annex IX to GP;
- EMEP/EEA Emission Inventory Guidebook 2009 (EEA 2009), updated June 2010, 4b Appendix – an example of emissions estimation for a 100 head farm. The case calculation data was used to define the contribution of ammonia emission from the farm (in Table 8.8 – "reference NH_3-N emissions" and "reduction NH_3-N", kg cow^{-1} $year^{-1}$);
- statistical data on dairy production in the Leningrad Oblast in 2009: the average annual per cow milk yield was 6,993 kg, the self cost of 1 kg of produced milk was 0.28 €, with the average rouble to euro exchange rate being 40;
- expert estimates of the contract price of farm construction, including manure handling systems, in the North-West of the Russian Federation.

In the process of assessment all the estimated indicators were reduced to specific values per one animal and also per 1 kg of produced milk. Table 8.8 shows all compared indicators per one animal; contractions of emission control options are those as currently implemented in GAINS.

As can be seen from Table 8.8, in case of implemented ammonia abatement measures for various emission sources the range of milk self cost supplement calculated by the IIASA data is from 1 to 7 %. With the average break-even level of milk production in Leningrad Oblast being from 14 to 20 %, this is an appreciable value. In case the cost of all emission abatement measures is 359 €/head, the self cost of milk grows by 18 %. So the key issue for the introduction of ammonia abatement into agricultural practice is to update ecological legislation in terms of financial provision for ammonia abatement.

The cost estimations of separate emission sources are shown in Table 8.9.

According to the Tables 8.8 and 8.9 emission abatement cost from GAINS may be considered suitable and for the most part acceptable for provisional assessment of accession cost of the Russian Federation to GP in terms of ammonia abatement on dairy farms.

Table 8.8 Assessment of ammonia abatement cost for a dairy farm, currently included in the GAINS model under conditions of RF PEMA

	Emissions			Cost		Assessment
	Emission reduction requirements, %	Reference NH_3-N emissions, kg cow^{-1} $year^{-1}$	Reduction NH_3-N emissions, kg cow^{-1} $year^{-1}$	Cost of abated NH_3-N, €/kg NH_3 (GAINS)	Cost of abated NH_3-N, €/cow (GAINS)	Self cost supplement of 1 kg of milk, % (RF PEMA)
LNF feeding strategies	10 %	28.6	2.8	14.7	42.1	2.2
SA building improvement	25 %	9.2	52.3	56.4	130.1	6.9
CS_Low storage	60	4.4	2.6	15.5	41.1	2.1
CS high covered storage	80	4.4	3.5	28.5	100.7	5.1
LNA_low land application	30	10	3.0	29.0	86.7	1
LNA_high land application	60	10	6.0	14.5	86.7	1

Table 8.9 The estimated cost of ammonia abatement used in GAINS compared with the expert estimations for RF PEMA

	Estimated object	Expert estimation for RF PEMA	Estimated indicators used in GAINS
LNF feeding strategies	Produced milk on dairy farm	Self cost of 1 kg of produced milk is 0.24€	Inputs in emission abatement through feeding strategies are 0.00615€/kg of milk
SA building improvement	Dairy barn	Capital inputs per 1 animal place are 4536€	Inputs in ammonia abatement through the building improvement per one animal are 130€
CS_Low storage	Manure storage for 5,000 m^3	Manure storage cost is from 55,000€ to 150,000€ The cover cost is from 50,000€ to 100,000€.	Inputs in ammonia abatement through a cover construction are 40,242€
LNA_high land application	Farm for 800 head	The overall investments in the system of manure land application is from 390,000 to 510,000€	Cost of ammonia abatement in land application is 67,280€.

8.4.3   Storage

The only one emission control option for manure storages currently implemented in GAINS is a cover construction. For RF PEMA conditions on many farms new storing facilities are yet to be constructed to mitigate emissions while storing manure. According to our calculations using GAINS prices the construction of a storage cover on a 400 head farm for will cost 40,242 €, with the storing capacity being around 5,000 cubic meters of manure. In RF PEMA the market cost of a concrete storage construction for 5,000 cubic meters is in the range from 55,000 € to 150,000 €, the cost of cover construction is from 50,000€ to 10,000€.

Land application of manure. According to available commercial quotations the overall investments into the system of land manure application on an 800 head farm for is from 390,000 to 510,000€. Revised estimation of ammonia emission abatement measures for this category according to GAINS is 67,280€.

8.4.4   Cattle Housing

According to Annex IX to GP a Party shall use for *all new or largely rebuilt* animal houses cattle housing systems, those which can reduce ammonia emissions by at least 25% compared with the reference. The "Guidance Document for Preventing and Abating Ammonia Emissions from Agricultural Sources" (Bittman et al. 2014) includes two reference systems for cattle housing – cubicle house (*reference technology 1*) with 12 kg/cow place/year ammonia emission and tied system (*reference technology 2*) with 4.8 kg/cow place/year ammonia emission. Only one option – "grooved floor" is considered as Category 1 technique with 25 % emission reduction as compared with reference cubicle house. Tied systems have less emission, but they are not favored for animal welfare reasons. Changes in the building design to meet new animal welfare legislation increase NH_3 emissions. In RF PEMA the reconstruction of barns is in progress involving the transfer from the tied to the loose cow keeping. So the following variants could be considered in the process of cattle housing cost estimation:

- in *rebuilt* barns with the transfer from tied stall to cubicle housing emissions will increase, there is no "abated" NH_3;
- *new building* designs have to introduce NH_3 mitigation measures without additional investment. Expert estimations show (Table 8.11) that construction investments for a barn in RF are currently 4,536 € per animal; estimated cost of abatement measures according to GAINS is 130 € per animal;
- there is no abatement requirements for *existing* cattle buildings.

At the same time any existing systems (both new and rebuilt ones) seem to be able to reduce emissions, at least through better management.

As can be seen from the above, for RF PEMA the cost value of ammonia abatement for cattle housing can be neglected in the cost estimations of emission mitigation on regional level. It seems that integrated research is needed to assess the cattle housing systems on big and medium farms within the general chain of milk production from the feeding system to manure soil application.

8.4.5 Technologies of Cattle Manure Processing into Solid Organic Fertilizers

Slurry-based cattle housing systems under conditions of RF PEMA involve a number of problems associated with possible long-lasting frosts, excessive soil moisture content and capital-intensive manure storing facilities. The weak points of bedding-free housing are somewhat worse welfare of animals and substantially – 1.5 to two times – higher volume of slurry against the excrements output. This requires bigger capacity of manure storing facilities and greater transportation efforts during the high peak of field works. On the big farms this results in higher capital costs. Besides in most cases this practice does not provide elimination of pathogenic micro-organisms and bacteria.

One of the ways to reduce the nutrients loss and, consequently, air emissions, is introduction of technologies for accelerated preparation of high quality organic fertilizers from animal and poultry manure, including the use of bioreactors.

Introduction of such technological solutions is especially topical on big animal farms with substantial manure output and the lack of land for raw manure application. Some fragments of comparative analysis of various technology solutions of organic fertilizer production from animal and poultry manure are shown on Figs. 8.6 and 8.7 (see as well Maximov et al. 2014).

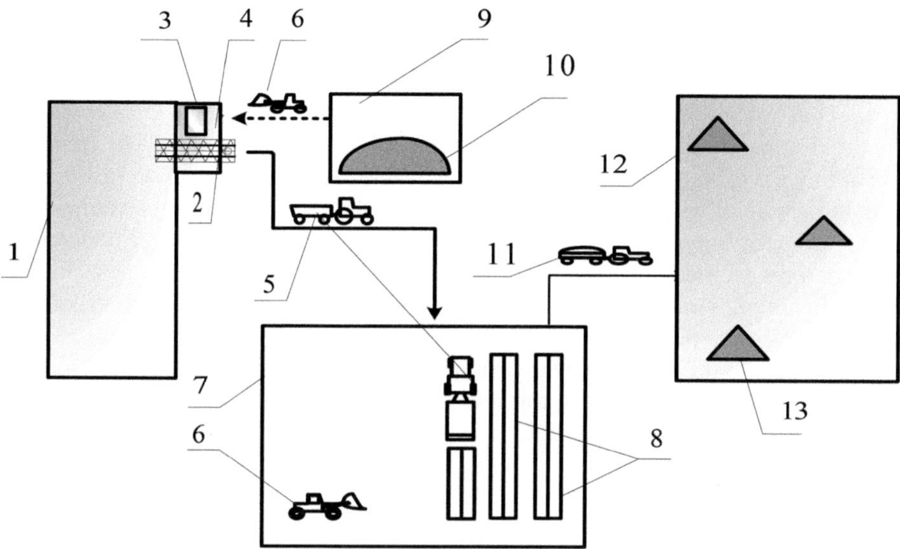

Fig. 8.6 Composting technology for the mix of manure with some moisture-absorbing materials on the farm batch-ground: *1* barn; *2* auger conveyor; *3* a moisture-absorbing material meter; *4* mixing station; *5* tractor with trailer for piling; *6* front-side forklift; *7* composting ground; *8* compost piles; *9* moisture-absorbing material storing site; *10* moisture-absorbing material; *11* vehicle for ready compost transportation; *12* field; *13* field-side piles of ready compost

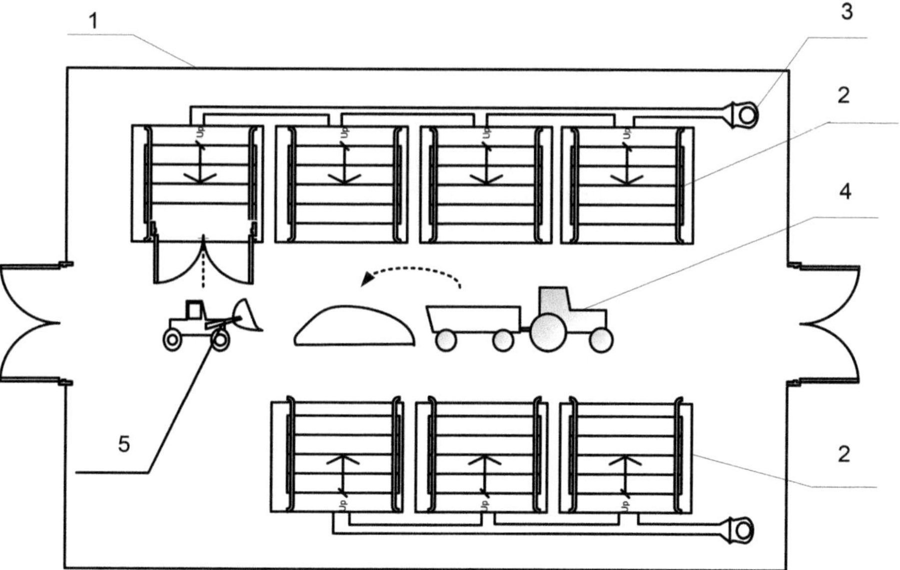

Fig. 8.7 Composting technology for the mix of manure with some moisture-absorbing materials in stationary chamber-type bioreactors: *1* facility for making composts; *2* fermentation chambers; *3* ventilator; *4* tractor with trailer; *5* front-side forklift

The major strong point of this technology is its low metal and power consumption but it involves the need for special rooms to accommodate bioreactors. As shown in Fig. 8.7, manure from the cross passage in a barn comes to a mixing screw conveyor 2, where the moisture absorbing material is supplied by a meter 3. The meter 3 is installed in an additional building of the mixing station. The moisture absorbing material is kept on a special site 9. The mixing screw is also loading the mix to the tractor with trailer for piling 5, which delivers the mix to the composting site 7. Composting on the open site lasts for about 3 months and the ready compost is loaded by a front forklift 6 into a vehicle 11, is transported to the field 12 and stored in big piles up to application time. Technical and economic characteristics of this technique are presented in Table 8.10.

According to this technology the composted mix, which is prepared on a mixing station (as 4 on Fig. 8.7), is transported to the composting department 1 by tractor 4. The composting department is located in a separate building, not necessarily heated, where fermenting chambers are placed. They have a punched bottom for air supply by a ventilator 3 and the gates to load the starting material with a front forklift 5. The composting in the fully loaded bioreactor lasts for 7 days. Technical and economic characteristics of the composting technology in a chamber fermentor are shown in Table 8.11.

8.4.6   *Conclusions*

Emission abatement cost as currently included in the GAINS model may be considered suitable and for the most part acceptable for provisional assessment of accession cost of the Russian Federation to GP in terms of ammonia abatement on dairy farms. Cost value of ammonia abatement for cattle housing can be neglected in the cost estimations of emission mitigation on regional level for RF PEMA.

Development of procedure of economic cost estimation for the conditions of the Russian Federation should be a part of integrated research on environmental pollution abatement from agricultural sources including updating of emission factors from all sources, initial data acquisition on manure handling technologies on various farms, etc.

When estimating the value of joining the GP for Russian Federation additional evaluation is needed of costs of organizational activities to update legislation and to monitor ammonia emission abatement measures on the farm and region levels.

Table 8.10 Technical and economic indices of manure composting technology (86 % moisture content) using a moisture-absorbing material on a farm batchground

Indicator	Unit	Livestock and type of moisture-absorbing material								
		200			400			600		
		Straw	*Peat*	*Poultry manure*	*Straw*	*Peat*	*Poultry manure*	*Straw*	*Peat*	*Poultry manure*
Moisture-absorbing material	t/year	950	6,030	6,030	1,900	12,060	12,060	2,850	18,090	18,090
Capital input	Thousand €	93.75	165.7	165.7	170	301.5	301.5	2356	460	460
Operation cost	Thousand €	45	142.5	76.5	66	252.9	114.5	84	379.2	171.5
Unit cost	€/t	8.55	13.8	7.4	6.275	12.2	5.55	5.325	12.2	5.5
Unit cost with account for transportation and land application	€/t	12.45	17.3	10.9	9.85	14.5	7.8	8.6	14.9	8.0

Table 8.11 Technical and economic indices of manure composting technology (86 % moisture content) using a moisture-absorbing material in stationary chamber-type bioreactors

Indicator	Unit	Livestock and type of moisture-absorbing material								
		200			400			600		
		Straw	*Peat*	*Poultry manure*	*Straw*	*Peat*	*Poultry manure*	*Straw*	*Peat*	*Poultry manure*
Moisture-absorbing material	t/year	950	6,030	6,030	1,900	12,060	12,060	2,850	18,090	18,090
Capital input	Thousand €	141.3	245	245	247.5	550	550	422.5	807.5	807.5
Operation cost	Thousand €	53.9	154.8	91.3	154.8	286.5	133.2	103.4	382.5	175.1
Unit cost	€/t	10.3	15.0	8.8	14.7	13	6.5	6.6	12.3	5.7
Unit cost with account for transportation and land application	€/t	14.2	18.5	12.4	18.3	15.3	8.7	9.9	15	8.1

8.5 Costs of Adoption of Low Ammonia Emission Slurry Application Methods on Grassland in Ireland

Stanley T.J. Lalor (⊠)
Grassland AGRO, Dock Rd., Limerick, Ireland

Formerly Teagasc, Johnstown Research Centre, Wexford, Ireland

8.5.1 Introduction

The emphasis on maximising the nitrogen fertilizer replacement value (NFRV) of cattle slurry has been revived in Ireland in recent years for a number of reasons. Nitrogen (N) fertilizer prices have increased substantially in recent years, resulting in farmers seeking to make better use of N resources in slurry to offset N fertilizer inputs. This has coincided with the introduction of legislative restrictions in 2006, and updated in 2010, to comply with the EU Nitrates and Water Framework Directives that control the quantities of fertilizers that can be applied to crops. This legislation also specifies the NFRV that must be assumed for cattle slurry applications (Anon 2010). There has also been a continued emphasis on reducing national ammonia (NH_3) emissions. Approximately 30 % of NH_3 emissions from Irish agriculture is attributable to landspreading of cattle slurry (Hyde et al. 2003). While Ireland is currently compliant with current NH_3 emission targets, the requirement to comply with future targets for reduced NH_3 emissions may affect future slurry management practices. The combination of these factors has resulted in farmers becoming more aware of the fertilizer benefits of cattle slurry and improving the NFRV is seen as a key driver of both improving nutrient use efficiency, and decreasing the contribution of landspreading to national NH_3 emissions.

Slurry application method, and in particular its effect on slurry placement, is considered a key determinant of the NFRV of slurry (Schröder 2005). Application methods that reduce gaseous losses of N as NH_3 have the potential to increase the NFRV of slurry, since the N not lost to the atmosphere is retained in the soil and may be utilised by the crop. The trailing shoe (TS) application method increased the NFRV of cattle slurry by 0.1 kg kg^{-1} total slurry N applied compared to conventional splashplate or broadcast (SP) application in grassland experiments in Ireland (Lalor et al. 2011).

At present in Ireland, almost all (97 %) of the cattle slurry application to grassland is performed using the SP-method (Hennessy et al. 2011). Historically, the most common timing of slurry application was after grass silage harvest in the summer period in the months May to July (Hyde and Carton 2005). However, in recent years, the proportion of slurry applied in the spring period from mid January to April has increased from 34 % in 2003 (Hyde et al. 2006) to 52 % in 2009

(Hennessy et al. 2011) as farmers seek to maximise NFRV by applying slurry in cooler weather conditions.

The environmental benefits of low-emission slurry application methods such as band spreading (using trailing hose (TH) or TS) and injection for reducing the gaseous emissions of NH_3 from landspreading of animal slurries are well established. However, the implementation of these technologies is often limited by the increased purchase and running costs associated with this machinery compared with the SP application method. In some European countries, this obstacle to technology adoption has been overcome by enforcing legislation. Since such legislation is not in place in Ireland, high rates of adoption will be dependant on measurable economic advantages to individual farmers.

8.5.2 Potential for Low Emission Application Methods in Ireland

The TH, TS and shallow injection (SI) methods are the most common low emission application methods available to grassland farmers. The reductions in NH_3 emissions associated with low emission application methods compared to SP have been shown to vary between a number of experiments reported. Within the review by Webb et al. (2010), the mean emission abatement from slurry applied to grassland, calculated as the mean % reductions in emissions compared to SP across a range of studies, were 35 %, 64 % and 80 % with TH, TS and SI, respectively. However, the range around these mean values was high. Studies from the UK have shown abatement levels lower than these mean values. Smith et al. (2000) measured reductions of 39, 43 and 57 %, and Misselbrook et al. (2002) measured reductions of 26, 57 and 73 % compared to SP for the three methods, respectively. Experiments conducted in Ireland measured a mean reduction in emissions of 36 % with TS compared with SP (Dowling et al. 2008). Current guidelines in the Integrated Pollution Prevention and Control (IPPC) Best Available Techniques Reference (BREF) document for intensive rearing of poultry and pigs suggests emission reductions compared to SP of 30, 40 %, and 60 % for TH, TS and SI (open slot) for pig slurry application to grassland (Anon 2003). This potential range in emission abatement needs to be considered when calculating costs and benefits of low emission application methods.

The objective of this analysis was to evaluate the economic implications of adopting the low emission application methods in Ireland, including both the costs and benefits to the farmer. Costs were calculated as the net additional costs of adopting low emission application methods per unit of slurry volume applied and per unit of NH_3-N abated. The analysis also examined the sensitivity of the calculated costs to variation in a range of input variables such as potential abatement levels achievable, and costs of various inputs that contribute to the net cost of low emission application method adoption.

8.5.3 Methodology

8.5.3.1 Estimating Costs

The analysis was conducted following the approach outlined for calculating cost associated with the application of emissions reduction techniques in the BREF document (Anon 2003). This approach estimates the 'unit' cost of techniques, which is defined as the "annual increase in costs that a typical farmer will bear as a result of introducing the technique". The increase in costs in this case means that only the additional costs incurred due to the adoption of the technique should be included. Therefore, in reality, these increased costs are incurred in addition to the current cost of continuing to apply slurry using the current reference method. The following equation was used for calculating the unit cost:

$$C_u = \frac{A_C + A_R + A_L + A_F - A_S}{V},$$ (8.1)

where C_u is the unit cost of the technique ($€$ m^{-3}); A_C is the annualized cost of additional capital ($€$ year^{-1}); A_R is the annual cost of additional repairs associated with the technique ($€$ year^{-1}); A_L is the annual additional labour costs ($€$ year^{-1}); A_F is the annual additional fuel costs ($€$ year^{-1}); A_S is the annual savings and/or value of production benefits arising as a result of the technique ($€$ year^{-1}); and V is the total volume of slurry applied using the technique each year (m^3 year^{-1}).

The value of A_C was calculated as the sum of the annual cost of all the capital investment required. Where separate pieces of investment are required, such as in this case where additional tractor power may be required in addition to the new application equipment, the annual cost of each capital investment was calculated and summed to give the total A_C. Therefore, for landspreading equipment where additional tractor power is also required, the A_C was calculated using the equation:

$$A_C = \left(C_t \times \frac{r_t(1+r_t)^{n_t}}{(1+r_t)^{n_t} - 1} \right) + \left(C_e \times \frac{r_e(1+r_e)^{n_e}}{(1+r_e)^{n_e} - 1} \right),$$ (8.2)

where C_t and C_e are the additional capital investment costs of the tractor and application equipment, respectively ($€$); r_t and r_e are the interest rates (expressed as a decimal of 1) for the tractor and application equipment, respectively; and n_t and n_e are the terms of the investment for the tractor and application equipment, respectively (y). While the interest rate is likely to be equal for both the tractor and the application machinery, the term of investment may vary. The cost of additional tractor power was calculated using the equation:

$$C_t = (P_e - P_o).C_p$$ (8.3)

where P_e and P_o are the tractor power requirements to operate the reference equipment and the low emission application equipment, respectively (kWh); and C_p is the capital cost of the tractor per unit increase in power ($€$ kWh^{-1}).

The value of A_R was calculated on the basis that the additional repair cost can be calculated as a percentage of the additional capital cost, using the equation:

$$A_R = C_t.rm_t + C_e.rm_e, \tag{8.4}$$

where rm_t and rm_e are the annual repair cost rate of the additional capital cost of the tractor (C_t) and application equipment (C_e), respectively (expressed as a decimal of 1).

A change in labour costs may arise due to the application technique having a different work rate than the reference method. Hence the number of hours work required may change due to increased hours required to apply the same volume of slurry. Labour costs may also change due to the new application machinery requiring a more skilled and highly paid operator. The value of A_L was calculated as the sum of the labour cost for additional hours that may be required to apply the same volume of slurry at a slower work rate, and the additional labour cost associated with paying an operator a higher rate for all hours worked because of the increased operator skill required., using the equation:

$$A_L = L_e.(H_e - H_o) + H_o.(L_e - L_o), \tag{8.5}$$

where L_e and L_o are the hourly labour costs assumed with the low emission application method and with the reference method, respectively ($€$ h^{-1}); and H_e and H_o are the hours of labour required each year with the low emission application method and with the reference method, respectively (h year^{-1}). The values of H_o and H_e can be calculated using the equation:

$$H_{o,e} = \frac{V}{R_{o,e}}, \tag{8.6}$$

where R_o and R_e are the slurry application rate with the reference equipment and the low emission application equipment, respectively (m^3 h^{-1}). The value of R_e was estimated by applying a coefficient to the value assumed for R_o to account for changes in spreading work rate based on differences in the bout width of the machine. This approach assumed that the time in the tanker load cycle that was spent filling and travelling between the field and the store was constant with all methods. It was also assumed that the tractor forward speed during the time spent spreading in the field was constant across application methods. Therefore, the difference in work rate between the application methods was assumed to be only affected by the time spent emptying the tanker. The narrower the working width of the machine, the longer it takes to empty the tanker. Therefore, R_e was calculated using the following equation:

$$R_e = \frac{R_o}{1 - T_s + \left(T_s.\dfrac{w_o}{w_e}\right)}, \tag{8.7}$$

where T_s was the proportion of the tanker load cycle time spent in the applying slurry in the field; W_o was the working width of the reference equipment (m); and W_e was the working width of the low emission application equipment (m).

Additional fuel costs may be incurred due to the low emission application for two reasons. Firstly, an increased power requirement of the tractor will result in higher fuel requirements for the hours worked that would have been worked with the reference method. Secondly, additional fuel will be required due to the additional hours due to the decrease in work rate with the low emission method. The value of A_F was calculated using the following equation:

$$A_F = C_f.\left(F_p. H_o.(P_e - P_o) + F_p.P_e.(H_e - H_o)\right), \tag{8.8}$$

where C_f was the cost of fuel ($€\ L^{-1}$); and F_p was the hourly fuel consumption per kWh of tractor power ($L\ h^{-1}\ kWh^{-1}$).

The term A_S was calculated based on the potential for the low emission application technique to result in mineral N fertilizer cost savings. Other potential benefits of low emission application methods compared to the reference SP application method could also be argued for inclusion such as the fertilizer benefits of more uniform application, or the reduction of odour emissions or pasture contamination. However, in this study, only the N fertilizer benefit was considered. It was assumed that NH_3-N not volatilised could replace mineral fertilizer N requirements on a 1:1 basis. The value of A_S was calculated using the following equation:

$$A_S = N.V.T.\left(\frac{E_o}{100}.\frac{E_e}{100}\right), \tag{8.9}$$

where N was the cost of mineral fertilizer N ($€\ kg^{-1}$); T was the total ammoniacal N in slurry ($kg\ m^{-3}$); E_o was the NH_3 emission factor for the reference method, expressed as loss of NH_3-N as a percentage of the TAN applied (%); and E_e was the NH_3 emission abatement potential of the low emission application method (%).

The cost of each technique per kg of NH_3-N emission abated was also calculated using the following equation:

$$C_{NH3} = \frac{C_u}{T.\left(\frac{E_o}{100}.\frac{E_e}{100}\right)}, \tag{8.10}$$

where C_{NH3} was the additional cost of adopting the low emission application method per kg of NH_3-N emission abated ($€\ kg^{-1}$).

8.5.3.2 Assumptions Adopted for Comparing Costs of Application Methods

The C_u and C_{NH3} for each of the low emission application methods of TH, TS and SI were calculated relative to a reference method of SP application. A number of assumptions were made for the parameters in Equations 8.1 to 8.10. The assumed values of these parameters and the rationale for these assumptions are listed in Table 8.12.

Table 8.12 Assumed values of parameters required for cost calculations, and the rationale and justification of each assumption adopted

Parameter	Unit	Assumed value				Rationale and justification
		SP	TH	TS	SI	
V	m^3 $year^{-1}$	10,000				Assumed as an average annual workload for machine operated by a contractor.
r_t		0.07				Average current interest rate for farm finance.
n_t	year	10				Typical life span of medium to high power tractor.
C_e	€	–	12,000	20,000	25,000	Typical additional prices in Ireland for low emission application machinery compared with SP tanker of equal size, including additional hydraulic and electrical fittings and chopping systems.
r_e		0.07				Typical interest rate on medium term borrowing for farm machinery.
n_e	year	7				Typical life span of application equipment.
P_o	kWh	75	–	–	–	Typical power requirement for a 9 m^{-3} SP tanker. Progressively higher tractor power requirement is assumed with each low emission application method due to increased weight and contact with soil.
P_e	kWh	–	85	100	120	
C_p	€ kWh^{-1}	930				Based on comparison of tractor price listings (Anon 2012).
rm_t		0.08				BREF guidelines suggest a value of 5–8 % for tractors (Anon 2003).
rm_e		–	0.10			BREF guidelines suggest a value of 3-6 % on slurry spreaders. However, a higher value was assumed in this case due to expected high mainte-nance due to moving parts and soil contact (Anon 2003).
L_o	€ h^{-1}	12	–	–	–	Higher labour costs were assumed for the low emission application methods due to the requirement for more skilled operator due to the increase in machine complexity and value
L_e	€ h^{-1}	–	15	15	15	
R_o	m^3 h^{-1}	30	–	–	–	Typical hourly work rate for a 9 m^3 tanker (3.3 loads per hour).
T_s		0.25	–	–	–	Typical proportion of load spreading cycle that is spent in the field.

(continued)

Table 8.12 (continued)

Parameter	Unit	Assumed value				Rationale and justification
		SP	TH	TS	SI	
W_o	m	10	–	–	–	SP spread width can vary considerably. An average width of 10 m is assumed.
W_e	m	–	6	6	4	Widths assumed are typical of commonly available units suitable for applications to grassland.
C_f	€ L^{-1}	0.90				Typical price of agricultural diesel in Ireland in February 2012.
F_p	l kWh^{-1}	0.30				Fuel requirement per kWh of power is typically in the range 0.25–0.35 L kWh^{-1} (Kim et al. 2005).
N	€ kg^{-1}	1.20				Typical price of mineral N fertilizer based on price in Ireland in February 2012.
T	kg m^{-3}	1.8				Typical total N concentration in cattle slurry in Ireland is 3.6 kg m^{-3} (Coulter 2004). Approximately 50 % of the total N is assumed to be present in the form of NH$_3$-N (DEFRA 2010).
E_o	%	55	–	–	–	Mean emissions of NH$_3$-N as a % of TAN following SP application as measured in Irish studies (Dowling et al. 2008).
E_e	%	–	30	35	70	Emission abatement efficiencies of 30, 60 and 70 % are assumed for TH, TS and SI, respectively, compared to application with SP in UNECE Guidance document (UNECE 2007). Respective average emission abatement of 35, 65 and 70 % are reported in the literature (Webb et al. 2010). Studies in Ireland measured emission reduction of 36 % with TS compared with SP (Dowling et al. 2008).

SP splashplate (reference method), *TH* trailing hose, *TS* trailing shoe, *SI* shallow injection. One value is shown where assumptions are equal for all application methods

8.5.3.3 Sensitivity Analysis

Given that the values assumed for many of the parameters required are based on typical and current estimates of various parameters, a sensitivity analysis was also conducted to examine the influence of changes in these factors over time on the cost estimates of the application machinery. The sensitivity analysis was conducted on a single factor basis by calculating the value of C_u and C_{NH3} by adjusting the value of one parameter while holding all other parameters constant.

The parameters considered for sensitivity analysis were the emission abatement efficiency of the low emission application method (E_e) in the range of 20–90 %; the cost of mineral fertilizer N (N) in the range of €0.70 kg^{-1} to €1.50 kg^{-1};manure volume (V) in the range of 500–20,000 m^3 year^{-1}; the tractor power requirement for the low emission application equipment (P_e) in the range of 75–150 kWh; additional capital cost of the application equipment (C_e) in the range of €5,000–€40,000; the hourly application rate of the reference SP method (R_o) in the range of 10–40 m^3 h^{-1}; interest rate for both tractor and equipment ($r_{t,e}$) in the range of 0.04–0.10; the repair cost rate for the tractor and equipment ($rm_{t,e}$) in the range of 0.03–0.15; and the cost of fuel (C_f) in the range of €0.50 L^{-1} to €1.20 L^{-1}.

8.5.4   Results

8.5.4.1   Costs of Application Methods

The calculated values of C_u and C_{NH3} for each of the low emission application methods are shown in Table 8.13. The TH method had the lowest C_u while the SI method had the highest. However, the TS method had a higher C_{NH3} value than the SI method. This was mainly due to the SI method having a higher assumed NH$_3$ emission abatement potential, and therefore the higher unit cost of SI was offset by a higher level of NH$_3$ abatement when compared with the TS method.

The contribution of capital costs (A_c), repairs and maintenance (A_R), labour (A_L), fuel (A_F) and savings (A_S) to the overall value of C_u is shown in Fig. 8.8. The total units cost of adoption of the low emission application equipment were €0.95 m^{-3}, €1.65 m^{-3} and €2.74 m^{-3} for TH, TS and SI, respectively. The differences between the total costs and the C_u of each method were due to savings in mineral N fertilizer

Table 8.13 Additional units cost (C_u) and cost per kg NH$_3$ abated (C_{NH3}) with trailing hose (TH), trailing shoe (TS) and shallow injection (SI) compared with the reference application method of splashplate. Calculations were based on the parameter values assumed in Table 6.15

Application method	C_u (€ m^{-3})	C_{NH3} (€ kg^{-1}) NH$_3$ abated
TH	€ 0.59	€ 2.00
TS	€ 1.23	€ 3.55
SI	€ 1.91	€ 2.76

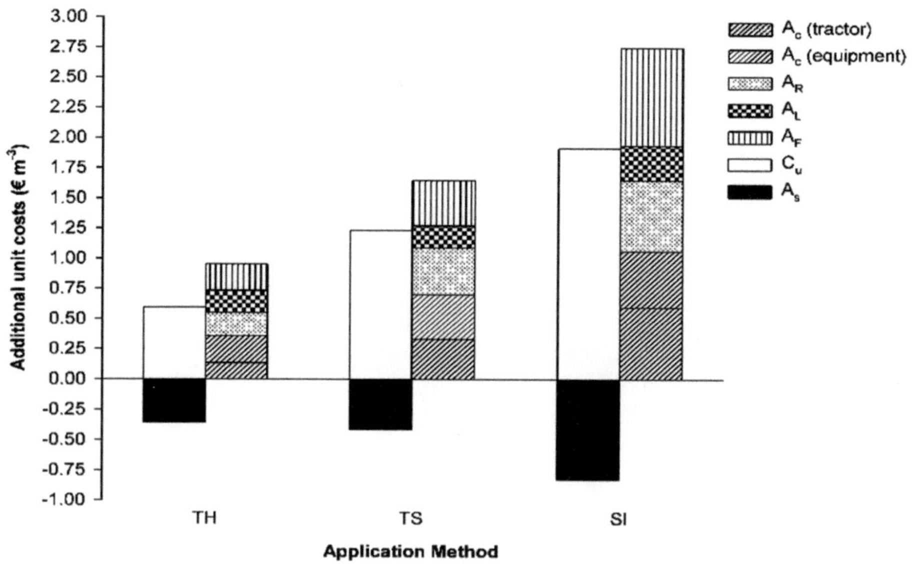

Fig. 8.8 Contribution of the additional costs of capital (A_c), repairs and maintenance (A_R), labour (A_L) and fuel (A_F), and the value of savings in mineral N fertilizer (A_S) to the overall unit cost (C_U) of the trailing hose (*TH*), trailing shoe (*TS*), and shallow injection (*SI*) application methods. Capital costs are divided into costs associated with the requirement for additional tractor power (A_c (tractor)) and costs associated with purchasing equipment (A_c (equipment))

due to the reduced NH_3-N emissions compared with the reference method (A_S). These savings offset 38 %, 25 % and 20 % of the total additional unit costs of TH, TS and SI adoption, respectively.

Capital costs (A_c) accounted for the largest percentage of total costs for all methods, being 37 %, 43 % and 39 % for TH, TS and SI, respectively. The percentage of capital costs due to the equipment was higher with TH (63 % of A_c) than with TS (53 % of A_c), which was higher than SI (44 % of A_c).

Repairs and maintenance costs (A_R) accounted for 20 %, 23 % and 21 % of the total costs for TH, TS and SI methods, respectively. Labour costs (A_L) accounted for the smallest proportion of increased costs for all methods, being 19 %, 10 % and 11 % for the TH, TS and SI methods, respectively. Fuel costs (A_F) accounted for 23 %, 23 % and 30 % of the total costs for TH, TS and SI methods, respectively.

8.5.4.2 Sensitivity Analysis

Additional Unit Cost (C_u)

The effects of varying the assumptions of a number of the cost calculation input variables on C_u are shown in Fig. 8.9. Across the ranges of all the cost calculation inputs examined, the TH method consistently had the lowest C_u, while SI had the highest. The TS method was intermediate in the case of all variables examined.

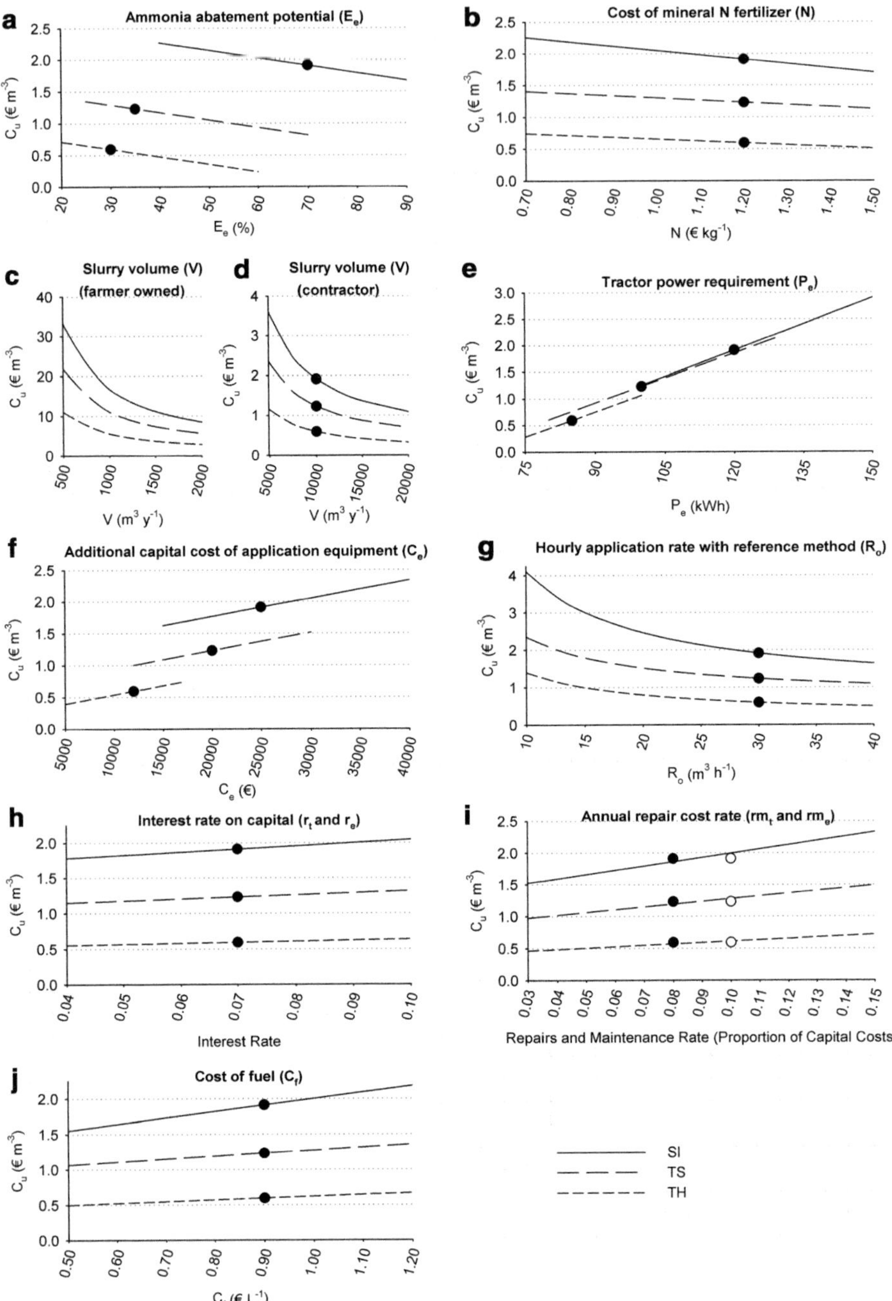

Fig. 8.9 Sensitivity of the additional unit cost (C_u) of adopting trailing hose (*TH*), trailing shoe (*TS*) and shallow injection (*SI*) to variation in the values assumed for a number of cost calculation input variables. *Solid shaded circles* indicate the assumed value and corresponding C_u for each variable. In the case of annual repair cost rate (i), the *solid circles indicate* assumed value of rm_t and the *open circles* indicate the assumed values of rm_e

The effect of varying the NH_3 abatement potential (E_e) of each application method on C_u is shown in Fig. 8.9a. The total range of E_e included in the analysis was 20–90 %. However, the ranges were restricted within each method to 20–60 % with TH, 25–70 % with TS, and 40–90 % with SI. This distinction between methods was made to reflect reasonable extremes of the E_e of each method based on previous studies (Webb et al. 2010; Dowling et al. 2008). Within the range of E_e included for each method, the C_u ranged from €2.27 m^{-3} to €1.67 m^{-3} with SI, from €1.35 m^{-3} to €0.81 m^{-3} with TS, and from €0.71 m^{-3} to €0.24 m^{-3} with TH. The effect of E_e was linear. A change in E_e of 10 % resulted in a change in the C_u of €0.119 m^{-3} with all methods.

The effect of varying the cost of mineral N fertilizer (N) on C_u is shown in Fig. 8.9b. The effect of varying N was more significant with the SI method than with the TH method, reflecting the higher fertilizer N savings with SI due to the higher NH_3 emission abatement potential. Within the range of N from €0.70 kg^{-1} to €1.50 kg^{-1} included, the C_u ranged from €2.26 m^{-3} to €1.70 m^{-3} with the SI method, from €1.40 m^{-3} to €1.13 m^{-3} with the TS method, and from €0.74 m^{-3} to €0.50 m^{-3} with the TH method. The effect of N was linear, with a change in N of €0.1 kg^{-1} resulting in an inverse change in C_u of €0.069 m^{-3}, €0.035 m^{-3}, and €0.030 m^{-3} with SI, TS and TH methods, respectively. In order for savings in fertilizer N to fully offset the additional costs of the equipment, (i.e. to achieve a value of C_u of €0.00 m^{-3}) the value of N of €3.96 kg^{-1}, €4.75 kg^{-1} and €3.20 kg^{-1} would be required with SI, TS and TH methods, respectively.

The volume of slurry applied annually with each machine (V) had a large effect on C_u. The range of V with a typical farmer-owned machine is shown in Fig. 8.9c.

The range of 500–2,000 m^3 year^{-1} is approximately equivalent to the slurry produced from a herd of approximately 40–150 dairy cows plus followers over a winter period of 18 weeks (Anon 2010). Within this range of V, the C_u ranged from €33.16 m^{-3} to €8.49 m^{-3} with the SI method, from €21.91 m^{-3} to €5.58 m^{-3} with the TS method, and from €11.03 m^{-3} to €2.79 m^{-3} with the TH method. The range of 5,000–20,000 m^3 year^{-1} shown in Fig. 8.9d is more typical of the slurry volume applied annually by a contractor. Within this range of V, the C_u ranged from €3.55 m^{-3} to €1.09 m^{-3} with the SI method, from €2.32 m^{-3} to €0.69 m^{-3} with the TS method, and from €1.14 m^{-3} to €0.32 m^{-3} with the TH method.

The effect of varying the tractor power requirement (P_e) on C_u is shown in Fig. 8.9e. The total range of P_e included in the analysis was 75–150 kWh. However, the ranges were restricted within each method to 75–100 kWh with TH, 80–130 kWh with TS, and 100–150 kWh with SI. Within the range of P_e included for each method, the C_u ranged from €1.25 m^{-3} to €2.90 m^{-3} with the SI method, from €0.61 m^{-3} to €2.17 m^{-3} with the TS method, and from €0.28 m^{-3} to €1.06 m^{-3} with the TH method. The effect of P_e was linear, with a change in P_e of 1 kWh resulting in a change in the C_u of approximately €0.031 m^{-3} with all methods.

The effect of varying the cost of application equipment (C_e) on C_u is shown in Fig. 8.9f. The total range of costs included in the analysis was €5,000 to €40,000. However, the ranges were restricted within each method to €5,000 to €17,000 with TH, €12,000 to €30,000 with TS, and €15,000 to €40,000 with SI. This distinction between methods was applied in order to reflect reasonable extremes of the C_e of each method for machine working widths assumed. Within the range of C_e included for each method, the C_u ranged from €1.62 m^{-3} to €2.34 m^{-3} with the SI method, from €1.00 m^{-3} to €1.52 m^{-3} with the TS method, and from €0.39 m^{-3} to €0.74 m^{-3} with the TH method. The effect of C_e was linear, with a change in C_e of €1,000 resulting in a change in the C_u of €0.029 m^{-3} with all methods.

The effect of varying the assumption of the hourly application rate with the reference method (R_o) on C_u is shown in Fig. 8.9g. The value of R_o has a large effect on C_u, particularly at lower hourly application rates that are typical where slurry has to be transported longer distances between the slurry store and the field. Decreasing the value of R_o from the baseline assumption of 30 m^3 h^{-1} to 10 m^3 h^{-1} increased the C_u to €4.11 m^{-3}, €2.35 m^{-3} and €1.40 m^{-3} with SI, TS and TH, respectively. Increasing the value of R_o to 40 m^3 h^{-1} decreased C_u to €1.64 m^{-3}, €1.09 m^{-3} and €0.49 m^{-3} with SI, TS and TH, respectively.

The effect of varying interest rate (r_t and r_e) on C_u is shown in Fig. 8.9h. The effect of varying r_t and r_e was more significant with the SI method than with the TH method, reflecting the higher capital investment costs with SI. Within the range of interest rates from 0.04 to 0.10 included, the C_u ranged from €1.78 m^{-3} to €2.05 m^{-3} with the SI method, from €1.15 m^{-3} to €1.32 m^{-3} with the TS method, and from €0.55 m^{-3} to €0.64 m^{-3} with the TH method. The effect of interest rate was approximately linear, with a change in the interest rate of 0.01 resulting in a change in the C_u of €0.044 m^{-3}, €0.028 m^{-3}, and €0.014 m^{-3} with SI, TS and TH methods, respectively.

The effect of varying the repairs and maintenance rate (rm_t and rm_e) on C_u is shown in Fig. 8.9i. The effect of varying rm_t and rm_e was more significant with the SI method than with the TH method, reflecting the higher capital investment costs with SI. Within the range of rm_t and rm_e from 0.03 to 0.15 included, the C_u ranged from €1.53 m^{-3} to €2.33 m^{-3} with the SI method, from €0.97 m^{-3} to €1.49 m^{-3} with the TS method, and from €0.46 m^{-3} to €0.72 m^{-3} with the TH method. The effect of repairs and maintenance rate was linear, with a change in the repairs and maintenance rate of 0.01 resulting in a change in C_u of €0.067 m^{-3}, €0.043 m^{-3}, and €0.040 m^{-3} with SI, TS and TH methods, respectively.

The effect of varying the cost of fuel (C_f) on C_u is shown in Fig. 8.9j. The effect of varying C_f was more significant with the SI method than with the TH method, reflecting the higher fuel requirements of this method due to higher power requirement and reduced work rate. Within the range of C_f from €0.50 L^{-1} to €1.20 L^{-1} included, the C_u ranged from €1.55 m^{-3} to €2.18 m^{-3} with the SI method, from €1.06 m^{-3} to €1.36 m^{-3} with the TS method, and from €0.50 m^{-3} to €0.67 m^{-3} with the TH method. The effect of C_f was linear, with a change in C_f of €0.1 L^{-1} resulting in a change in C_u of €0.090 m^{-3}, €0.042 m^{-3}, and €0.024 m^{-3} with SI, TS and TH methods, respectively.

Additional Unit Cost Per kg of NH$_3$-N Abated

The effect of varying a number of the assumptions on C_{NH3} is shown in Fig. 8.10. In contrast with C_u where the SI method had consistently higher costs, the TS method

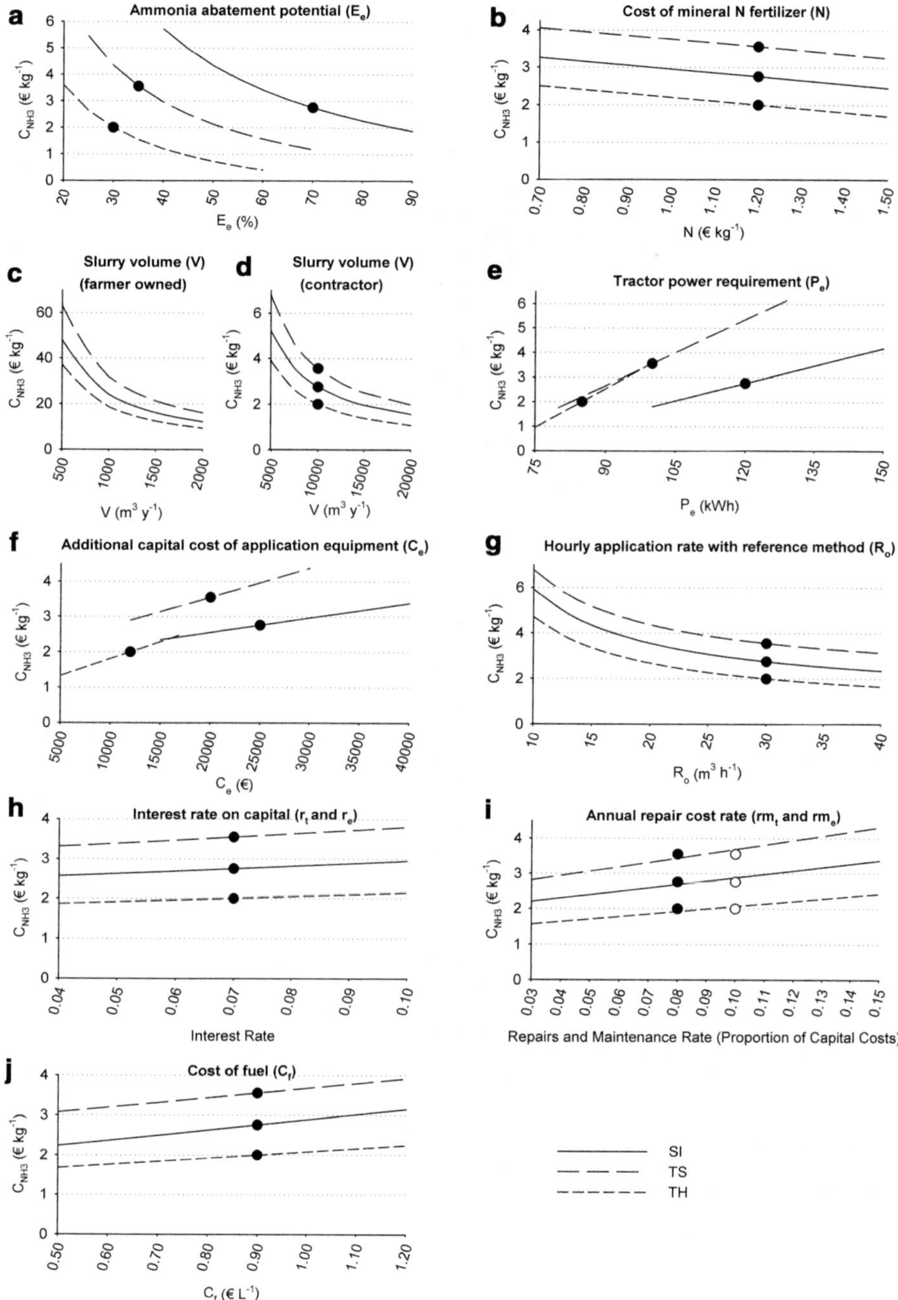

Fig. 8.10 Sensitivity of the additional unit cost per kg NH_3-N abated (C_{NH3}) of adopting trailing hose (*TH*), trailing shoe (*TS*) and shallow injection (*SI*), to variation in the values assumed for a number of cost calculation input variables. *Solid shaded circles* indicate the assumed value and corresponding C_u for each variable. In the case of annual repair cost rate (i), the *solid circles indicate* assumed value of rm_t and the *open circles* indicate the assumed values of rm_e

C_{NH3} is the highest cost method, and is most sensitive to variation in the cost calculation variables.

The only variable that showed exception to this trend was the NH_3 abatement potential (E_e), where the SI resulted in the highest C_{NH3} of all machines at equal levels of E_e (Fig. 8.10a). Within the range of E_e included for each method, the C_{NH3} ranged from €5.72 kg^{-1} to €1.88 kg^{-1} with SI, from €5.45 kg^{-1} to €1.18 kg^{-1} with TS, and from €3.60 kg^{-1} to €0.40 kg^{-1} with TH. Unlike with C_u, the effect of E_e on C_{NH3} was not linear, with the sensitivity to change increasing with decreasing values of E_e.

The effect of varying the cost of mineral N fertilizer (N) on C_{NH3} is shown in Fig. 8.10b. The effect of varying N was similar with all methods since the NH_3-N abated by each method corresponds to fertilizer N savings. Within the range of N from €0.70 kg^{-1} to €1.50 kg^{-1} included, the C_{NH3} ranged from €4.05 kg^{-1} to €3.25 kg^{-1} with the TS method, from €3.26 kg^{-1} to €2.46 kg^{-1} with the SI method, and from €2.50 kg^{-1} to €1.70 kg^{-1} with the TH method. The effect of N was linear, with a change in N of €0.1 kg^{-1} resulting in an inverse change in C_{NH3} of €0.1 kg^{-1} with all methods.

The volume of slurry applied annually with each machine (V) had a large effect on C_{NH3}. The range of V with a typical farmer-owned machine is shown in Fig. 8.10c. Within this range of V, the C_{NH3} ranged from €63.22 kg^{-1} to €16.11 kg^{-1} with the TS method, from €47.84 kg^{-1} to €12.25 kg^{-1} with the SI method, and from €37.15 kg^{-1} to €9.40 kg^{-1} with the TH method. Within the range of 5,000–20,000 m^3 year^{-1} more typical to a contractor (Fig. 8.10d), the C_{NH3} ranged from €6.69 kg^{-1} to €1.98 kg^{-1} with the TS method, from €5.13 kg^{-1} to €1.57 kg^{-1} with the SI method, and from €3.85 kg^{-1} to €1.07 kg^{-1} with the TH method.

The effect of varying the tractor power requirement (P_e) on C_{NH3} is shown in Fig. 8.10e. The C_{NH3} was more sensitive to a changes in P_e in the case of the TH and TS methods compared to SI, since with SI, the increased cost associated with higher power were offset to a greater extent by fertilizer N saved due to the higher assumption of E_e. Within the range of P_e included for each method, the C_{NH3} ranged from €1.75 kg^{-1} to €6.25 kg^{-1} with the TS method, from €1.80 kg^{-1} to €4.19 kg^{-1} with the SI method, and from €0.95 kg^{-1} to €3.57 kg^{-1} with the TH method. The effect of P_e was linear, with a change in P_e of 1 kWh resulting in a change in the C_{NH3} of €0.090 kg^{-1}, €0.048 kg^{-1} and €0.105 kg^{-1} with TS, SI and TH methods, respectively.

The effect of varying the cost of application equipment (C_e) on C_{NH3} is shown in Fig. 8.10f. Within the range of C_e included for each method, the C_{NH3} ranged from €2.89 kg^{-1} to €4.38 kg^{-1} with the TS method, from €2.34 kg^{-1} to €3.37 kg^{-1} with the SI method, and from €1.33 kg^{-1} to €2.48 kg^{-1} with the TH method. The effect of C_e was linear, with a change in C_e of €1,000 resulting in a change in the C_{NH3} of €0.082 kg^{-1}, €0.041 kg^{-1} and €0.096 kg^{-1} with TS, SI and TH methods, respectively.

The effect of varying the assumption of the hourly application rate with the reference method (R_o) on C_{NH3} is shown in Fig. 8.10g. The value of R_o has a large effect on C_{NH3}, particularly at lower hourly application rates. Decreasing the value of R_o from the baseline assumption of 30 m^3 h^{-1} to 10 m^3 h^{-1} increased the C_{NH3} to €6.77 kg^{-1}, €5.92 kg^{-1} and €4.70 kg^{-1} with TS, SI and TH, respectively. Increasing the value of R_o to 40 m^3 h^{-1} decreased C_{NH3} to €3.15 kg^{-1}, €2.36 kg^{-1} and €1.66 kg^{-1} with TS, SI and TH, respectively.

The effect of varying interest rate (r_t and r_e) on C_{NH3} is shown in Fig. 8.10h. Within the range of interest rates from 0.04 to 0.10 included, the C_{NH3} ranged from €3.31 kg^{-1} to €3.80 kg^{-1} with the TS method, from €2.57 kg^{-1} to €2.95 kg^{-1} with the SI method, and from €1.86 kg^{-1} to €2.14 kg^{-1} with the TH method. The effect of interest rate was approximately linear, with a change in the interest rate of 0.01 resulting in a change in the C_{NH3} of €0.081 kg^{-1}, €0.063 kg^{-1}, and €0.047 kg^{-1} with TS, SI and TH methods, respectively.

The effect of varying the repairs and maintenance rate (rm_t and rm_e) on C_{NH3} is shown in Fig. 8.10i. Within the range of rm_t and rm_e from 0.03 to 0.15 included, the C_{NH3} ranged from €2.81 kg^{-1} to €4.31 kg^{-1} with the TS method, from €2.20 kg^{-1} to €3.36 kg^{-1} with the SI method, and from €1.56 kg^{-1} to €2.42 kg^{-1} with the TH method. The effect of repairs and maintenance rate was linear, with a change in the repairs and maintenance rate of 0.01 resulting in a change in C_{NH3} of €0.125 kg^{-1}, €0.096 kg^{-1}, and €0.072 kg^{-1} with TS, SI and TH methods, respectively.

The effect of varying the cost of fuel (C_f) on C_{NH3} is shown in Fig. 8.10j. Within the range of C_f from €0.50 L^{-1} to €1.20 L^{-1} included, the C_{NH3} ranged from €3.07 kg^{-1} to €3.91 kg^{-1} with the TS method, from €2.24 kg^{-1} to €3.15 kg^{-1} with the SI method, and from €1.67 kg^{-1} to €2.24 kg^{-1} with the TH method. The effect of C_f was linear, with a change in C_f of €0.1 L^{-1} resulting in a change in C_{NH3} of €0.121 kg^{-1}, €0.130 kg^{-1}, and €0.081 kg^{-1} with TS, SI and TH methods, respectively.

8.5.5 Discussion

The TH application method had the lowest costs both in terms of C_u and C_{NH3}. However, the method with the highest costs depended on the metric used for comparison of the TS and SI methods. The high C_{NH3} of TS was partly due to the low value of E_e (35 %) assumed in this analysis. While this assumed level of abatement is consistent with the findings of Irish research (Dowling et al. 2008), it is lower than higher values of up to 60–65 % that might be assumed based on other data sources (Anon 2003; Webb et al. 2010). The sensitivity analysis showed that the C_u (Fig. 8.9a) and C_{NH3} (Fig. 8.10a) would have been reduced to €0.93 m^{-3} and €1.57 kg^{-1}, respectively, if a value of E_e of 60 % had been assumed for TS. In this scenario, the TS would have been the lowest cost option based on C_{NH3}. However, the assumption of the lower value of E_e for TS in an Irish context is justified based on data from Irish studies (Dowling et al. 2008).

The estimated additional unit costs are highly dependent on the assumptions used for the range of factors that contribute to costs. Of the factors that were isolated in the sensitivity analysis, C_u was most sensitive to changes in V and R_o, while C_{NH3} was also highly sensitive to changes in E_e. In the case of V, applying higher volumes of slurry has the effect of spreading the total costs of application over a larger volume of slurry, and over a larger quantity of NH_3-N emission abatement. For slurry volumes typical of farmer owned machines (Fig. 8.10c), both the C_u and C_{NH3} are increased by factors of between approximately 4 and 18 with all three low emission methods compared with the baseline scenario assumption of V of 10,000 m^3 $year^{-1}$. Approximately 50 % of slurry in Ireland is applied using farmer-owned SP equipment (Hennessy et al. 2011). An increase in the cost of slurry application of these proportions would restrict the level to which these application methods could be adopted by operators other than contractors. The explanation of the sensitivity to the assumed value of R_o is similar to that for V, whereby the lower hourly application rates result in higher fuel and labour costs per unit volume of slurry applied or per unit of NH_3-N abated. The C_{NH3} was also sensitive to the effect of E_e, particularly at lower values of E_e where the marginal effect of change in C_{NH3} was greater than at higher values.

The sensitivity of C_u and C_{NH3} to the effect of varying the additional capital costs inferred by P_e and C_e highlight the importance of machine design and performance that reduce the investment cost in capital, and the power requirement for their operation. The contribution of additional costs capital costs for the tractor to the total additional capital costs (A_c) (Fig. 8.8) also indicates the importance of considering the additional capital cost of the tractor in addition to the application equipment where incentives such as grant aid on capital investment in equipment are being designed to promote the adoption of low emission equipment.

Cost savings with reduced mineral N fertilizer inputs due to NH_3-N emission abatement is often viewed as a means of offsetting the cost of low emission application method adoption. However, the results of this analysis show that there was a net additional cost of adoption after mineral N fertilizer savings were included, even at the higher range of the values of N included. Current agronomic advice in Ireland assumes that larger savings on fertilizer nitrogen can be made by applying slurry to grassland in the spring (February to April) period, rather than in the summer (June and July). The NFRV of slurry applied with SP in summer (May–July) is assumed to be 0.12 kg kg^{-1}, whereas the NFRV increases to 0.21 kg kg^{-1} for application in spring (February–April). Low emission application methods are assumed to increase the NFRV by 0.10 kg kg^{-1} in both spring and summer (Coulter and Lalor 2008; Lalor et al. 2011). Nutrient advice in the UK also assumes a higher NFRV for spring application (0.25–0.45 kg kg^{-1}) compared to summer (0.20–0.35 kg kg^{-1}). The increase in NFRV with bandspreading is assumed to be 0.05 kg kg^{-1} (DEFRA 2010). While these estimates of the NFRV were not adopted directly in this study, they correspond closely with the quantities of NH_3-N abated in the calculations of this study. Based on the assumed value of T in this study of 1.8 kg m^{-3}, the NH_3-N conserved was 0.297 and 0.347 kg m^{-3} with TH and TS methods, respectively. This equated to an increase in NFRV due to the application method of 0.08 and 0.10 kg kg^{-1} with TH and TS, respectively. These are in agreement with the effects of TS on NFRV cited above. The corresponding increase in NFRV with SI based on this study was 0.19 kg kg^{-1}.

The main restriction to SP application in spring is the requirement for suitable soil trafficability conditions to coincide with short grass covers so that herbage contamination can be minimised. The low emission application methods minimise grass contamination by applying slurry in lines rather than on the entire grass canopy. Therefore, they allow greater flexibility of application timing by facilitating application on taller swards (Laws et al. 2002). This results in more spreadland being available for slurry application on the days in spring when weather conditions allow traffic. There is potential for greater savings on fertilizer N costs through adoption of low emission application technology, as a greater proportion of slurry may be applied in the spring when the nitrogen fertilizer replacement value can be maximised (Lalor and Schulte 2008). Of the low emission application methods, the TH and TS methods are considered to be the most suitable for Irish grassland, as they avoid potential problems with slurry injection in Irish soils due to variability in stone content, texture, drainage and topography. The TS may also infer additional benefits over TH by reducing the contamination of herbage with slurry, as the shoe coulter is designed to improve the precision of slurry placement at the base of the sward.

Where additional NFRV benefits due to flexibility in application timing allowing application in spring are also inferred by the adoption of low emission application equipment, the net costs would be reduced as greater mineral N fertilizer cost savings could be achieved (Lalor 2008). Where application in spring can be facilitated, the NFRV is increased by approximately 0.10 kg kg^{-1}. This equates to an additional cost saving of €0.43 m^{-3} of slurry. Assuming that this increased flexibility application timing and NFRV benefit is achievable with all methods, the additional cost saving would reduce the C_u to €0.16 m^{-3}, €0.80 m^{-3} and €1.48 m^{-3} with TH, TS and SI, respectively. However, Lalor and Schulte (2008) demonstrated that this benefit is more likely with TS than with TH or SI since the TS was considered the most effective machine at reducing sward contamination with slurry.

8.5.6 *Conclusions*

The TH method of slurry application was the most cost effective of the low application methods based on the assumptions adopted in this study. The SI method had the highest costs per unit of slurry volume applied, while TS had the highest cost per kg of NH_3-N abated. However, this conclusion was based on assuming a level of NH_3-N emission abatement with TS specific to Irish conditions that is lower than that suggested in other literature sources. The benefit of mineral N fertilizer savings due to NH_3-N emission abatement was not sufficient to offset the total cost of adoption, even when additional benefits of improved flexibility in application timing were taken into account. The sensitivity analysis showed that the factors that had greatest impact on the cost were the assumed NH_3-N abatement potentials, the volume of slurry being applied annually with each machine, and the hourly work rate of the equipment. The capital costs of increased tractor power contributed a significantly to the total capital cost of adoption of low emission equipment.

8.6 Costs Incurred by the Trailing Hose Technique and by Slurry Dilution with Water Under Swiss Frame Conditions

Martin Raaflaub
School for Agricultural, Forestry and Food Sciences (HAFL),
Bern University of Applied Sciences (BFH), Länggasse 85,
CH-3052 Zollikofen, Switzerland

Harald Menzi (✉)
School for Agricultural, Forestry and Food Sciences (HAFL),
Bern University of Applied Sciences (BFH), Länggasse 85,
CH-3052 Zollikofen, Switzerland

Agroscope, Tioleyre 4, 1725 Posieux, Switzerland

8.6.1 Introduction

Frame conditions regarding economic aspects and farm structure differ considerably between Switzerland and other European countries, even between Switzerland and its neighbours. The capacity utilization of agricultural machinery tends to be lower in Switzerland, namely due to smaller farm size and the higher part of sloped surfaces. Also machinery purchasing prices are higher, possibly due to lesser competition outside the EU common market. Salaries are higher than in neighboring countries while interest rates are lower. Thus, even though the methodology of abatement cost calculation is consistent with the methodology of Webb et al. showed in the present document, the resulting abatement costs can differ considerably.

In the present document, specific cost data are presented for the measures "trailing hose technique" and "dilution with water", as these measures have a large dissemination in Switzerland and ample field data are available.

8.6.2 Methodology, Data Sources, Assumptions

Base data on farm machinery costs and performance where utilized from the document "Maschinenkosten 2009/10" by Gazzarin and Albisser (2009). The method and base parameters used by these authors and in the calculations behind this contribution are to a great extent equal to the method described by Webb et al. (2015) (Chapter 6). The main divergences concern:

- **Interest rate**: 1.60 % (interest paid for Swiss Federal Bonds, contract period 10 years, average of the year 2011), resp. 4 % (data cited from Gazzarin and Albisser 2009)

- **Amortization period**: variable, according to the utilization intensity of the machine, but always 10 years or more
- **Labour cost**: CHF 28.-/h

The costs of the abatement measures are calculated as the difference to a reference method representing the current good agricultural practice, i.e. vacuum tankers with splash plate. The value of the N retained by the abatement method is not included, thus the results obtained are labelled "gross costs".

8.6.3 Trailing Huse Technique: Description and Results

Costs related to the slurry tanker: Machinery experts recommend only pump tankers to be fitted with trailing hoses, mainly as the incidence of clogging is reduced, whereas when using splash plates (reference) the cheaper vacuum tanker is considered fully adequate. Also, reinforcements in the tanker structure and a dislocation of the axle are often necessary, particularly with small tankers, due to the added weight of the hose distributor.

Traction costs: Traction costs increase due to the extra power uptake of the chopper pump (ca. 5 kW), to the higher dragging force requirement and the need of using a heavier tractor due to the extra weight.

Time consumed: It is assumed that the trailing hoses techniques requires no additional time compared to the reference splash plate technique, as the time used to fold and unfold the spreader and to eliminate eventual clogging are considered insignificant.

Results: The costs of the trailing hose technique as compared to splash plate technique and its components are presented in Tables 8.14 and 8.15.

The extra costs incurred by the trailing hose technique show a considerable decrease with increasing tanker size, particularly for tankers with a capacity of 8,000 L or more.

The lower additional investment for the 12,000 L tanker compared to the 8,000 L option is mainly due to differences in price and characteristics of the base slurry tanker with splash plate. Apparently, the structure of the big tanker requires lesser reinforcement than for the smaller model.

Considerably lower additional costs are incurred if the trailing hose spreading technique is used with an umbilical system instead of a tanker. One has to be aware that the umbilical system can only be used for plots in the vicinity of the livestock operation. However, according to a survey on manure management performed in 2007 over 30 % of the Swiss slurry was spread with umbilical systems (Kupper, personal communication, 2011). When comparing the Swiss data in Table 8.14. with the data presented by Webb et al. (2015, chapter 6 in this volume) and data published by KTBL (2005), the considerably higher price of the trailing hose spreader equipment in Switzerland is particularly striking. A comparison with earlier Swiss price data shows a considerable price increase between 2004 and 2009, the reasons of which are hitherto unknown.

Table 8.14 Additional costs of the trailing hose technique as compared to splash plate for different slurry spreading configurations

Spreading configuration and performance	Tanker 5,000 l, 20 m^3/h	Tanker 8,000 l, 26 m^3/h	Tanker 12,000 l, 31 m^3/h	Umbilical system, 60 m^3/h	Umbilical system, 60 m^3/h
Capacity utilization	1,000 m^3/year	4,000 m^3/year	6,000 m^3/year	1,000 m^3/year	4,000 m^3/year
Additional investment in CHF	36,500	48,000	47,000	18,100	18,100
Fixed costs (interest, depreciation, insurance, shelter) in CHF/year	3,161	4,157	4,070	1,917	1,917
Fixed costs in CHF/ m^3	3.16	1.04	0.68	1.92	0.48
Variable costs (maintenance, additional traction costs) in CHF/m^3	0.20	0.18	0.16	0.23	0.23
Total costs in CHF/m^3	**3.36**	**1.22**	**0.84**	**2.15**	**0.71**

Source: Gazzarin and Albisser (2009), own calculations

Table 8.15 Costs of adding water to slurry before application (in CHF/m^3 added water)

Tanker volume (litres)	Tractor	Tanker/spreader	Labour	Total costs	Marginal costs
vacuum tanker with splash plate					
3,000	1.87	2.07	1.87	5.81	3.16
6,000	1.54	1.41	1.27	4.22	2.22
12,000	1.59	1.29	0.90	3.78	2.00
pump tanker with splash plate					
3,000	1.87	3.33	1.87	7.07	3.88
6,000	1.54	1.96	1.27	4.77	2.51
12,000	1.59	1.65	0.90	4.14	2.14
Pump tanker with trailing hose					
6,000	1.59	2.36	1.27	5.23	2.62
12,000	1.90	1.97	0.90	4.78	2.41
Umbilical system with splash plate					
	0.47	0.33	0.47	1.27	0.82

Source: Gazzarin and Albisser (2009), own calculations

8.6.4   Dilution of Slurry with Water: Description and Results

While the efficiency regarding emission abatement of slurry dilution is variable depending on the quality of the slurry (type of slurry, initial concentration levels, etc.) the costs per added volume of water remain constant. Thus, no assumptions are made relating to qualitative parameters of the slurry.

It is assumed that water is added to the slurry pit only when spare storage volume is available. The water is not filled directly into the slurry tanker but added into the slurry pit before starting the spreading operation. Thus, no extra costs for storage and tanker filling installations are included. Furthermore, no price for the water added is included, assuming that the water is either collected rain water or taken from a water source owned by the farmer. Solely the costs of transporting and spreading the additional volumes of liquid are being calculated. It is assumed that fixed time expenses (i.e. preparation, washing) remain unchanged.

Assuming the case when slurry dilution is only done when spare machine capacity is available, the additional marginal costs are presented. The marginal costs solely include the labour costs and variable machinery costs.

The measure of additional dilution of slurry with water is relatively costly when spreading slurry with a tanker, even when water is available free of charge and when only marginal costs instead of total costs are being considered. However, when slurry is spread by an umbilical system, the costs of diluting with water are moderate. Additional water is therefore propagated in Switzerland for farms that have their own water and use an umbilical system for slurry spreading. It must be considered that additional slurry dilution is effective to reduce emissions especially if the diluted slurry is applied at the same rate of nitrogen per hectare and not at the same m^3 per hectare (Menzi et al. 1998) because the latter would mean additional emitting surface (Fig. 8.11).

Fig. 8.11 Trailing hose spreader combined with an umbilical system

8.7 Estimated Cost of Abating Volatilized Ammonia from Urea by Urease Inhibitors in the EU

Alberto Sanz-Cobena (✉)
ETSI Agrónomos, Technical University of Madrid,
Ciudad Universitaria, 28040 Madrid, Spain

Antonio Vallejo
College of Agriculture, Technical University of Madrid, Ciudad Universitaria.
Avenida Complutense s/n 28040, Madrid, Spain

Tom Misselbrook
Rothamsted Research, North Wyke, Okehampton,
Devon EX20 2SB, UK

Brian Wade
Agrotain International, AGROTAIN International LLC,
1 Angelica Street, St. Louis, MO 63147, USA

8.7.1 Introduction

Urea is the predominant source of inorganic nitrogen (N) fertilizer used in agriculture throughout the world (Harrison and Webb 2001). However, large ammonia (NH_3) losses are associated with its use. Besides the decreased agronomic effectiveness of the applied fertilizer, volatilized NH_3 is involved in the processes of acidification and eutrophication of natural ecosystems and the formation of air-borne fine particulate matter, which can negatively affect human health (Sutton et al. 1993; Sommer and Hutchings 2001). According to the advisory code of good agricultural practices by the Expert Group on Ammonia Abatement (UNECE 2001), mitigation of NH_3 emissions from urea application can be achieved by (a) specific measures based on incorporating it into the soil, (b) applying it during appropriate weather conditions, (c) its substitution for different forms of N fertilizer (e.g. ammonium nitrate) and (d) the use of urease inhibitors (UIs). The first two options are mainly focussed on enhancing the contact between ammonium (NH_4^+) and the soil colloid in such a way that the concentration of NH_4^+ and NH_3 would be decreased in the soil solution and hence decrease the potential for volatilization. The costs associated with these strategies will depend on those of the machinery and infrastructure needed (e.g. additional capital/running costs of equipment to directly place fertilizer in the soil compared with surface application) and the effectiveness of the strategy. The third option relies on the difference in the average emission factor between urea and the alternative N fertiliser. Differences in costs between the forms of fertilizer have to be taken into account in addition to any crop yield impacts. The fourth option is a biochemical method which retards the hydrolysis of urea by inhibiting the urease enzyme in the soil (Gill et al. 1997). This slower hydrolysis rate is associated with a smaller increase in pH around the urea granule and hence a lower potential for NH_3 volatilization. Among the various types of UIs which have been identified and tested, N-(n-butyl) thiophosphoric triamide (NBPT) has been found significantly effective at relatively low concentrations under both laboratory (Carmona et al. 1990; Gill et al. 1997) and field conditions (Sanz-Cobena et al. 2008; Zaman et al. 2009). However, little information has been published on the cost of NH_3 abatement by using urease inhibitors. The objective of this paper was to address the cost-effectiveness of using Agrotain®, the only UI (NBPT-based) currently commercially available on the European market. For this purpose, a cost: benefit analysis was conducted based on the ECETOC data, which provides NH_3 volatilization emission estimates from urea fertilizers for 17 Western European countries (ECETOC 1994).

8.7.2 Cost-Benefit Analysis

We considered that urea is used primarily as a surface-applied "top dressing" and therefore liable to volatilization loss; the cost calculations assume all urea used is treated with urease inhibitor. Additionally, benefit calculations examine the value of mitigated N and the value of additional grain production possible from the mitigated N.

Cost and benefits of mitigation were calculated as follows:

$$\text{MITICOST}_{co}\left(\text{€ kg}^{-1}\text{N}_{miti}\right) = \left(\text{UCONSUM}_{co}*\text{AGTCOST}\right)$$

$$*\left(\text{UCONSUM}_{co}*\text{SURFAPP}_{co}*\text{EF}_{co}*\text{MITIEFF}\right)^{-1}$$

$$\text{MITIBENU}_{co}\left(\text{€ kg}^{-1}\text{N}_{miti}\right) = \text{UCONSUM}_{co}*\text{SURFAPP}_{co}*\text{EF}_{co}*\text{MITIEFF}*\text{UPRICE}$$

$$\text{MITIBENGR}_{co}\left(\text{€ kg}^{-1}\text{N}_{miti}\right) = \left(\text{UCONSUM}_{co}*\text{SURFAPP}_{co}*\text{EF}_{co}*\text{MITIEFF}\right.$$

$$\left.*\text{NUE}*\text{GRNCONT}^{-1}\right) \quad *\left(\text{UCONSUM}_{co}*\text{SURFAPP}_{co}*\text{EE}_{co}*\text{MITIEFF}\right)^{-1}$$

Where:

MITICOST_{co}	Mitigation cost per country
N_{miti}	N mitigated through the treatment of urea with a urease inhibitor
MITIBENU_{co}	Mitigation benefit calculated in value of N mitigated per country
MITIBENGR_{co}	Mitigation benefit calculated as additional grain derived from mitigated N per country
UCONSUM_{co}	Consumption of solid urea fertilizer per country
AGTCOST	price of Agrotain product, taxes and costs for treating urea
SURFAPP_{co}	percentage of urea that is surface applied (90 % default value)
EF_{co}	emission factor of ammonia volatilization adopted from ECETOC
MITIEFF	mitigation effectiveness of urease inhibitor (70 % default value)
UPRICE	prevailing farmer price of solid urea in Europe
NUE	nitrogen use efficiency (50 % default value) (Trenkel 2010)
GRNCONT	grain nitrogen content (2 % N in grain default value)
GRPRICE	price of winter wheat on Chicago Board of Trade, January 2011

Urea consumption data by member state was adopted from the fertilizer industry association database for the cropping year 2007 (IFA 2010). Commodity prices fluctuate greatly within and across seasons, so prices were fixed based on the general market prices at the time of writing. Urea was fixed at 289 € t^{-1}, winter wheat at USD 250 t^{-1} and cost of Agrotain product (only urease inhibitors registered under EC fertilizer legislation) and application to urea was fixed at USD 50 t^{-1}. Urease inhibitor (NBPT) mitigation effectiveness was fixed at 70 %, the mean value reported from a 16-site research project in the UK (Chambers and Dampney 2009). This assumption should be considered with care. The 70 % effectiveness of by NBPT (Agrotain) based on Chambers and Dampney (2009) is higher than that found in other studies carried out outside UK, and was associated with a range from 25 to 100 %. Sanz-Cobena et al. (2008) found a mitigation of 50 % when using Agrotain in a sunflower crop fertilized with urea under Mediterranean conditions. Local variability is expected to exist in the effectiveness of the inhibitor, probably due to specific soil and weather conditions.

The analysis excluded NH_3 mitigation with UAN fertilizer (Urea-Ammonium-Nitrate or UAN typically contains 50 % urea-N, 50 % ammonium nitrate-N). Although urease inhibitors are used with UAN, the use of UAN across Europe is relatively low (<15 % of total N) with the exception of France which consumes approximately half of the total consumption in Europe (IFA 2010).

8.7.3 Results and Discussion

Calculated mitigation cost with a urease inhibitor across the member states averaged 0.76 € kg N_{miti}^{-1}. Costs were relatively lower in member states with higher NH_3 emission factors because the cost was divided over more units of mitigated N (Table 8.16). Mitigation costs at the extremes of the Chambers and Dampney data showed costs of 2.13 € kg N_{miti}^{-1} assuming 25 % average urease inhibitor effectiveness and 0.53 € kg N_{miti}^{-1} assuming 100 % average urease inhibitor effectiveness.

8.7.4 Conclusions

As fertilizer N is an input to crop production, any "additional" N gained through mitigation could be utilized in the production system to save input costs or increase revenue to offset the cost of mitigation.

The 'efficiency option' would utilize mitigated nitrogen to reduce overall application rates with the expectation of maintaining equivalent crop yields. The above calculations allow 11–14 % N rate reductions (e.g. mitigating 70 % of the 15–20 % urea-N volatilized). The reduction of N rate would decrease total nitrogen costs for crop production and partially offset the higher cost of urea treated with urease inhibitor. Also, reduced N rates may have additional environmental benefits as the total load of reactive N in the environment is reduced.

The 'productivity option' would maintain normal urea-N application rate with the expectation of higher crop yields. When the mitigated nitrogen is converted to grain production, the grain value would be 8.93 € kg N_{miti}^{-1}. The cost of mitigation could be completely neutralized if only 9 % of the mitigated nitrogen is converted into grain. The "value multiplier" of grain production would provide a financial incentive for farmers to adapt urease inhibitors as a technological mitigation strategy.

Table 8.16 Quantity and cost-benefit analysis (CBA) of mitigating ammonia volatilization loss from urea fertilizer by treatment with a urease inhibitor

Country	Urea consumed	N volatilized	N mitigated	UI mitigation cost	Mitigated N value	Potential grain value
	kg urea-N	kg NH_3-N	kg NH_3-N	€/kg N_{miti}	€/kg N_{miti}	€/kg N_{miti}
AU	2,300,000	310,500	217,350	0.81	0.61	8.93
BE	460,000	62,100	43,470	0.81	0.61	8.93
CH	2,760,000	372,600	260,820	0.81	0.61	8.93
DE	138,460,000	18,692,100	13,084,470	0.81	0.61	8.93
DK	4,600,000	621,000	434,700	0.81	0.61	8.93
FN	920,000	124,200	86,940	0.81	0.61	8.93
FR	144,440,000	19,499,400	13,649,580	0.81	0.61	8.93
GR	1,840,000	331,200	231,840	0.60	0.61	8.93
IR	11,960,000	1,614,600	1,130,220	0.81	0.61	8.93
IT	190,440,000	25,709,400	17,996,580	0.81	0.61	8.93
LU	460,000	62,100	43,470	0.81	0.61	8.93
NL	460,000	62,100	43,470	0.81	0.61	8.93
NO	460	62	43	0.81	0.61	8.93
PT	4,600,000	621,000	434,700	0.81	0.61	8.93
SP	113,160,000	20,368,800	14,258,160	0.60	0.61	8.93
SW	460	62	43	0.81	0.61	8.93
UK	32,200,000	4,347,000	3,042,900	0.81	0.61	8.93
TOTAL	**649,060,920**	**92,798,224**	**64,958,757**			
AVERAGE				**0.76**	**0.61**	**8.93**

8.8 Potential N₂O Reduction Associated to the Use of Urease Inhibitors in Spain

Diego Abalos and Alberto Sanz-Cobena (✉)
ETSI Agrónomos, Technical University of Madrid,
Ciudad Universitaria, 28040 Madrid, Spain

Antonio Vallejo
College of Agriculture, Technical University of Madrid,
Ciudad Universitaria, Avenida Complutense s/n, 28040 Madrid, Spain

Eduardo Aguilera
Universidad Pablo de Olavide, Ctra. de Utrera, km. 1, 41013 Sevilla, Spain

Luis Lassaletta
Ecotoxicology of Air Pollution, CIEMAT, Madrid, Spain

8.8.1 Introduction

Productivity of agricultural ecosystems is highly dependent on nitrogen (N) inputs from fertilizer application. Nevertheless, the use of fertilizers represents a threat to environmental quality since they are considered one of the main sources of atmospheric ammonia (NH_3) and the greenhouse gas (GHG) nitrous oxide (N_2O) (IPCC 2007). Emissions of N_2O from fertilizer application contribute 58 % of anthropogenic N_2O emissions according to the IPCC estimates (IPCC 2007), whereas total NH_3 volatilized from ureic and ammonium based fertilizers is close to 17 % of anthropogenic emissions (Buijsman et al. 1987). These N losses are responsible for an important reduction in the effectiveness of the applied fertilizer and therefore represent an economic concern. This is specially the case for urea, the predominant source of inorganic N fertilizer used in agriculture throughout the world (Harrison and Webb 2001).

Originally developed to reduce NH_3 volatilization from urea fertilization, urease inhibitors have received additional attention in the last years due to their potential capacity to reduce N_2O emissions. Among the various types of urease inhibitors that have been identified and tested, N-(n-butyl) thiophosphoric triamide (NBPT) is the most widely used (Abalos et al. 2014) and it has already been found to be highly effective at mitigating NH_3 volatilization at relatively low concentrations under both laboratory (Gill et al. 1997) and field conditions (Sanz-Cobena et al. 2008).

8.8.2 Results and Discussion

Recent studies in Spain have been focussed on understanding the degree of success of NBPT in reducing N_2O emissions for Mediterranean agricultural systems (Fig. 8.12). The results previously found elsewhere showed no consistent effect which was probably due to differences in management and environmental factors

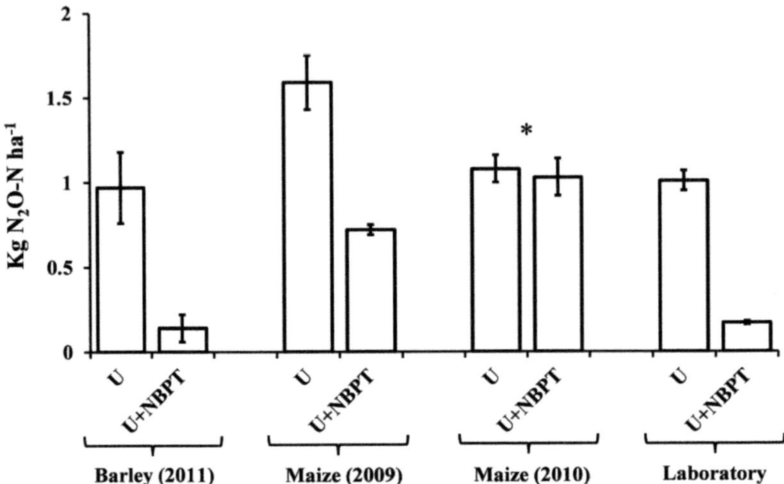

Fig. 8.12 Cumulative N_2O emissions for Barley 2011 (Abalos et al. 2012), Maize 2009 and 2010 (Sanz-Cobena et al. 2012) and Laboratory experiments (Sanz-Cobena et al. 2014). Error bars indicate standard errors. *Note that Maize (2010) is the only experiment without significant effects and carried out under denitrifying conditions

between studies (Akiyama et al. 2010). Based on that, experiments in Spain were built on the hypothesis that NBPT reduces N_2O emissions at soil moisture contents close to field capacity (favorable for nitrification) but not at saturation (favorable for denitrification). An initial 2-year field experiment was carried out in irrigated maize in order to evaluate the inhibitor's effectiveness under two contrasting irrigation regimes (Sanz-Cobena et al. 2012). In the first year (2009), irrigation was controlled the first two weeks after fertilizing in order to enhance nitrifying conditions whereas the second year (2010) was characterized by an intense irrigation and so denitrification was expected to be the main process responsible for N_2O production. Nitrous oxide emissions were effectively reduced by NBPT under nitrification-favoring conditions (55 % reduction) and no significant effect was found when denitrification was the dominant pathway of N_2O production. Then, a field study with a barley crop was undertaken in order to confirm if the previous results remain true for rainfed systems under low rainfall Mediterranean conditions (Abalos et al. 2012). Again, NBPT was effective mitigating these emissions during periods of low rainfall leading to soil moisture contents close to or below field capacity (86 % reduction).

Subsequently, a laboratory study was carried out with the aim of further elucidating the mechanisms by which the inhibitor is able to affect N_2O forming processes (Sanz-Cobena et al. 2014). The study showed that the most likely mechanism through which NBPT affects nitrification is by affecting the NH_4^+ pool. Reducing the size of the NH_4^+ pool may reduce nitrification rates in such a way that the production of N_2O is also affected. It could be argued that reduced NH_4^+ concentrations via lower nitrification rates may decrease NO_3^- supply thereby affecting N_2O emissions from denitrification. However, this effect was found to be not significant during this set of experiments, which partially explains why some researchers did not found N_2O reductions when applying NBPT (e.g. Menéndez et al. 2009).

8.8.3   *Conclusions*

A recent meta-analysis has shown that NBPT can be recommended in order to increase both crop yields and N use efficiency (Abalos et al. 2014). Besides these benefits and its proven efficiency to reduce NH_3 volatilization, research conducted in Spain shows that the inhibitor may also play a significant role guiding EU mitigation policy for N_2O emissions. The economic viability of this product will thus, also depend on initiatives such as the implementation of Emission Trading Schemes including greenhouse gas emissions from agriculture. However, the applicability of this mitigation strategy is restricted to specific agricultural systems where soil moisture content is below saturation. For low rainfall Mediterranean conditions, NBPT use to reduce N_2O emissions may be suitable for rainfed crops and systems where water is applied matching crop needs.

References

Abalos D, Sanz-Cobena A, Misselbrook T, Vallejo A (2012) Effectiveness of urease inhibition on the abatement of ammonia, nitrous oxide and nitric oxide emissions in a non-irrigated Mediterranean barley field. Chemosphere 89:310–318

Abalos D, Jeffery S, Sanz-Cobena A, Guardia G, Vallejo A (2014) Meta-analysis of the effect of urease and nitrification inhibitors on crop productivity and nitrogen use efficiency. Agric Ecosyst Environ 189:136–144

Acutis M, Provolo G, Bertoncini G (2009) An expert system for the nitrate issue in Lombardian agriculture. In: Grignani C, Acutis M, Zavattaro L, Bechini L, Bertora C, Marino Gallina P, Sacco D (eds) Proceedings of the 16th nitrogen workshop: connecting different scales of nitrogen use in agriculture. Turin, Italy, pp 465–466

Akiyama H, Yan X, Yagi K (2010) Evaluation of effectiveness of enhanced efficiency fertilizers as mitigation options for N_2O and NO emissions from agricultural soils: meta-analysis. Global Change Biol 16:1837–1846

Anon (2003) Integrated Pollution Prevention and Control (IPPC). Reference document on best available techniques for intensive rearing of poultry and pigs. European Commission

Anon (2010) European Communities (Good Agricultural Practice for Protection of Waters) Regulations 2010. SI 610 of 2010. Department of Environment, Heritage and Local Government, The Stationary Office, Dublin, Ireland

Anon (2012) Tractor guide 2012, Irish Farmers Monthly. IFP Media, Dublin, Ireland

BAFU und BLW (2012) Baulicher Umweltschutz in der Landwirtschaft. Ein Modul der Vollzugshilfe Umweltschutz in der Landwirtschaft. Bundesamt für Umwelt, Bern. Umwelt-Vollzug Nr. 1101. http://www.bafu.admin.ch/publikationen/publikation/01581/index.html?lang=de

Bassanino M, Grignani C, Sacco D, Allisiardi E (2007) Nitrogen balances at the crop and farm-gate scale in livestock farms in Italy. Agric Ecosyst Environ 122(3):282–294

Bassanino M, Sacco D, Zavattaro L, Grignania C (2011) Nutrient balance as a sustainability indicator of different agro-environments in Italy. Ecol Indic 11(2):715–723

Bechini L, Castoldi N (2006) Calculating the soil surface nitrogen balance at regional scale: example application and critical evaluation of tools and data. Ital J Agron 1:665–676

Bechini L, Castoldi N (2009) On-farm monitoring of economic and environmental performances of cropping systems: results of a 2-year study at the field scale in northern Italy. Ecol Indic 9:1096–1113

Bittman S, Dedina M, Howard CM, Oenema O, Sutton MA (eds) (2014) Options for ammonia mitigation: guidance from the UNECE task force on reactive nitrogen. Centre for Ecology and Hydrology, Edinburgh

Buijsman E, Mass HFM, Asman WAH (1987) Anthropogenic NH_3 emissions in Europe. Atmos Environ 21:1009–1022

Carmona G, Christianson CB, Byrnes BH (1990) Temperature and low concentration effects of the urease inhibitor N-(n-butyl) thiophosphoric triamide (NBPT) on ammonia volatilisation from urea. Soil Biol Biochem 22:933–937

Chambers BJ, Dampney PMR (2009) Nitrogen efficiency and ammonia emissions from urea-based and ammonium nitrate fertilisers. International Fertiliser Society Proceedings No. 657, York, UK

Coulter BS (ed) (2004) Nutrient and trace element advice for grassland, tillage, vegetable and fruit crops. Teagasc, Wexford

Coulter BS, Lalor S (eds) (2008) Major and micro nutrient advice for productive agricultural crops, 3rd edn. Teagasc, Wexford

CRPA (2011) SIM.BA-N: Simplified Balance –N. Centro Ricerche Produzioni Animali (CRPA), Reggio Emilia. http://www.crpa.it/nqcontent.cfm?a_id=1150&tt=crpa_www&sp=optiman. Last access 15 Sept 2014

CRPA (2014) Achieving good water quality status in intensive animal production areas. http://aqua.crpa.it/. Accessed 15 Sept 2014

DEFRA (2010) Fertiliser Manual (RB 209), 8th edn. The Stationery Office, Norwich

Dowling C, Curran T, Lanigan GJ (2008) The effect of application technique and climate conditions on ammonia emissions from cattle slurry. In: 13th International RAMIRAN proceedings, Albena, 4pp. ISBN 978-954-9067671-6-3

ECETOC (1994) Ammonia emission to air in western Europe. Technical report 62. European Centre for Ecotoxicology and Toxicology of Chemicals, Avenue E Van Nieuwenhuyse 4, Brussels, Belgium

EMEP/EEA emission inventory guidebook – EEA (2009) Updated June 2010.4.B Animal husbandry and manure management Appendix B. http://www.eea.europa.eu/publications/emep-eea-emission-inventory-guidebook-2009/part-b-sectoral-guidance-chapters/4-agriculture/4-b/4-b-appendixb.xls. Accessed 15 Sept 2014

Fumagalli M (2009) Indicator-based and modelling approaches for the integrated evaluation and improvement of agronomic, economic and environmental performances of farming and cropping systems in northern Italy. Ph.D. thesis, University of Milano, Italy

Fumagalli M, Acutis M, Mazzetto F, Vidotto F, Sali G, Bechini L (2011) An analysis of agricultural sustainability of cropping systems in arable and dairy farms in an intensively cultivated plain. Eur J Agron 34(2):71–82

Gazzarin C, Albisser G (2009) Maschinenkosten 2009/2010. ART-Bericht Nr. 717, Forschungsanstalt Agroscope Reckenholz-Tänikon, Ettenhausen, Switzerland. http://www.agroscope.ch/publikationen/07703/07707/07711/index.html?lang=en&sb_pubsearch=1&sb_pubsearch=1&sort[3_10]=0&dir[3_10]=asc&page[3_10]=2

Gill JS, Bijay-Singh, Khind CS, Yadvinder-Singh (1997) Efficiency of N-(n-butyl) thiophosphoric triamide in retarding hydrolysis of urea and ammonia volatilization losses in a flooded sandy loam soil amended with organic materials. Nutr Cycl Agroecosyst 53:203–207

Harrison R, Webb J (2001) A review of the effect of N fertilizer type on gaseous emissions. Adv Agron 73:65–108

Hennessy T, Buckley C, Cushion M, Kinsella A, Moran B (2011) National farm survey of manure application and storage practices on Irish farms. Agricultural Economics and Farm Surveys Dept., Teagasc, Wexford

Hyde B, Carton OT (2005) Manure management facilities on farms and their relevance to efficient nutrient use. Fertiliser Association of Ireland, Winter Scientific Meeting. UCD, Dublin, pp 27–43

Hyde BP, Carton OT, O'Toole P, Misselbrook TH (2003) A new inventory of ammonia emissions from Irish agriculture. Atmos Environ 37:55–62

Hyde B, Carton OT, Murphy WE (2006) Farm facilities survey – Ireland 2003. Teagasc, Wexford

International Fertilizer Industry Association (2010) IFA fertilizer database for 2007/08. Slow- and controlled-release and stabilized fertilisers: an option for enhancing nutrient efficiency in agriculture, Paris

IPCC (2007) Climate change, synthesis report of the fourth assessment report of IPCC, Geneva, Switzerland

ISTAT (2013) I prodotti agroalimentari di qualità DOP, IGP e STG. http://www.istat.it/it/archivio/98939/. Accessed 15 Sept 2014

ISTAT, several years, Agricultural dabatabase from Istituto Nazionale di Statistica from ISTAT (Istituto Nazionale di Statistica). http://agri.istat.it/. Last access 15 Sept 2014.

Kim KU, Bashford LL, Sampson BT (2005) Improvement of tractor performance. Appl Eng Agric 21(6):949–954

Kuratorium für Technik und Bauwesen in der Landwirtschaft KTBL (2005) KTBL-Datensammlung Betriebsplanung 2004/05, KTBL-Schriftenvertrieb im Landwirtschaftsverlag Münster, 2004; ISBN 3-7843-2178-X

Lalor S (2008) Economic costs and benefits of adoption of the trailing shoe slurry application method on grassland farms in Ireland. In: Koutev V (ed) 13th RAMIRAN international conference, Albena, Bulgaria, pp 75–80

Lalor STJ, Schulte RPO (2008) Low ammonia emission application methods can increase the opportunity for application of cattle slurry to grassland in spring in Ireland. Grass Forage Sci 63:531–544

Lalor STJ, Schröder JJ, Lantinga EA, Oenema O, Kirwan L, Schulte RPO (2011) Nitrogen fertilizer replacement value of cattle slurry in grassland as affected by method and timing of application. J Environ Qual 40(2):362–373

Laws JA, Smith KA, Jackson DR, Pain BF (2002) Effects of slurry application method and timing on grass silage quality. J Agric Sci 139:371–384

Mantovi P (2007) L'efficienza dell'azoto in azienda si può migliorare. L'Informatore Agrario 34:2–5

Maximov DA et al (2014) Selecting manure management technologies to reduce ammonia emissions from big livestock farms in the Northwestern Federal District of Russia. In: Van der Hoek KW, Kozlova NP (eds) Ammonia workshop 2012 Saint Petersburg. Abating ammonia emissions in the UNECE and EECCA region. Семинар по аммиаку 2012, Санкт Петербург. Снижение выбросов аммиака в регионах ЕЭК ООН и ВЕКЦА.RIVM Report 680181001/SZNIIMESH Report. Bilthoven, The Netherlands, p 185. http://www.rivm.nl/Documenten_en_publicaties/Wetenschappelijk/Rapporten/2014/juni/Ammonia_workshop_2012_Saint_Petersburg_Abating_ammonia_emission_in_UNECE_and_EECCA_region. Accessed 15/09/2014

Menéndez S, Merino P, Pinto M, Estavillo JM (2009) Effect of N-(n-butyl) thiophosphoric triamide and 3,4-dimethylpyrazole phosphate on gaseous emissions from grasslands under different soil water contents. J Environ Qual 38:27–35

Menzi H, Frick R, Kaufmann R (1997) Ammoniak-Emissionen in der Schweiz: Ausmaß und technische Beurteilung des Reduktionspotentials. Schriftenreihe der FAL 26. Eidgenössische Forschungsanstalt für Agrarökologie und Landbau, Zürich Reckenholz, Switzerland

Menzi H, Katz P, Fahrni M, Neftel A, Frick R (1998) A simple empirical model based on regression analysis to estimate ammonia emissions after manure application. Atmos Environ 32:301–307

Misselbrook TH, Smith KA, Johnson RA, Pain BF (2002) Slurry application techniques to reduce ammonia emissions: results of some UK field-scale experiments. Biosyst Eng 81(3):313–321

Penkhues B, Plans and cost estimates for layer hen housing systems. Big Dutchman GmbH, Vechta (unpublished)

Raison C, Pflimlin A (2006) Des groupes d'éleveurs améliorent leurs pratiques environmentales. Cap Élevage 10:14–15

Regione Lombardia (2008) Gestione dell'azoto sostenibile a scala aziendale: itinerari tecnici Progetto GAZOSA. A cura di Acutis M, Bechini L, Fumagalli M, Perego A, Carozzi M, Bernardoni E, Brenna S, Pastori M, Mazzetto F, Sali G, Francesco V. Quaderni della ricerca. Regione Lombardia N 94 Ottobre 2008

Roggero PP, Ceccon P, Mori M, Toderi M (2009) Innovative tools and strategies to design sustainable cropping systems for the Italian nitrate vulnerable zones. In: Grignani C, Acutis M, Zavattaro L, Bechini L, Bertora C, Marino Gallina P, Sacco D (eds) Proceedings of the 16th nitrogen workshop: connecting different scales of nitrogen use in agriculture. Turin, Italy, pp 343–344

Sacco D, Bassanino M, Grignani C (2003) Developing a regional agronomic information system for estimating nutrient balances at a larger scale. Eur J Agron 20(1–2):199–210

Sanz-Cobena A, Misselbrook TH, Arce A, Mingot JI, Diez JA, Vallejo A (2008) An inhibitor of urease activity effectively reduces ammonia emissions from soil treated with urea under Mediterranean conditions. Agric Ecosyst Environ 126:243–249

Sanz-Cobena A, Sánchez-Martín L, García-Torres L, Vallejo A (2012) Gaseous emissions of N_2O and NO and $NO_3{}^-$ leaching from urea applied with urease and nitrification inhibitors to a maize (Zea mays) crop. Agric Ecosyst Environ 149:64–73

Sanz-Cobena A, Abalos D, Meijide A, Sanchez-Martin L, Vallejo A (2014) Soil moisture determines the effectiveness of two urease inhibitors to decrease N_2O emission. Mitig Adapt Strateg Glob Chang. doi:10.1007/s11027-014-9548-5

Schröder JJ (2005) Manure as a suitable component of precise nitrogen nutrition. Proceedings No. 574, International Fertiliser Society, York, UK

Simon JC, Grignani C, Jacquet A, Le Corre L, Pagès J (2000) Typologie des bilans d'azote de divers types d'exploitation agricole: recherche d'indicateurs de fonctionllement. Agronomie 20:175–195

Smith KA, Jackson DR, Misselbrook TH, Pain BF, Johnson RA (2000) Reduction of ammonia emission by slurry application techniques. J Agric Eng Res 77(3):277–287

Sommer SG, Hutchings NJ (2001) Ammonia emissions from field applied manure and its reduction. Eur J Agron 15:1–15

Stoll P, Kessler J, Gutzwiller A, Chaubert C, Gafner J, Bracher A, Jost M, Pfirter HP, Wenk C (2004) Fütterungsempfehlungen und Nährwerttabellen für Schweine, 3rd edn. Eidgenössische Forschungsanstalt für Nutztiere und Milchwirtschaft (ALP), LMZ, Zollikofen, pp 55–60. http://www.agroscope.admin.ch/futtermitteldatenbank/04835/index.html?lang=de

Sutton MA, Pitcairn CER, Fowler D (1993) The exchange of ammonia between the atmosphere and plant communities. Adv Ecol Res 24:301–393

Trenkel ME (2010) Slow- and controlled-release and stabilized fertilizers. An option for enhancing nutrient use efficiency in agriculture. International Fertilizer Industry Association, Paris

UNECE (2001) UNECE framework code for good agricultural practices for reducing ammonia. Expert Group on Ammonia Abatement. United Nations Economic Commission for Europe, Geneva. http://www.clrtap-tfrn.org/sites/clrtaptfrn.org/files/documents/EPMAN%20Documents/eb.air.wg.5.2001.7.e.pdf. Accessed 18 Oct 2014

UNECE (2007) Guidance document on control techniques for preventing and abating emissions of ammonia. ECE/EB.AIR/WG.5/2007/13. United Nations Economic Commissions for Europe (UNECE), Geneva, Switzerland

Ventura M, Scandellari F, Ventura F, Guzzon B, Rossi Pisa P, Tagliavini M (2008) Nitrogen balance and losses through drainage waters in an agricultural watershed of the Po Valley (Italy). Eur J Agron 29:108–115

Webb J, Pain B, Bittman S, Morgan J (2010) The impacts of manure application methods on emissions of ammonia, nitrous oxide and on crop response - a review. Agric Ecosyst Environ 137:39–46

Webb J, Morgan J, Pain B (2015) Cost of ammonia emission abatement from manure spreading and fertilizer application. In: Reis S, Howard CM, Sutton MA (eds) Costs of ammonia abatement and the climate co-benefits. Springer Publishers, Amsterdam

Zaman M, Saggar S, Blennerhassett JD, Singh J (2009) Effect of urease and nitrification inhibitors on N transformation, gaseous emissions of ammonia and nitrous oxide, pasture yield and N uptake in grazed pasture system. Soil Biol Biochem 41:1270–1280

Chapter 9
Estimating Costs and Potential for Reduction of Ammonia Emissions from Agriculture in the GAINS Model

Zbigniew Klimont and Wilfried Winiwarter

Abstract The revision of the IIASA's integrated assessment model GAINS allowed for adoption of new cost estimates for ammonia abatement as developed in the framework of the Task Force on Reactive Nitrogen (TFRN). Here we describe key features and methods of the agricultural module of the GAINS model, discuss update of ammonia reduction cost, and recent applications of the revised model in discussion of the future European ecosystem and air quality. As ammonia is predominantly released from agriculture, also abatement needs to prioritize the same economic sector. The revision of GAINS also accommodates for the TFRN's concept to focus measures towards larger installations and avoid calling for emission reductions on small farms. The overall lower cost of ammonia abatement influence optimization towards applying more strict ammonia abatement, which will decrease overall abatement cost but not necessarily costs for ammonia abatement (as these are considered to be applied more readily). Finally, new analysis of reduction targets following optimized versus more even distribution of reduction measures, shows potential for synergies that could help in finalizing agreement on how future emission reduction targets could be achieved.

Keywords Ammonia emissions • Emission abatement • GAINS model • Greenhouse gases • Cost-effectiveness assessment

9.1 Introduction

Emissions of ammonia in Europe play an increasing role in the observed effects of air pollution, e.g., acidification, euthrophication, contribution to secondary particulate matter and associated health impacts. However, most of the international environmental agreements so far refrained from ambitious targets for reduction of ammonia emissions. Discussion of the revision of the Gothenburg Protocol

Z. Klimont (✉) • W. Winiwarter
International Institute for Applied Systems Analysis (IIASA), Schlossplatz 1,
A-2361 Laxenburg, Austria
e-mail: klimont@iiasa.ac.at; winiwarter@iiasa.ac.at

S. Reis et al. (eds.), *Costs of Ammonia Abatement and the Climate Co-Benefits*,
DOI 10.1007/978-94-017-9722-1_9

(UNECE 1999) and availability of the new data on ammonia abatement costs (which are being compiled in this volume) as well as reduction potential resulted in updates of the national assessments and the GAINS model. The latter is central to scenarios developed for discussion under the CLRTAP (e.g. Amann et al. 2010, 2011b). Here we summarize the current methodology and the data used for calculating costs and the reduction potential of measures included in the GAINS model to control ammonia emissions, and discuss key results obtained in preparation for the May 2012 revision of the Gothenburg Protocol revision (Reis et al. 2012; UNECE 2012).

9.2 The GAINS Model

9.2.1 General Model Concept

The Greenhouse gas – Air pollution INteractions and Synergies (GAINS) model is a tool to estimate the environmental effects of air pollution under consideration of greenhouse gas emissions. It allows assessing, at the level of economic sectors and individual countries, options to reduce emissions and the costs of their implementation with regard to their effect in terms of reducing ecosystem and human health impacts (Amann et al. 2011a).

GAINS operates under a multi-gas regime. Emissions of trace gases included cover greenhouse gases under the Kyoto protocol (CO_2, CH_4, N_2O, F-gases) as well as air pollutants (SO_2, NO_x, NMVOC, CO, NH_3 and several particulate matter species). Interventions in the emissions of one component, implemented as a technology change which affects the emission factors associated with a certain activity, may then also cause intended or unintended side effects on the emissions of one or more other components. The introduction of such control technology is always associated with certain costs.

Dispersion and transformation of trace constituents in the atmosphere is implemented in GAINS via source-receptor matrices, which are themselves the results of repeated long-term runs of atmospheric chemistry-transport models. Likewise, environmental impacts are quantified as a result of parameterized ecosystems or human health response. Using external information on energy and other activities, it estimates emissions and effects for every 5 years over the period 1990–2030.

GAINS has been used in a number of policy related exercises and described in detail in connection with these endeavors, specifically, this refers to its use in the CAFE programme, the NEC process (Amann et al. 2007) and for the EU climate policy (Höglund Isaksson et al. 2009). Further documentation as well as the model itself can be accessed at http://gains.iiasa.ac.at.

9.2.2 Agricultural Module of the GAINS Model

9.2.2.1 Basic Characteristics and Ammonia Emission Calculation

The GAINS ammonia module has been developed based on the work of Klaassen (1991a, b). Updated documentation, extending on the agricultural interactions, have been published by Brink et al. (2001a, b) and by Klimont and Brink (2004). These reports and papers describe the detailed structure used and the underlying data sources. For activity data, GAINS contains a number of scenarios where various sources have been used, including national projections as well as work of the international organizations like the Food and Agriculture Organization of the United Nations (FAO), the European Fertilizer Manufacturers Association (EFMA), the International Fertilizer Industry Association (IFA), the Organisation for Economic Co-operation and Development (OECD) and international modeling groups (CAPRI model: Britz and Witzke 2008). The historical data relies on statistical information (e.g., Eurostat, FAO, EFMA, IFA) validated by national experts during several consultation processes linked to the preparations of CLRTAP Protocols, EU Directives (specifically the National Emission Ceilings directive), and the Clean Air for Europe (CAFE) programme.

In its current implementation, the model distinguishes four key emission stages for ammonia released from animal manure.[1] In a mass-conservation approach, any measure that keeps ammonia from evaporating will keep it available for the next stage, such that an emission reduction in one stage may lead to an increase in the following stage. These stages are "housing", "storage", "application", and "grazing" (obviously, grazing is an own pathway somewhat independent of the three other stages). Emission factors and a set of abatement measures are defined for each of the stages. Most recently, this approach has been extended to discriminate more stages and include in a consistent manner NH_3, N_2O, and CH_4 emissions as well as nitrate losses (Asman et al. 2011, and Tier 2 approach in Klimont and Brink 2004). However, this extension has not been implemented in the on-line model yet.

The current approach to assess emissions thus can be described as presented by Klimont and Brink (2004):

$$EL_{j,l} = \sum_i L_{i,j} \sum_k \sum_{s=1}^{4} \left[ef_{i,j,l,s}\left(1 - \eta_{i,k,l,s}\right)X_{i,j,k,l} \right] \qquad (9.1)$$

where:

EL ammonia emissions from livestock farming [kt NH_3/year];
i,j,k,l livestock category, year, abatement technique, country;
s emission stage (four stages)

---

[1] For nitrogen mineral fertilizers emissions are calculated by multiplying applied amounts with emission factors specific to region and fertilizer type.

L animal population [thousand heads];
ef emission factor [kg NH_3 / animal per year];
η reduction efficiency of abatement technique;
X implementation rate of the abatement technique

In the above equation, the emission factors of each stage are influenced by the N-losses at previous stages (Eqs. 9.2a, 9.2b, 9.2c, and 9.2d).

$$ef_1 = Nx_1 v_1 \qquad (9.2a)$$

$$ef_2 = Nx_1 \left(1 - v_1\right) v_2 \qquad (9.2b)$$

$$ef_3 = Nx_1 \left(1 - v_1 - \left(1 - v_1\right) v_2\right) v_3 \qquad (9.2c)$$

$$ef_4 = Nx_4 v_4 \qquad (9.2d)$$

where:

$ef_{1,2,3,4}$ NH_3-nitrogen loss at distinguished emission stages, i.e., housing (1), storage (2), application (3), and grazing (4),

$Nx_{1,4}$ N excretion during housing (1) and grazing (4),

$v_{1,2,3,4}$ N volatilization rates at distinguished emission stages

The parameters in Eqs. 9.1 and 9.2 can be viewed and downloaded from the on-line version of the GAINS model (http://gains.iiasa.ac.at).

9.2.2.2 Activity Categories and Emission Control Options in GAINS

In order to reflect the significant differences in national practices of animal husbandry, GAINS not only differentiates livestock into major categories, but also distinguishes between animals kept on liquid (slurry) and solid manure systems (often referred to as farmyard manure or FYM) while for mineral fertilizers distinguishes urea from other N-fertilizers:

- Livestock categories

    - Dairy cows (distinguishing liquid and solid manure systems)
    - Other cattle (distinguishing liquid and solid manure systems)
    - Pigs (distinguishing liquid and solid manure systems)
    - Sheep and goats
    - Horses, donkeys and mules
    - Laying hens
    - Other poultry
    - Fur animals
    - Camels
    - Buffaloes

- Mineral N-fertilizers

  – Urea
  – Other

In the last two decades a number of reduction measures have been developed and applied in several countries. In GAINS, the measures are grouped into key categories of abatement techniques (see Table 9.1 for a listing of feasible combinations). Individual abatement technologies address specific stages of the process chain (see Table 9.2), but will also influence the emission factors of subsequent stages according to Eq. (9.2). A detailed description of these options has been presented by Klimont and Brink (2004).

Low nitrogen feed describes a method of dietary changes, where a lower protein (nitrogen) content of animal feed leads to reduced nitrogen excretion.

Low emission housing covers a number of options that prevent ammonia emissions from animal housing, basically reducing the surface area and exposure time of manure in the animal house. This includes flushing systems or other means of immediate transport of manure into storage.

Air purification includes options which treat the air ventilated from animal housing. As discussed in the guidance document to the Annex IX of the Gothenburg Protocol (UNECE 2007), the treatment of exhaust air by acid scrubbers or biotrickling filters has proven to be practical and effective for large scale operations in the Netherlands, Germany and Denmark. Thus the GAINS database has been updated to consider the recent shift away from biofilters (for which the previous cost data had been developed) to acid scrubber systems.

Covered storage means a reduction of exposure of stored manure to air. We distinguish between low efficiency systems (e.g. floating foils or polysterene) and high efficiency systems that would allow more efficient separation from the atmosphere (using concrete, corrugated iron or polyester caps)..

Low ammonia application refers to distributing manure to agricultural fields in a way to minimize surface exposure, by placing it underneath a soil or plant cover layer instead of spreading it over the surface (broadcasting). Low efficiency methods include slit injection, trailing shoe, slurry dilution, and band spreading for liquid slurry, and incorporation of solid manure by ploughing into the soil the day after application. As high efficiency methods we understand immediate incorporation by ploughing (within 4 h after application), deep and shallow injection of liquid manure and immediate incorporation by ploughing (within 12 h after application) of solid manure.

Improved application or substitution of urea refers to either appropriate timing and dose of application or to the substitution of urea (and ammonium carbonate, if relevant) by other chemical forms of fertilizers that are less easily releasing ammonia, e .g. ammonium nitrate.

Further, the combinations of the above options are defined and respective emission factors and costs are calculated. The defined combinations attempt to mimic a realistic combination of options applied at different stages (see Table 9.1).

Table 9.1 Emission control options for livestock, as currently implemented in GAINS

| Animal category | FEED | HOUSING | | STORAGE | APPLICATION | TOTAL NUMBER OF OPTIONS |
| | Low nitrogen feed | Low emission housing | Air purification | Covered storage | Low ammonia application | (including combinations) |
	(LNF)	(SA)	(BF)	(CS)	(LNA)	
Dairy cows	x	x		x	x	18
Other cattle		x		x	x	9
Pigs	x	x	x	x	x	31
Laying hens	x	x	x	x	x	20
Other poultry[a]	x	x	x	x	x	21
Sheep					x	2
						101
Total number of measures which include the given option	45	18	30	32	58	

[a]Includes also poultry manure incineration

Table 9.2 GAINS control options for ammonia emission by stage

		Affected stages			
Control	GAINS abbrev.	1	2	3	4
Low nitrogen feed	LNF	x	x	x	x
Low emission housing	SA	x	x	x	
Air purification	BF	x			
Covered storage	CS		x	x	
Low ammonia application	LNA			x	
Improved application or substitution of urea	SUB			x[a]	

[a]For application of N-mineral fertilizers, emissions are calculated only after application

9.3 Ammonia Emissions Control Costs in GAINS

9.3.1 General Concept

The basic intention of a cost evaluation in the GAINS model is to identify the value to society of the resources diverted in order to reduce emissions of a specific compound. In practice, these values are approximated by estimating costs at the production level rather than prices to the consumers. Therefore, any mark-ups charged over production costs by, e.g., food industry or the retail markets do not represent actual resource use and are ignored. Certainly, there will be transfers of money with impacts on the distribution of income or on the competitiveness of the market, but these should be removed from a consideration of the efficiency of a resource. Any taxes added to production costs are similarly ignored as transfers.

As in the cost modules for other pollutants, a central assumption in the GAINS ammonia module is the existence of a free market for abatement equipment across Europe that is accessible to all countries at the same conditions. Thus, the capital investments for a certain technology can be specified as being independent of the country. Likewise, certain elements of operating costs will principally be identical for all countries, here subsumed as either 'fixed operating costs' or some control measure specific characteristics included in 'variable operating costs'. Simultaneously, the calculation method takes into account several country-specific parameters that characterize the situation in a given country or region in order to assess the 'variable operating costs', for instance, labor, energy, water, disposal costs, etc.

Thus, the expenditures on emission controls are differentiated into three categories, although for some technologies not all categories are relevant:

- investments,
- fixed operating costs (costs of maintenance, insurance, administrative overhead), and
- variable operating costs (e.g., energy, water, labor costs, feed and fertilizer price, costs of waste disposal, etc.).

Considering the above, costs per unit of activity, i.e., number of life animals, or tons of fertilizer use, are calculated. Furthermore, taking into account the abatement efficiency of the specific measure, unit costs per unit of removed pollutant (NH_3) can be estimated.

The following sections introduce the cost calculation principles used in GAINS and explain the construction of the cost curves that can be further used in the optimization module of the GAINS model. To illustrate the methodology, examples of cost calculations are given. Cost parameters derive from actual available data on costs for known circumstances. Using the generalization principles of the equations below, values of all parameters used to calculate country-specific costs and the national cost curves have been derived, which have been presented in detail by Klimont and Winiwarter (2011) for all European countries. Moreover, they are also accessible from the on-line implementation of the GAINS model (http://gains.iiasa. ac.at). The elements of the method are in general compatible with other modules of the model as described, e.g., by Klimont et al. (2002).

9.3.2 Investment Cost

Investments cover the expenditure accumulated until the start-up of an abatement technology. These costs include, e.g., delivery of the installation, construction, civil works, ducting, engineering and consulting, license fees, land requirement (purchase) and capital. The GAINS model uses investment functions where these cost components are aggregated into one term.

The investment costs for individual control measures are defined as a function of the size of an installation. In its generic form, the total investment costs T consist of a constant and a size-dependent part, the latter typically characterized by the average farm size ss expressed as the average number of animal places on a farm for a specific livestock category. This linear approach may be transformed to express a specific investment I per animal place. The form of either of these functions is described by its fixed and variable coefficients, ci^f and ci^v.

$$T_{i,j,k} = ss_{i,l} \cdot ci^f_{i,k} + ci^v_{i,k} \tag{9.3a}$$

$$I_{i,k,l} = ci^f_{i,k} + \frac{ci^v_{i,k}}{ss_{i,l}} \tag{9.3b}$$

where

ci^f, ci^v investment function coefficients (Klimont and Winiwarter 2011; Table A1)

ss average farm size (Klimont and Winiwarter 2011; Table A2)

i,k,l livestock category, abatement technique, country

Note that the "average farm size" relates only to the larger farms in a country and excludes very small (subsistence or hobby) farms from the analysis, for which measures are not considered practical or would be associated with excessive costs.

A slightly different function has been developed to estimate investment costs for storage options, as typically costs depend on the volume of manure stored *ManVol* rather than on the number of animal places. Conversion between these parameters may be performed using the national specific parameters determining the livestock and country specific average size of the store, which in GAINS is calculated considering typical storage time, annual manure production, and the number of production cycles.

$$T_{i,k,l} = ManVol_{i,l} \cdot ci^f_{i,k}{}^* + ci^v_{i,k} \tag{9.4a}$$

Conversion of this equation may be performed via

$$ManVol = ss \cdot \frac{st}{12} \cdot mp \cdot ar \tag{9.4b}$$

With the parameter 12 (number of months per year) factored into the coefficient ci^f, this conversion yields the investments per animal place:

$$I_{i,k,l} = ci^f_{i,k} \cdot st_{i,l} \cdot mp_{i,l} \cdot ar_{i,l} + \frac{ci^v_{i,k}}{ss_{i,l}} \tag{9.4c}$$

where

st storage time (Klimont and Winiwarter 2011; Table A4)
mp manure 'production' of a single animal per year (Klimont and Winiwarter 2011; Table A3)
ar production cycles per year (Klimont and Winiwarter 2011; Table A5)

Costs calculated this way refer to the total manure produced, both inside housing and during grazing. While manure excreted during grazing would not need to be collected in stores, which would reduce the requirement for retrofitting capacity (and costs), dimensioning of such installations has to be done for the period it is used full time. Thus GAINS cost calculations assume capacities as of full-time use of storage.

The number of production cycles per year ar allows converting between the number of animals produced, as typically presented in production statistics, and the number of animal places which are strong determinants in costs of measures. Manure production mp is given for a single animal, e.g. for the lifetime of a pig that is fattened over a 4-month-period, but yearly for longer-living animals like dairy cows.

Coefficients ci^f, ci^v are derived from actual cost data (see Klaassen 1991b; Klimont and Winiwarter 2011, see Table 9.3 for examples) as a result of a regression calculation performed on the linearized expression (Eqs. 9.3a and 9.4a,

Table 9.3 Examples of cost data for manure storage options (high efficiency measures)

Capacity [m^3 manure]	Total investment [€ 2005]	Comment
29	3,591	Own estimates from Dutch data (Klaassen 1991b)
76	5,130	Own estimates from Dutch data (Klaassen 1991b)
159	6,669	Own estimates from Dutch data (Klaassen 1991b)
900	23,498	UK: rigid lids (pers. comm. D. Cowell)
340	2,985	Swiss (wood)
340	5,364	Swiss (tent)
233	7,934	Dutch data – originals (see Klaassen 1991b)
475	9,744	Dutch data – originals (see Klaassen 1991b)
950	13,217	Dutch data – originals (see Klaassen 1991b)
1,425	15,763	Dutch data – originals (see Klaassen 1991b)
1,900	43,639	Denmark (concrete roof)
1,900	19,795	Denmark (PVC tent)

respectively). They represent the costs for a cover (lid), assuming an existing manure tank. Figure 9.1 presents this regression calculation as performed for high efficiency measures in manure storage. The inversion into size-specific costs (here by manure storage capacity) is shown in Fig. 9.2, both for the sample points and the regression. Both figures indicate the considerable scatter of available cost data, and their representation in the cost function.

The general costs per amount of manure produced can be translated into a typical example of investment functions depending on specific parameters. Average farm size (expressed as number of animals per farm) and the time manure typically remains in storage in a specific country are strong determinants to costs. The example of pig manure (Fig. 9.3) applies the equations presented above to calculate costs vs. farm size, for two different values of storage time (all other parameters constant). The influence of storage time on the size of storage tank needed can be visualized this way as a function of the tank-size dependent investment costs.

A comparison of the results derived from the GAINS equations with costs data collected for the UK (Ryan 2004) shows, on the one hand, the large scatter of available cost data, but also indicates that the results derived from GAINS are fully compatible with that range (Fig. 9.4).

The investment costs are annualized over the technical lifetime of the installation lt by using the interest rate q (as %/100); GAINS allows for using different interest rates although for all the calculations performed within the Gothenburg Protocol, NEC, and CAFE related work an agreed social interest rate of 4 % was used:

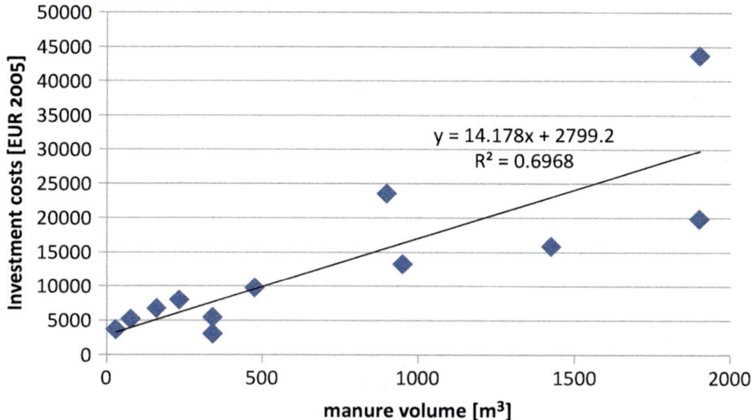

Fig. 9.1 Regression function to derive cost coefficients (costs expressed in € 2005) for high efficiency measures in manure storage

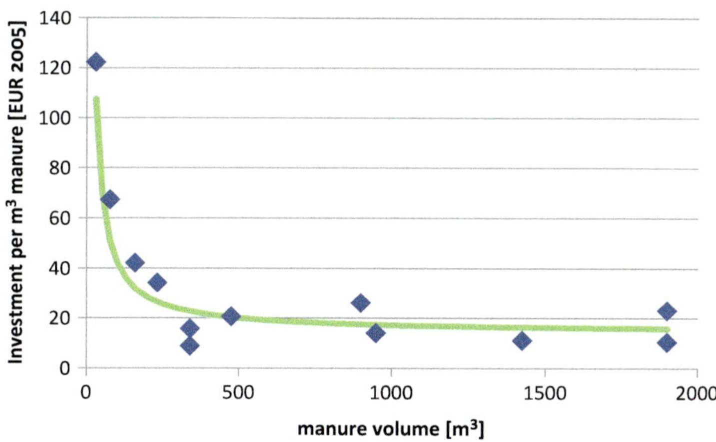

Fig. 9.2 Size-dependent investment costs for high efficiency measures to abate ammonia from manure storage. The inverted regression function (*line*) indicates large costs for small units (costs expressed in € 2005)

$$I_{i,k,l}^{an} = I_{i,k,l} \cdot \frac{(1+q)^{lt_k} \cdot q}{(1+q)^{lt_k} - 1} \tag{9.5}$$

where

i,k,l livestock category, abatement technique, country
lt lifetime of abatement technique (Klimont and Winiwarter 2011; Table A1)
q interest rate (e.g., $0.04 = 4\ \%$)

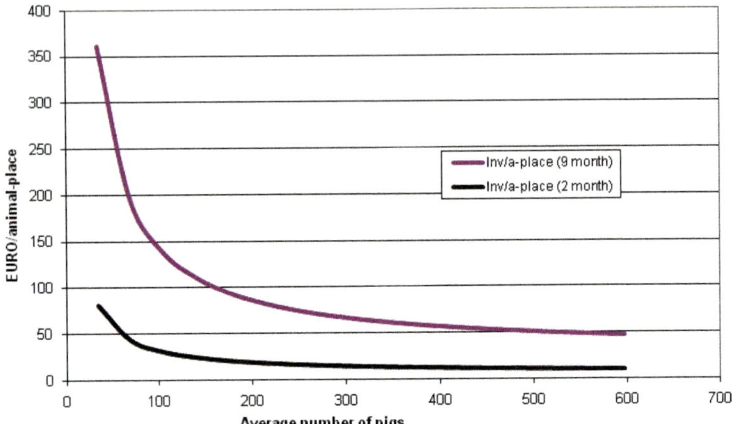

Fig. 9.3 GAINS investment functions for storage of pig manure (per animal place) for different storage capacity required (storage time)

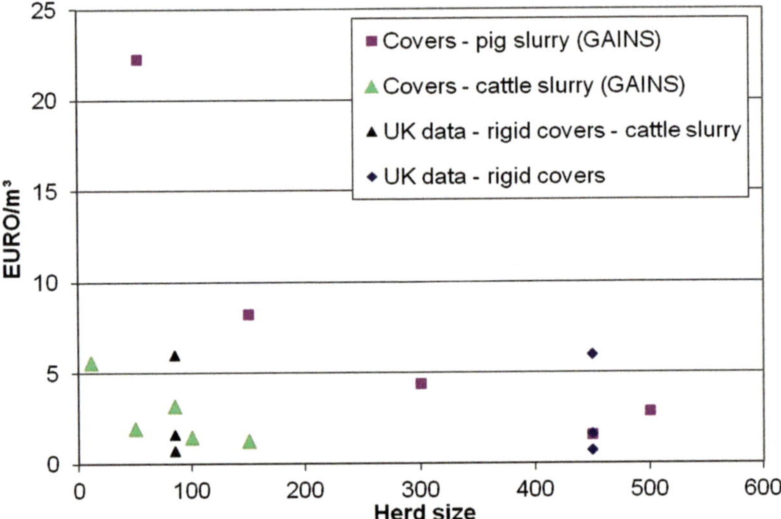

Fig. 9.4 Comparison of costs for storage covers between GAINS and UK data

All parameters used to derive investment costs are listed in the Annex of the underlying IIASA Interim Report (Klimont and Winiwarter 2011; Tables A1–A5) and are available from the on-line application of GAINS.

9.3.3 Operating Costs

The annual fixed expenditures OM^{fix} cover the costs of repairs, maintenance and administrative overhead per animal place. These cost items are not related to the actual use of the installation. As a rough estimate for annual fixed expenditures, a standard percentage fk of the total investments is used:

$$OM^{fix}_{i,k,l} = I_{i,k,l} \cdot fk_{i,k} \tag{9.6}$$

where

i,k,l livestock category, abatement technique, country
fk percentage of investment costs (Klimont and Winiwarter 2011; Table A7)

In turn, the variable operating costs OM^{var} are related to the actual operation of an installation and take into account additional costs incurred beyond the "no control" baseline situation (the reference technology) due to additional supplies needed. These supplies are given per animal produced and year:

- additional labor demand,
- increased energy demand for operating the device (e.g., for the fans and pumps), either as gas or electricity,
- animal feed,
- water, or
- waste disposal.

The variable operating costs are calculated from the quantity Q needed (demand) of certain extra supply p for a given control technology k, and from its (country-specific) price c.

$$OM^{var}_{i,k,l} = \sum_{p} Q_{i,k,p} c_{i,k,l,p} \tag{9.7}$$

where

p parameter type (additional energy, labour, waste disposal, etc.)
i,k,l livestock category, abatement technique, country
Q quantity of p (Klimont and Winiwarter 2011; Table A6)
c unit price of a given p (Klimont and Winiwarter 2011; Table A8)

While the equations above are used in GAINS generally, a somewhat adapted version is needed to derive the costs of low ammonia application options. In this adaptation, costs (per cubic meter of manure) are derived from constant parameters as a function of the manure application rate Q^{mh}. Cost parameters are specific for grassland and arable land, requiring separate treatment:

$$C^{mg}_{k',l} = ci^{fg}_{k'} - ci^{vg}_{k'} \cdot Q^{mh}_{k',l} \qquad (9.8a)$$

$$C^{ma}_{k',l} = ci^{fa}_{k'} - ci^{va}_{k'} \cdot Q^{mh}_{k',l} \qquad (9.8b)$$

where

k', l	abatement technique (low or high efficiency; applied to grassland or arable land), country
C^{mg}, C^{ma}	cost of option k' per m³; grassland, arable land
ci^{fg}, ci^{vg}	cost coefficients for a specific option k' used on grassland (Klimont and Winiwarter 2011; Table A9)
ci^{fa}, ci^{va}	cost coefficients for a specific option k' used on arable land (Klimont and Winiwarter 2011; Table A9)
Q^{mh}	manure application rate per hectare for option k' (Klimont and Winiwarter 2011; Table A8)

The total annual costs of the low ammonia application measures are calculated using the country-specific share of manure applied on grassland S^{mg}. At the same time we convert the costs to costs expressed per animal produced using country- and livestock category-specific manure production rates mp. Here only the indoor share needs to be considered, as low ammonia application only applies to manure collected during the housing period:

$$OM^{var}_{i,k,l} = \left(S^{mg}_{i,l} \cdot C^{mg}_{i,k,l} + (1 - S^{mg}_{i,l}) \cdot C^{ma}_{i,k,l}\right) \cdot mp_{i,l} \cdot \frac{Nx_{1,i,l}}{Nx_{1,i,l} + Nx_{4,i,l}}$$

$$(9.9a)$$

where

i,k,l	livestock category, abatement technique (low or high efficiency), country
S^{mg}	share of manure applied to grassland (the rest of manure is considered to be applied on agricultural land) (Klimont and Winiwarter 2011; Table A10)
mp	manure 'production' of a single animal per year (Klimont and Winiwarter 2011; Table A3)
$Nx_{1,4}$	N excretion during housing (1) and grazing (4), considered proportional to the respective manure production shares

All the individual parameters needed for performing all calculations are presented by Klimont and Winiwarter 2011 (see their Annex, Tables A6–A10). The fact that solid manure typically is not applied at grassland at all can be handled in appropriately choosing the S^{mg} parameter as zero.

Low ammonia application, in avoiding loss of ammonia to the atmosphere, furthermore supplies soils with nitrogen. Any ammonia nitrogen not emitted may be considered extra fertilizer that contributes to savings in mineral fertilizer application, which in turn can be calculated as cost reduction using fertilizer costs available as commodity costs:

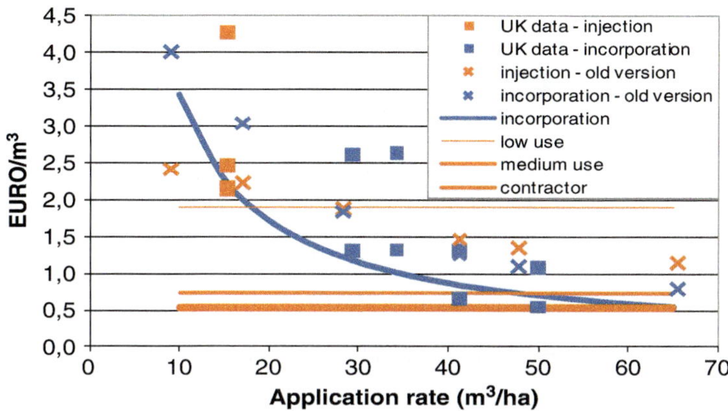

Fig. 9.5 Comparison of costs for slurry injection and incorporation of slurry and manures

$$Nsav_{i,k,l} = ef_{i,l,3} \cdot \eta_{i,k,l,3} \cdot c_{fert,l} \cdot {}^{14}\!/_{17} \tag{9.9b}$$

where

i,k,l	livestock category, abatement technique, country
$Nsav$	saved fertilizer costs
ef_3	unabated emission factor (as in Eq. (9.2))
η_3	removal efficiency (as in Eq. (9.1) for stage 3, application)
c_{fert}	fertilizer costs (Eq. (9.7): Klimont and Winiwarter 2011; Table A8)
14/17	stoichiometric factor (N content in ammonia)

An example for calculation of operating costs is presented in Fig. 9.5; size matters – in this case it is the application rate of manure that influences the costs of the abatement option. A comparison with the UK assessment (Ryan 2004) is presented in Fig. 9.5. The UK cost data, while having a considerable spread themselves, are within the same range as those calculated in GAINS.

9.3.4 Calculation of Unit Costs

Considering the cost elements discussed above, the unit costs ca of specific measures can be calculated. Unit costs in GAINS are expressed per activity unit, i.e., the annual average number of live animals, and, respectively, the mineral fertilizer use.

The unit costs ca are derived by adding[2] the annualized investment costs, the fixed operation costs and the variable operation costs times the intensity of their

---

[2] Depending on the actual technology, some of the cost components might be irrelevant.

application (number of production cycles) and considering savings in mineral fertilizer due to ammonia buried in soil during application. A conversion from number of animals produced and animal places to the average number of live animals at any given time is provided by the number of production cycles *ar* and capacity utilization factor *sb*:

$$ca_{i,k,l} = \frac{I^{an}_{i,k,l} + OM^{fix}_{i,k,l} + OM^{var}_{i,k,l} \cdot ar_{i,l}}{sb_{i,l}} - Nsav_{i,k,l} \qquad (9.10)$$

where

i,k,l	livestock category, abatement technique, country
ca	unit costs per live animal
ar	production cycles per year (Klimont and Winiwarter 2011; Table A5)
sb	capacity utilization factor (Klimont and Winiwarter 2011; Table A11)
Nsav	saved fertilizer costs

An alternate way of cost notation is to express costs per unit of abated emissions. In a multi-pollutant environment as in GAINS this notation is of limited value, but when comparing costs for abatement of a specific compound it may be very useful.

$$cn_{i,k,l} = \frac{ca_{i,k,l}}{ef_{i,l} \cdot \eta_{i,k,l}} \qquad (9.11)$$

where

η_k	removal efficiency of option k
$ef_{i,l}$	emission factor for livestock category i and country l, assuming no abatement is in place (unabated emission factor per live animal)

Data on production cycles and capacity utilization are presented by Klimont and Winiwarter 2011 (see their Annex Tables A5 and A11); emission factors and removal efficiencies are essential parameters of emission calculation and are available in the GAINS on-line application (http://gains.iiasa.ac.at).

9.4 Marginal Costs and Development of Cost Curves

Unit costs, as calculated in the previous section, do not necessarily provide information about the cost efficiency of measures. Information about cost efficiency is essential when discussing the future strategies, specifically considering their reduction potential and associated costs. Very often a concept of marginal cost curve is applied to serve such purpose. Here we explain how marginal costs and a cost curve are calculated in GAINS.

Costs as presented in previous section refer to a change in abatement relative to the base case, the no-control option that should be representative of the reference

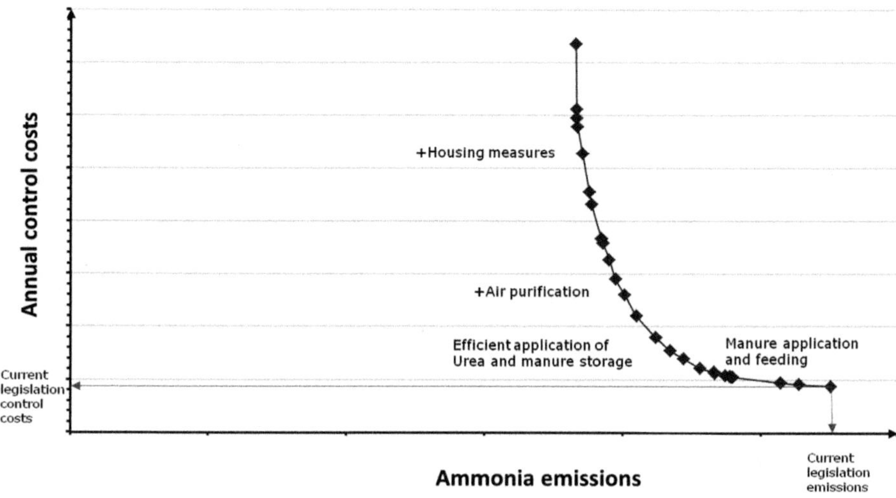

Fig. 9.6 Ammonia cost curve: example of a typical cost curve

technology in a given country. Marginal costs relate the extra costs for an additional measure to the extra abatement of that measure (compared to the abatement of the less effective option). GAINS uses the concept of marginal costs for ranking the available abatement options, according to their cost effectiveness, into the so-called "national cost curves" (see example of an idealized cost curve in Fig. 9.6).

If, for a given emission source (category), a number of control options are available, these options are sorted by their cost efficiency. The marginal costs mc for control option k are calculated by comparing to parameters of option k-1, which is the next less effective one:

$$mc_k = \frac{cn_k\eta_k - cn_{k-1}\eta_{k-1}}{\eta_k - \eta_{k-1}} \qquad (9.12)$$

where

cn_k cost efficiency for option k
η_k removal efficiency of option k

The marginal costs relate to incremental costs for additional emission reduction. Sorting emission reduction options by increasing marginal costs results in cost-optimal combination of measures for a given emission reduction target. In a first step, all available capacity of the cheapest option (least marginal cost) is taken; the next step applies to the second cheapest option and so forth. Multiplying, for each step, the available capacity with the emission savings per unit (removal efficiency × emission factor) yields the saved emissions, available capacity times marginal costs is the total annual costs. A cost curve can be constructed by stepwise subtracting the respective emission savings from the total emissions before abatement, and by adding the costs of each of the options taken. A more detailed discussion of cost curves has been presented by Klimont et al. (2002).

A cost curve indicates the potential for further abatement and allows for estimation of associated costs as well as indicates which options are necessary and cost-effective to achieve the required reduction. In the example presented in Fig. 9.6, the starting point is reflected by the highest emissions on the right hand side, i.e., before any of the further measures are taken into account. Cost curves start out with the most cost effective measures and often considerable emission reductions are achieved at relatively low costs. The actual shape of the curve will depend on actual situation in a given country, including already implemented measures and further abatement potential. For example, if all the cheap measures have been installed, the curve would be typically much steeper than that shown in Fig. 9.6, i.e., looking more like the left most quarter of the chart where the most expensive measures (in terms of marginal costs) cause a considerable increase of overall costs at fairly little reductions.

9.5 Implementation Limits (Applicability) of Measures

It is important to consider the constraints associated with application of control measures. These constraints may be of very different nature, including soil conditions (stoniness, slope), farm practices and size, local regulation, technical limitations, etc. Such constraints are referred often as applicability parameter and are included in GAINS as specific to country/region and livestock category. This way, GAINS recognizes that measures can only be applied to a certain extent (given as a percentage of the total activity) – further implementation will not be considered (Klimont and Winiwarter 2011; Table A12). A realistic assessment of these constraints is essential to provide valid information about the total reduction potential as they will determine how far the cost curve extends, i.e. to which degree emissions can be abated.

Implementation of ammonia abatement measures depends, in practice, on the size of farms. GAINS accounts for that by using the average farm sizes as a parameter driving the costs of certain measures (specifically in housing and in storage of manure). In some countries where a large number of subsistence farmers, often coupled with extra income on a paid job ("hobby farm"), represents significant share of farms, country average numbers remained small and in consequence cost estimated in GAINS grew very high. This procedure distorted cost estimates even where measures would have been possible, in large farms that may still cover a sizable fraction of the animal population in the respective country.

In order to better match the actual situation, and also request measures to be implemented where they are most useful, we now consistently and for all countries exclude farms smaller than 15 LSU from the evaluation of farm sizes. This threshold is very practicable as provided in the available statistics. It virtually has no effect for countries that are dominated by "industrial type" farms (Netherlands,

Denmark, Czech Republic), but will lead to a significant change in countries that have a sizeable share of "hobby farmers" and subsistence farmers, as Poland, Romania or Bulgaria. Exclusion of small farms <15 LSU in determining the average farm size of these countries will lead to a much more realistic cost estimate of measures that actually can be introduced at larger farms. In a similar way, the UNECE Task Force on Reactive Nitrogen (TFRN) suggests to exempt small farms from recommended measures (ECE/EB.AIR/WG.5/2011/3; Jan 28, 2011).

Estimating costs on the basis of only large and medium sized farms in consequence means that no measures would occur on small farms. In the past, we have used the results of consultations with country experts and results of specific questionnaires to derive applicabilities of agricultural measures (distinguishing between liquid and solid manure systems). Here, we extend and update the applicabilities according to the following rules:

- The percentage of animals living (by GAINS animal category) on farms larger than 15 LSU (Eurostat database, see discussion below) was determined.
- This percentage was multiplied to the percentage of applicability already implemented, assuming that applicabilities from previous assessments (e.g., due to climatic, topographical or geological conditions in a country) need to be applied uniformly to each size of farms.
- Specific consideration was given to animal categories for which liquid and solid systems are calculated separately in GAINS. Using the respective shares, we assumed a separation strictly by farm size, such that the largest farm on solid system is still a little bit smaller than the smallest farm on liquid system – obviously the boundary between these types to be very different by country, as determined from the respective fractions implemented in GAINS. As a consequence, applicability of measures was extended to solid systems only if fully applied to liquid systems already.

$$A^*_{i,l} = A_{i,l} \cdot share_{i,l} \tag{9.13}$$

where

A^* Applicability excluding small farms
A Applicability (limitations according to other parameters than farm size)
i GAINS animal category
l country

$$share_{i,l} = nlarge_{i,l} / n_{i,l} \tag{9.14}$$

where

$nlarge$ number of animals on large farms
n total number of animals

with an exemption for separation into solid and liquid systems for liquid manure systems:

$$\left.\begin{array}{l} share_{i,l} = nlarge_{i^*,l}/n_{i^*,l} \\ share_{i,l} = 1 \end{array}\right\}\left\{\begin{array}{c} if\ \ share < Fl_{i^*,l} \\ else \end{array}\right. \qquad (9.14a)$$

for solid manure systems:

$$\left.\begin{array}{l} share_{i,l} = 0 \\ share_{i,l} = \dfrac{nlarge_{i^*,l}/n_{i^*,l} - Fl_{i^*,l}}{1 - Fl_{i^*,l}} \end{array}\right\}\left\{\begin{array}{c} if\ \ share\ for\ liquid\ \ system < 1 \\ else \end{array}\right. \qquad (9.14b)$$

where

i^* GAINS animal category, but not differentiating manure systems ("dairy cattle", "other cattle", "pigs")

Fl GAINS fraction of animals on liquid systems

It can be argued that feeding measures (LNF) and manure application by contractors (LNA) may likewise be applied on small farms (and at a cost comparable to those of large farms). As, on the one hand, this may require excess training of very small farmers and control of implementation, and on the other hand it may be argued that in practice the number of those small farms will strongly decrease in the future, we decided to keep the same thresholds (15 LSU per farm) also for these options as an applicability limit, unless a specific information for a particular country was made available by the national experts.

Statistical data to be used for this task are available from EUROSTAT data explorer   http://epp.eurostat.ec.europa.eu/portal/page/portal/agriculture/data/data base. The details required are found in table: *ef_ls_ovlsureg* – "Livestock: Number of farms and heads by livestock units (LSU) of farm and region". Note that farm sizes given in LSU comprise of all animals on this farm, not only the GAINS animal type to be used. We argue that, for the purpose of this exercise, overall manure production is needed for cost estimation of storage capacity, independent of how many animal categories there are in a farm.

Dividing animal numbers for each country/LSU-size class by the respective number of holdings allows deriving average animal numbers per holding for each class. Again dividing these average animal numbers by the respective utilization rate *sb* (Klimont and Winiwarter 2011; Table A11) yields the farm size *ss* in units of animal places. Calculating the weighted average (by animal number) of the classes larger than 15 LSU allows to assess the farm size an average animal is staying at, to be used as the "farm size" for the respective country. The resulting farm sizes (as animal places) are displayed by Klimont and Winiwarter (2011; Table A2). The number of animals in all classes larger than 15 LSU divided by total number of animals allows obtaining the shares of animal on large and medium sized farms, needed to identify applicability of measures.

9.6   Integration of TFRN Cost Data

Estimating cost coefficient in the GAINS model draws on the available data from real life applications. The TFRN collected and reviewed new cost data on ammonia abatement measures and we have integrated these in GAINS. A brief summary and comparison with the current GAINS datasets is provided below.

9.6.1   Costs of Low Nitrogen Feed

Following van Vuuren et al. (2015, this volume), the variability of feed costs basically depends on market fluctuations rather than change of local conditions. Prices of soybeans as alternative (low nitrogen) feed may be more expensive or even cheaper than conventional feeding. Average costs, according to these authors, are estimated at 0.5 €/kg NH_3-N abated (for the most ambitious and thus most expensive reduction target of 15 %, the same reduction efficiency is used in GAINS), excluding grazing animals. As phase feeding operations may be considered in place already for the farm sizes considered, GAINS will not require investment costs for this option; additional feed costs of 2 (cattle and pigs), 5 (poultry) and 8 (laying hens) €-cents per 100 kg feed, much lower than current GAINS implementation, explain the cost range in terms of ammonia abated for most countries.

9.6.2   Costs for Animal Housing

Most recent information on costs of low emission animal housing (Montalvo et al. 2015, this volume) support the assumptions currently used in GAINS.

Relevant for housing emission, however, are also chemical scrubbers cleaning exhaust air[3] (this measure is referred in GAINS as "BF" – previously referring to biofiltration). Scrubbers will not produce waste (thus amount of waste to be disposed can be set zero), and fixed investment costs are lower than assumed for biofilters. With costs of 30, 3 and 1.5 € per animal place for pigs, layers and other poultry (about half the previous GAINS values, all other parameters unchanged) we arrive at ammonia abatement costs for most countries near 10 €/kg NH_3-N as suggested by Montalvo et al. (2015, this volume).

---

[3] These scrubbers remove also particulate matter (PM) emissions.

9.6.3 Costs for Storage

GAINS implementation of costs for storage was not altered, as the expert discussions at the TFRN workshop (see a comprehensive overview prepared by VanderZaag et al. 2015, this volume) seemed to indicate general agreement in costs per m^3 storage capacity and year (<1 €/m^3 for low efficiency measures, ~40 % reduction; <5 €/m^3 for high efficiency measures, ~80 % reduction) as well as costs related to ammonia abatement (up to 2 and up to 4 €/kg NH_3-N abated: costs presented for the savings in this stage only, not for the whole chain) can be in principle reproduced, even if large national variability also causes considerable scatter. In order to be conservative, GAINS lower end of range tends to coincide with the upper end of data presented by VanderZaag et al. rather than their lower end (where zero costs as in "natural crust" are not considered by GAINS). The GAINS costs refer to the lid construction only (for high efficiency options), i.e., do not include costs of building the tank.

9.6.4 Costs for Spreading

New evidence presented at the workshop and discussed subsequently (Webb et al. 2015, this volume) demonstrates cost differences depending on the utilization of equipment. For large farm sizes or for contractors performing the work, investment costs will play significantly lower role, to the effect that contractors would operate clearly cheaper. Assuming a cost optimized way being taken (as is the principle in an economic model such as GAINS) small or medium size farms would simply not choose the more costly option of buying their equipment and instead rely on contractor work. Costs assessed may depend on labor costs and other country parameters, but we use here an overall result of 0.52 €/m^3 manure spread. Solid manure will be added to arable land only (immediate incorporation) at only slightly different costs (0.70 €/m^3) – see Webb et al. (2015, this volume) for the details on how these factors were assessed. Notably, the costs are applicable to manure from housing only, i.e. time animals stay inside housing is considered while grazing period is not. This is different to storage/housing, as for these processes size of installations might be adapted to seasons when animals are indoors over extended periods. Relevant costs relative to abatement are below 1 €/kg NH_3-N, and are lower for the high efficiency options.

9.7 Results and Discussion

Following the update of the cost data and applicability constraints in GAINS, the new cost coefficients and reduction potentials were used in the scenarios prepared within the work supporting the May 2012 revision of the Gothenburg Protocol; the details of the model runs have been described by Amann et al. (2011b).

The scenarios developed for that report optimize costs with respect to a number of environmental targets, to which ammonia reduction is just one element. Ammonia, together with NO_x, contributes to eutrophication; these two compounds as well as SO_2 are responsible for acidification. Human health impacts due to atmospheric PM pollution has been associated with primary and secondary particulate matter, where precursors like NO_x, SO_2 and ammonia play an important role in the formation of the latter.

In consequence, changes in each of the emitted compounds will lead to spillover effects in terms of optimization results. If reducing ammonia emissions becomes cheaper, the same overall effect may be achieved (at lower total costs) by reducing ammonia rather than, for example, reducing NO_x. Consequently, applying in this analysis the recently estimated lower ammonia control costs, as discussed in Sect. 8.6, tend to increase the requirements to reduce NH_3, compared to previous simulations.

Country-specific optimization results for ammonia reductions in the MID scenario (Amann et al. 2011b) are shown in Fig. 9.7. The figure shows relative importance of reductions of ammonia emissions by source category and region.

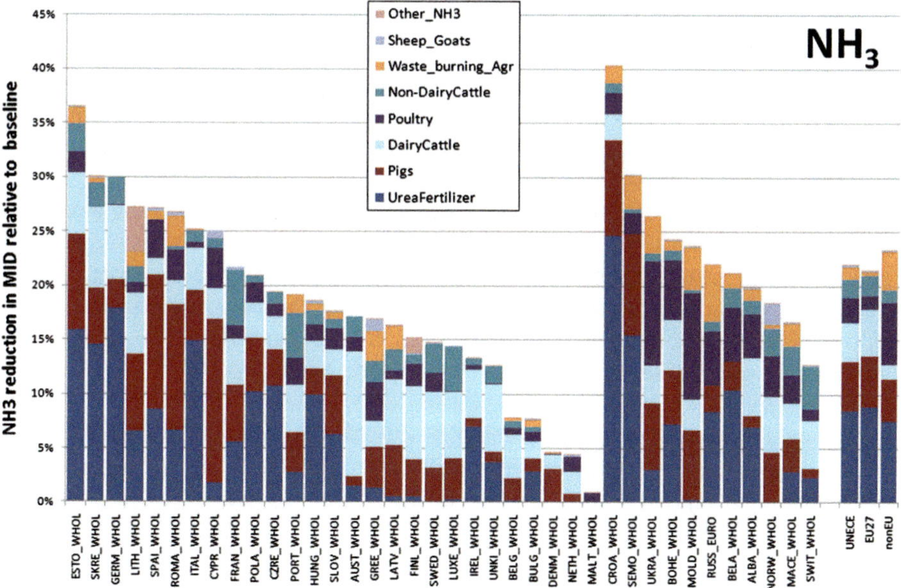

Fig. 9.7 Percentage of reduction according to the GAINS optimization (MID-scenario)

While nearly all categories are affected by reductions, there are important common elements in the proposed strategies. An important contribution originates from improved application or substitution of urea fertilizer. At the UNECE level this contributes nearly half of the required reduction clearly indicating cost-efficiency of this option. In countries where potential for urea substitution does not exist, typically a higher proportion of emissions is reduced in cattle sector. Nearly a quarter of the mitigation is expected in pig production sector and for a number of countries the cost-effective potential in this sector was estimated at more than 50 %.

Another interesting aspect of the discussed central scenario (MID) is to assess the emission stage providing the most significant reductions. Using the MID scenario from Amann et al. (2011b), and also determining the "distance to target" between baseline and the maximum feasible reduction (MFR) scenario, we present results in Table 9.4. The left part of the table shows the emission reductions (in per cent of baseline emissions) achieved due to the introduction of cost-effective measures. While large differences occur between countries (as shown in Fig. 9.7), reflecting national specificities like current agricultural practice, applicability of measures, extent of abatement needed, and potential of abatement for other compounds than ammonia, a common pattern remains that largest reductions seem to be achievable in the "application" stage. Exceptions to this pattern (Denmark) indicate a large degree of implementation in the baseline already, not allowing further measures to be taken. In addition to the emission reductions, Table 9.4 also provides information to which degree available reductions have been implemented in the "MID" scenario already – the right part of the table presents the reduction achieved in the MID scenario as a percentage of the reduction possible (difference between baseline and MFR). The high share of closure for application stage becomes evident, indicating that application of manure is recognized as the most effective as well as cost-effective measure in the optimization. Variation is significant between countries, and some caution needs to be taken when interpreting results. For example, cross-effects between stages will result in MFR application emissions becoming lower than those of the MID scenario, allowing for apparent reductions larger than 100 %.

Amann et al. (2011c) estimated the total additional[4] costs of reducing ammonia emissions in the EU-27 in the MID scenario at about 600 million €/year in 2020 (see also Fig. 9.8). As discussed earlier, the MID scenario represents a cost-effective solution where burden of reductions, and consequently penetration of additional measures, is unevenly distributed across countries. The TFRN has proposed an alternative set of strategies to reduce emissions in the UNECE; the principle was to develop different ambition level strategies that would ask for a more even distribution of efforts across the sectors and countries. Detailed discussion of the various ambition levels (typically referred to as A, B, C – A being the most ambitious) is provided in the draft Annex-IX for the Gothenburg Protocol and

---

[4] Additional costs over the baseline cost that was estimated at about 1.620 billion €/year in 2020 (Amann et al. 2011b).

Table 9.4 Implementation of measures of the central "MID" scenario by the individual emission stages of manure treatment

	% reduction				% closure to MFR			
	Housing	Storage	Application	Grazing	Housing	Storage	Application	Grazing
Austria	2 %	−3 %	43 %	2 %	6 %	−6 %	91 %	60 %
Belgium	4 %	3 %	13 %	4 %	41 %	14 %	94 %	100 %
Bulgaria	5 %	7 %	25 %	4 %	20 %	22 %	91 %	85 %
Cyprus	16 %	34 %	47 %	2 %	35 %	82 %	100 %	88 %
Czech Rep.	5 %	17 %	23 %	9 %	25 %	68 %	102 %	97 %
Denmark	4 %	18 %	3 %	3 %	30 %	70 %	108 %	100 %
Estonia	11 %	30 %	56 %	9 %	34 %	79 %	102 %	100 %
Finland	15 %	20 %	19 %	1 %	74 %	99 %	97 %	100 %
France	7 %	8 %	66 %	2 %	16 %	17 %	98 %	89 %
Germany	1 %	8 %	44 %	−1 %	8 %	25 %	114 %	−20 %
Greece	15 %	22 %	41 %	1 %	48 %	40 %	92 %	80 %
Hungary	9 %	12 %	28 %	7 %	19 %	33 %	97 %	100 %
Ireland	3 %	2 %	13 %	1 %	21 %	7 %	53 %	100 %
Italy	6 %	32 %	47 %	2 %	19 %	74 %	98 %	100 %
Latvia	9 %	18 %	32 %	7 %	44 %	78 %	90 %	96 %
Lithuania	28 %	29 %	46 %	10 %	83 %	62 %	103 %	100 %
Luxembourg	8 %	6 %	52 %	3 %	34 %	12 %	87 %	67 %
Malta	0 %	0 %	2 %	0 %	1 %	1 %	14 %	0 %
Netherlands	4 %	3 %	9 %	0 %	34 %	20 %	57 %	60 %
Poland	10 %	12 %	27 %	8 %	33 %	39 %	103 %	100 %
Portugal	1 %	1 %	49 %	0 %	3 %	3 %	90 %	37 %

(continued)

Table 9.4 (continued)

	% reduction				% closure to MFR			
	Housing	Storage	Application	Grazing	Housing	Storage	Application	Grazing
Romania	11 %	33 %	39 %	4 %	42 %	66 %	94 %	95 %
Slovakia	8 %	17 %	46 %	6 %	24 %	37 %	92 %	75 %
Slovenia	5 %	14 %	30 %	2 %	16 %	33 %	108 %	71 %
Spain	9 %	9 %	66 %	0 %	17 %	16 %	97 %	50 %
Sweden	6 %	20 %	45 %	1 %	22 %	47 %	109 %	67 %
UK	7 %	9 %	21 %	2 %	44 %	46 %	99 %	100 %

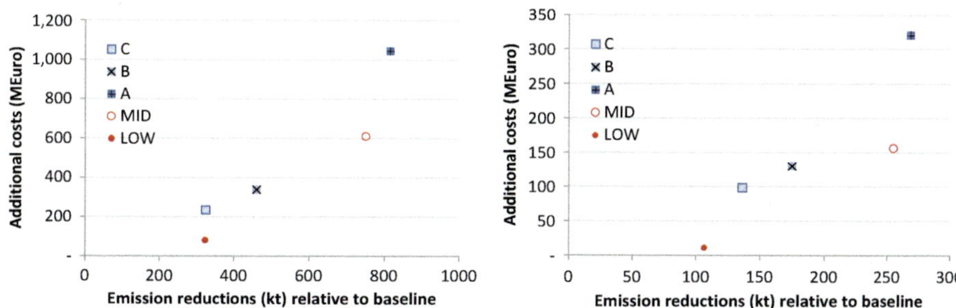

Fig. 9.8 Cost-effectiveness of the A, B, and C scenarios in comparison to the MID and LOW scenarios described in the Amann et al. (2011c). *Left*: EU27, *right*: non-EU countries

a summary is available in Wagner et al. (2012). We have implemented the A, B, and C strategies in the GAINS model and compared them to the LOW and MID optimized ambition scenarios. The discussion of the implementation and results are presented in Wagner et al. (2012) and summarized in Fig. 9.8.

In principle, the ambition C can be compared with the LOW and A with the MID scenario with respect to expected reduction of emissions, the cost vary significantly, i.e., the optimized scenarios are about half as expensive for similar reductions (Fig. 9.8). This difference is more apparent for the non-EU countries where there are less measures in the baseline and therefore larger scope for optimization. In spite of these differences, the analysis shows that an alternative way of strengthening reduction ambitions follows a similar pattern as the optimization approach. Since, the A, B, and C scenario approach assures that a Europe wide legislation could be introduced and therefore will be easier to agree upon and manage, the eventual agreement might look into the synergies that exist between the two sets analyzed in Wagner et al. (2012).

References

Amann M, Asman WAH, Bertok I, Cofala J, Heyes C, Klimont Z, Rafaj P, Schöpp W, Wagner F (2007) Cost-effective emission reductions to address the objectives of the thematic strategy on air pollution under different greenhouse gas constraints, NEC Scenario Analysis Report No. 5. IIASA, Laxenburg

Amann M, Bertok I, Cofala J, Heyes C, Klimont Z, Rafaj P, Schöpp W, Wagner F (2010) Scope for further environmental improvements in 2020 beyond the baseline projections. Background paper for the 47th session of the Working Group on strategies and review of the convention on long-range transboundary air pollution. CIAM report 1/2010, International Institute for Applied Systems Analysis (IIASA), Laxenburg

Amann M, Bertok I, Borken-Kleefeld J, Cofala J, Heyes C, Höglund-Isaksson L, Klimont Z, Nguyen B, Posch M, Rafaj P, Sandler R, Schöpp W, Wagner F, Winiwarter W (2011a) Cost-effective control of air quality and greenhouse gases in Europe: modeling and policy applications. Environ Model Software 26:1489–1501

Amann M, Bertok I, Borken-Kleefeld J, Cofala J, Heyes C, Höglund-Isaksson L, Klimont Z, Rafaj P, Schöpp W, Wagner F (2011b) Cost-effective emission reductions to improve air quality in Europe in 2020 – scenarios for the negotiations on the revision of the Gothenburg protocol under the convention on long-range transboundary air pollution. CIAM report 1/2011, International Institute for Applied Systems Analysis (IIASA), Laxenburg

Amann M, Bertok I, Borken-Kleefeld J, Cofala J, Heyes C, Höglund-Isaksson L, Klimont Z, Rafaj P, Schöpp W, Wagner F, Winiwarter W (2011c) An updated set of scenarios of cost-effective emission reductions for the revision of the Gothenburg protocol – background paper for the 49th session of the Working Group on strategies and review, Geneva, 12–15 Sept 2011. CIAM report No. 4/2011, Centre for Integrated Assessment Modelling (CIAM), International Institute for Applied Systems Analysis, Laxenburg

Asman WAH, Klimont Z, Winiwarter W (2011) A simplified model of nitrogen flows from manure management, IIASA Interim Report IR-11-030. IIASA, Laxenburg

Brink C, Kroeze C, Klimont Z (2001a) Ammonia abatement and its impact on emissions of nitrous oxide and methane – Part 1: method. Atmos Environ 35:6299–6312

Brink C, Kroeze C, Klimont Z (2001b) Ammonia abatement and its impact on emissions of nitrous oxide and methane – Part 2: application for Europe. Atmos Environ 35:6313–6325

Britz W, Witzke P (eds) (2008) CAPRI model documentation 2008: Version 2. University of Bonn, Germany. http://www.capri-model.org/docs/capri_documentation.pdf. Accessed 14 Aug 2014

Höglund Isaksson L, Winiwarter W, Tohka A (2009) Potentials and costs for mitigation of non-CO_2 greenhouse gases in annex 1 countries: version 2.0, IIASA Interim Report IR09-044. IIASA, Laxenburg

Klaassen G (1991a) Past and future emissions of ammonia in Europe, Status Report 9101 IIASA. IIASA, Laxenburg

Klaassen G (1991b) Costs of controlling ammonia emissions in Europe, Status Report 9102 IIASA. IIASA, Laxenburg

Klimont Z, Cofala J, Bertok I, Amann M, Heyes C, Gyarfas F (2002) Modeling particulate emissions in Europe. A framework to estimate reduction potential and control costs, IIASA Interim Report IR-02-076. IIASA, Laxenburg

Klimont Z, Brink C (2004) Modeling of emissions of air pollutants and greenhouse gases from agricultural sources in Europe, IIASA Interim Report IR-04-048. IIASA, Laxenburg

Klimont Z, Winiwarter W (2011) Integrated ammonia abatement – modelling of emission control potentials and costs in GAINS, IIASA Interim Report IR-11-027. IIASA, Laxenburg

Montalvo G, Pinero C, Herreroc M, Bigeriegod M, Prins W (2015) Chapter 4: Ammonia abatement by animal housing techniques. In: Reis S, Howard CM, Sutton MA (eds) Costs of ammonia abatement and the climate co-benefits. Springer, Dordrecht, pp 53–73

Reis S, Grennfelt P, Klimont Z, Amann M, ApSimon H, Hettelingh J-P, Holland M, LeGall A-C, Maas R, Posch M, Spranger T, Sutton MA, Williams M (2012) From acid rain to climate change. Science 338:1153–1154. doi:10.1126/science.1226514

Ryan M (2004) Ammonia emission abatement measures. Report and data set

Van Vuuren A, Pineiro C, van der Hoek KW, Oenema O (2015) Chapter 3: Economics of low nitrogen feeding strategies. In: Reis S, Howard CM, Sutton MA (eds) Costs of ammonia abatement and the climate co-benefits. Springer, Dordrecht, pp 35–51

VanderZaag A, Amon B, Bittman S, Kuczynski T (2015) Chapter 5: Ammonia abatement with manure storage and processing techniques. In: Reis S, Howard CM, Sutton MA (eds) Costs of ammonia abatement and the climate co-benefits. Springer, Dordrecht, pp 75–112

UNECE (1999) Protocol to abate acidification, eutrophication and ground-level ozone. UNECE Convention on long-range transboundary air pollution, Geneva. http://www.unece.org/env/lrtap/full%20text/1999%20Multi.E.Amended.2005.pdf. Accessed 18 Aug 2014

UNECE (2007) Guidance document on control techniques for preventing and abating emissions of ammonia. Paper ECE/EB.AIR/WG.5/2007/13, Geneva. http://www.unece.org/env/documents/2007/eb/wg5/WGSR40/ece.eb.air.wg.5.2007.13.e.pdf. Accessed 14 Aug 2014

UNECE (2012) Parties to UNECE air pollution convention approve new emission reduction commitments for main air pollutants by 2020. UNECE, Geneva. http://www.unece.org/index.php?id=29858

Wagner F, Winiwarter W, Klimont Z, Amann M, Sutton M (2012) Ammonia reductions and costs implied by the three ambition levels proposed in the Draft Annex IX to the Gothenburg Protocol, CIAM report 5/2011; version 1.2. IIASA, Laxenburg

Webb J, Morgan J, Pain B (2015) Chapter 6: Cost of ammonia emission abatement from manure spreading and fertilizer application. In: Reis S, Howard CM, Sutton MA (eds) Costs of ammonia abatement and the climate co-benefits. Springer, Dordrecht, pp 113–135

Chapter 10
Costs of Ammonia Abatement: Summary, Conclusions and Policy Context

Clare Howard, Mark A. Sutton, Oene Oenema, and Shabtai Bittman

Abstract This chapter summarises the information in the preceding chapters, which builds on the outcomes of an Expert Workshop held by the UNECE Task Force on Reactive Nitrogen in October 2010 in Paris, France, which examined the state-of-the-art regarding abatement measures for ammonia in agriculture. Cost information is provided by farm activity and abatement measure, including a discussion on integrated nitrogen management at the farm scale. The chapter also reports the conclusions of the Expert Workshop, noting the finding that in many cases the costs for the abatement techniques were cheaper than previously estimated. Wider policy contexts of the information are explored, including identifying priority measures for ammonia abatement and links to Annex IX of the Gothenburg Protocol.

Keywords Ammonia emissions • Emission abatement • International policy • UNECE • Convention on Long-range Transboundary Air Pollution

C. Howard (✉)
NERC Centre for Ecology & Hydrology (CEH), Bush Estate, Penicuik EH26 0QB, UK

School of GeoSciences, University of Edinburgh, Grant Institute, Kings Buildings Campus, West Mains Road, Edinburgh EH9 3JW, UK
e-mail: cbritt@ceh.ac.uk

M.A. Sutton
NERC Centre for Ecology & Hydrology (CEH), Bush Estate, Penicuik EH26 0QB, UK
e-mail: ms@ceh.ac.uk

O. Oenema
Environmental Sciences Group, Alterra, Wageningen University, P.O. Box 47, NL-6700 AA Wageningen, The Netherlands
e-mail: oene.oenema@wur.nl

S. Bittman
Agriculture and Agri-Food Canada, 6947 Highway 7, PO Box 1000, Agassiz, BC V0M 1A0, Canada
e-mail: shabtai.bittman@agr.gc.ca

© Springer Science+Business Media Dordrecht 2015

S. Reis et al. (eds.), *Costs of Ammonia Abatement and the Climate Co-Benefits*,
DOI 10.1007/978-94-017-9722-1_10

10.1 Introduction

Emissions of ammonia from agriculture lead to pollution at local, regional and transboundary scales. This is due to direct pollution from ammonia, which is transported and re-deposited (with local effects on biodiversity and plant health), but also through its role in the formation of particulate matter and in the wider context of the 'nitrogen cascade' (see Fig. 10.1). Once reactive nitrogen such as ammonia is released, it continues to be recycled into other forms of reactive nitrogen, 'cascading' through the environment, impacting on air, water and soil quality, contributing to particulate matter and nitrate production, impacts such as acidification and eutrophication, as well as interacting with the greenhouse gas balance through nitrous oxide (N_2O). As noted in Chap. 1, due to success in abating emissions of non-NH_3 pollutants, such as SO_2 and NO_x it is estimated that NH_3 will be the largest single contributor to each of acidification, eutrophication and secondary particulate matter formation in Europe by 2020. The impacts of ammonia emissions, such as on health and ecosystems remediation, all have a cost. Therefore policy discussions should be informed by the costs of technical options wherever possible. This volume seeks to address this need and the present chapter is a summary of that activity and its interaction with policy development in the United Nations Economic Commission for Europe (UNECE).

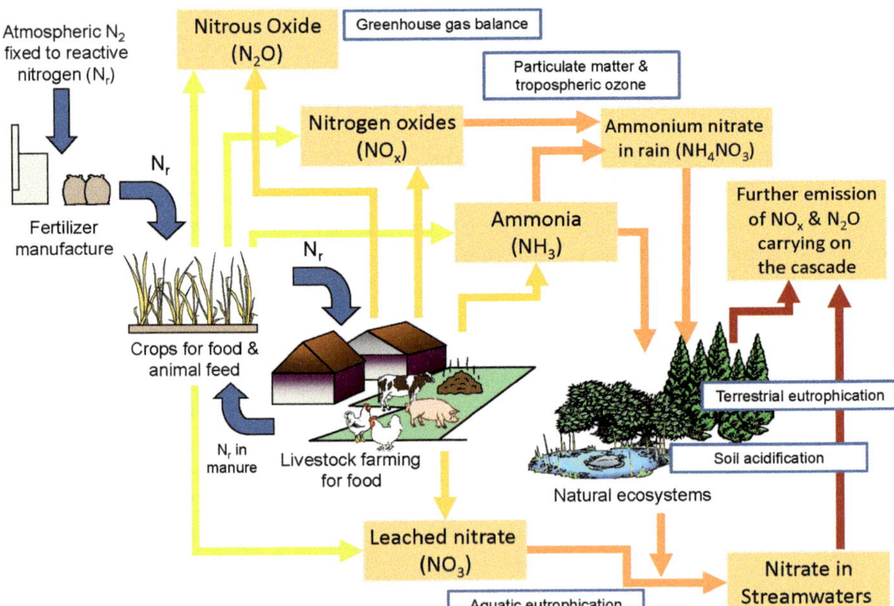

Fig. 10.1 Simplified view of the nitrogen cascade, indicating flows and impacts from the use of fertilisers and manures (Sutton et al. 2011)

The information in the preceding chapters builds on the outcomes of an Expert Workshop held by the UNECE Task Force on Reactive Nitrogen (see Box 10.1, TFRN) in October 2010 in Paris, France, which examined the state-of-the-art regarding abatement measures for ammonia in agriculture, their associated costs, co-benefits and implications for greenhouse gas emissions. Further development of the material was undertaken by the experts after the workshop and in later meetings of the Task Force on Reactive Nitrogen and its Expert Panels. As such this volume represents the culmination of these efforts to address the costs and co-benefits of measures to mitigate ammonia emissions from agriculture. It addresses all factors and sources of nitrogen in the farm system, from feeding and housing, through to manure and slurry storage, manure (including slurry) and fertiliser application (see Chap. 1, Fig. 1.3) as well as considering nitrogen management at the farm scale. Previous chapters have provided the detailed methodologies and cost calculations for these agricultural activities, along with national case studies.

This chapter summarises the costs information currently available (Sect. 10.2), reports the conclusions of the Expert Workshop (Sect. 10.3) and explores the wider policy context of the information presented. Finally priority measures are identified for ammonia abatement (Sect. 10.4) in relation to recent discussions within the Convention to revise Annex IX of the Gothenburg Protocol (Sect. 10.5). Although those discussions did not yet lead to a revised version of Annex IX (the 2012 negotiations foundered on what became an unresolvable difference in level of ambition regarding Annex IX), the discussion represents a starting point for future renewed negotiations of Annex IX within the Convention.

Box 10.1: Abating Ammonia, the Policy Context and the Role of the Task Force on Reactive Nitrogen (TFRN)

The 1999 'Gothenburg Protocol to Abate Acidification, Eutrophication and Ground-level Ozone' is established under the auspices of the Convention on Long Range Transboundary Air Pollution (CLRTAP) of the United Nations Economic Commission for Europe (UNECE), (UNECE 2014a). Annex IX of the Gothenburg Protocol lists 'Measures for the control of emissions of ammonia from agricultural sources' (UNECE 2013a, b).

The Executive Body of the LRTAP Convention (EB) 'is the meeting of the representatives of the Parties to the Convention. It is responsible for taking action to implement the fundamental principles of the Convention, reviewing the implementation of the Convention and setting up subsidiary bodies to carry out the work on implementation and development' (UNECE 2014b). The Working Group on Strategies and Review (WGSR) 'is the principal negotiating body for the Convention. It assists the Executive Body in policy-oriented matters including:

(a) Assessing scientific and technical activities relating to the preparation and revision of protocols;

(continued)

Box 10.1 (continued)

(b) Negotiating revisions to existing protocols and the preparation of new ones;
(c) Promoting the exchange of technology;
(d) Preparing proposals for any strategic development under the Convention' (see UNECE 2014c).

The Task Force on Reactive Nitrogen reports to WGSR and 'has the long-term goal of developing technical and scientific information, and options which can be used for strategy development across the UNECE to encourage coordination of air pollution policies on nitrogen in the context of the nitrogen cycle and which may be used by other bodies outside the Convention in consideration of other control measures' (TFRN 2014). In recent years it has been responsible for delivering options for the revision of Annex IX, and work on supporting documents includes the now adopted, revised 'Guidance document on preventing and abating ammonia emissions from agricultural sources' (UNECE 2014; Bittman et al. 2014) and a currently ongoing revision of the 'UNECE Framework Code for Good Agricultural Practice for Reducing Ammonia' (UNECE 2001).

As part of its goal to develop a wider perspective on the nitrogen cycle, TFRN organized the European Nitrogen Assessment (Sutton et al. 2011) and has since developed Guidance on Nitrogen Budgets (UNECE 2013c), addressed nitrogen climate interactions (Erisman and Bleeker 2011) and analyzed the relationship between nitrogen pollution and European food choices (Westhoek et al. 2014).

10.2 Summary Cost Tables

Cost information by farm activity and abatement measure is provided below. The values represent those developed from the Expert Workshop and from subsequent discussions. They also form the basis of the recently adopted UNECE 'Guidance document on preventing and abating ammonia emissions from agricultural sources' (herein referred to as 'Ammonia Guidance Document', UNECE 2014; Bittman et al. 2014). All costs information is provided in relation to reference methods as set out in the 'Ammonia Guidance Document' (UNECE 2014). For example, for low-emission manure spreading the reference method for comparison is free broadcast surface spreading. To allow for clear comparison the cost estimates are expressed where possible in € per kg of ammonia-nitrogen abated (€ per kg NH_3-N abated). In previous chapters other cost related metrics have been mentioned (such as € per animal place per year or € per cubic metre of manure applied to land). Whilst these metrics are not mentioned here, it is noted that they are often useful both for farmers to compare technical options with practical application in the field, and for use with the Greenhouse Gas and Air Pollution Interactions and Synergies (GAINS) model (see also Chap. 9).

10.2.1 Integrated Nitrogen Management at the Farm Scale

The basis of farm-scale integrated nitrogen management is the assessment of nitrogen surplus ($N_{surplus}$) and Nitrogen Use Efficiency (NUE). These two parameters are critical in addressing overall performance, where the objective is to reduce nitrogen surplus, thereby reducing the risk of pollution losses, and at the same time to increase NUE, providing the basis for savings by famers (see Chap. 2 for further details). Improvement in these values (reduction of surplus and increase in efficiency) should therefore decrease ammonia emissions through the saving of losses at various stages in the system (housing, feeding, etc.). Similarly, on mixed livestock farms, between 10 and 40 % of the $N_{surplus}$ is related to NH_3 emissions.

Nitrogen management also aims to identify and where possible avoid 'pollution swapping', between different N compounds and/or environmental compartments. In principle an overall improvement in NUE at the farm scale is reflective of abatement synergies where several forms of nitrogen pollution are reduced simultaneously. To optimize the N management at a farm level, an N input-output balance is needed, where all N inputs (such as feed and fertiliser) and outputs (such as products) are accounted for. The estimated costs of establishing integrated N management at the farm level are given in Table 10.1, where greater costs are associated with a higher ambition level.

The savings involved (and also co-benefits) are:

- Decrease in fertiliser costs
- Potential increases in crop quality

In terms of calculating the costs, the following is taken into account (see Chap. 2 for further details):

- Advisory services
- Analytical tests, e.g.:

  - Soil
  - Crop
  - Feed
  - Manure

Table 10.1 Estimated costs of integrated N management at a farm level

Description	Cost
Undertaking a farm N budget/balance	€200–500 per farm per year[a]
Cost associated with decreasing $N_{surplus}$ and increasing NUE through optimizing the use of N	-€1.0–€1.0 per kg N saved[b]

[a]Note that this does not include the costs associated with education, promotion and start-up for such services

[b]Highly depends on farm type and the initial situation. Net gains are expected when initial $N_{surplus}$ is relatively high, net costs are expected when decreasing $N_{surplus}$ and increasing NUE is associated with increased risk of yield losses

The economic cost of possible investments in the techniques which such farm balances and following advisory services might recommend for use are not included in the cost listed here. These are instead included by farm activity (such as feeding, land application, storage), below, to avoid double accounting.

Further information on the potentially achievable NUE and $N_{surplus}$ on differing farms, can be found in Chap. 2 and also the 'Ammonia Guidance Document' (UNECE 2014). It should also be noted that NUE should be managed with consideration of other nutrient efficiencies and other factors such as pest control.

10.2.2 Livestock Feeding Strategies

The economics of low nitrogen livestock feeding strategies was discussed in Chap. 3, where measures can lead to reductions in NH_3 emissions along the full chain, i.e. from excretion in housing, through storage to application on land. These strategies are more difficult to apply to grazing animals. However NH_3 emissions from pastures are much smaller than for housed animals (when considering the full manure chain), so that increased grazing period provides a means of reducing NH_3 emissions.

Several livestock feeding strategies can be implemented: phase or group feeding, low-protein feeding, supplementing by-pass protein (cattle) or amino acids (mono-gastrics), increasing the non-starch polysaccharide content of the feed and adding pH lowering substances (such as benzoic acid). However, calculating the specific costs per unit of N abated is complex as it depends on the initial animal feed composition and on the prices of the feed ingredients on the market (which can vary year to year). As well as cost savings, other potential benefits of not overfeeding protein are improved animal health and performance. Estimates and case studies with further detail for pigs, poultry and dairy cattle are provided in Chap. 3, but the overall economic cost of improved feeding strategies (as agreed on at the Expert Workshop) is roughly estimated at -0.5 to 0.5 € per kg NH_3-N saved, meaning that there are potential net gains (i.e. animal performance) and costs which will vary from farm to farm and year to year. The general relationship is that the cost to implement measures increases with the ambition level of NH_3 for emission abatement, mainly because of the need to provide essential amino acids via supplementary feeding. The need for proper management skills also increases with an increase in the ambition level of NH_3 emission abatement.

10.2.3 Livestock Housing

An overview of possible emission reduction from livestock housing and the estimated economic costs for major animal categories is given in Table 10.2. Note that more measures are available for new or largely rebuilt housing than for existing houses. The reference situation in each case is the most conventional housing

Table 10.2 Estimated costs for ammonia emission reduction techniques for animal housing, with their associated emission reduction levels[a]

Category	Emission reduction compared with the reference (%)[a]	Extra cost (€/kg NH$_3$-N reduced)
Existing pig and poultry housing on farms with > 2,000 fattening pigs or > 750 sows or > 40,000 poultry	20	0–3
New or largely rebuilt cattle housing	0–70	1–20
New or largely rebuilt pig housing	20–90	1–20
New and largely rebuilt broiler housing	20–90	1–15
New and largely rebuilt layer housing	20–90	1–9
New and largely rebuilt animal housing on farms for animals other than those already listed in this table	0–90	1–20

[a]Further information on reference situations can be found in UNECE (2014) (within Chapter 5 of the Ammonia Guidance Document, specifically paragraphs 63, 73, 78–81, 98–100, 108 & 115–116)

system, without any abatement techniques applied. Costs of abatement techniques in animal housing relate to the following:

(a) Depreciation of investments related to emission reduction measures (which includes the construction capital and associated labour costs over the life of the installation);
(b) Return on investments;
(c) Additional energy costs;
(d) Additional operation and maintenance costs.

Table 10.2 shows a wide range of costs for NH$_3$ emission reduction from livestock housing. This reflects the variety of mitigation techniques available, as well as the size and age of different housing systems the differing needs in each production phase and especially the effectiveness of the measures. Measures range in complexity from limiting water leaks associated with livestock drinking water to installing filters on exhaust fans. Higher costs can similarly reflect higher levels of emissions abatement (i.e. in the case of air scrubbers). The potential benefits of abatement measures that relate to increasing animal health and performance are very difficult to quantify and are therefore not always included in cost estimates. The wide cost ranges indicate that there is substantial market potential for reducing costs, especially when the NH$_3$ emission reductions are incorporated into planned redevelopment of livestock housing facilities.

Information on the methodology for such calculations, and more details on measures for the pig and cattle sectors (in specific national contexts) can be found in Chaps. 4 and 8. Note that there is a minimum farm size for applying measures because it is considered to be costly and ineffective to apply certain measures to housing in very small animal house operations. The wide ranges in both emission reduction percentages and in cost per kg NH$_3$-N reduced reflect that the emission reduction techniques have to be implemented in farm-specific ways.

Table 10.3 Estimated costs of ammonia emission reduction techniques for manure and slurry storage and their emission reduction levels

Techniques	Emission reduction (%)	Cost (€ per m^3 per year)	Cost (€ per kg NH_3-N saved)
Tight lid	>80	2–4	1–2.5
Plastic cover	>60	1.5–3	0.5–1.3
Floating cover	>40	1.5–3[a]	0.3–5[a]

[a]Not including crust; crusts form naturally on some manures and have no cost, but are difficult to predict

10.2.4 Slurry and Solid Manure Storage

There are several main actions which can be taken to abate ammonia emissions from manure storage systems:

(a) Decreasing the surface area of the storage to limit the area and time of manure contact with the atmosphere (i.e. through covers, encouraging natural crusting, deepening storage);
(b) Lower pH and the ammonium concentration;
(c) Minimizing disturbances.

Although generally applicable to both manure and slurry storage, it should be noted that the actions are easier to implement in the case of slurry storages than dung (solid manure) storages. To reduce emissions from solid manures the main options are (a) covering, such as with plastic sheeting and (b) in the case of poultry manure, keeping it dry prior to land application in order to reduce uric acid hydrolysis and ammonia formation. Costs relating to storage covers are given above. The references are an uncovered slurry store without a natural crust and an uncovered solid manure heap (Table 10.3).

10.2.5 Land Application of Slurry and Solid Manures

A summary of costs for low NH_3 emission application methods for both slurry and solid manures are provided in Table 10.4, where the reference is the broadcast surface spreading of slurry and solid manure. In the cases where a range of costs is specified this reflects variation in the NH_4 content of the slurry or manure, where a higher NH_4 content translates to a lower abatement cost, and to the quantity of manure that is spread. In this context, there is a co-benefit of low-emission manure storage, as this leaves more NH_4 in manure, so that low emission manure spreading becomes more cost effective.

Given the varying economies of scale, the mean costs of these techniques are likely to be in the lower half of the ranges when the application work is done by

Table 10.4 Estimated costs of ammonia emission reduction techniques for manure application and their emission reduction levels

Manure type	Application techniques	Emission reduction (%)	Cost (€ per kg NH_3-N saved)
Slurry	Injection	>60	−0.5–1.5
	Shallow injection	>60	−0.5–1.5
	Band application with trailing shoe	>30	−0.5–1.5
	Band application with trailing hose	>30	−0.5–1.5
	Slurry dilution	>30	−0.5–1.0
	Application Timing Management Systems (ATMS)	>30	0.0–2.0
	Direct incorporation following surface application	>30	−0.5–2.0
Solid manure	Direct incorporation	>30	−0.5–2.0

contractors or implemented on large farms or with shared equipment. Similarly, costs may be reduced in some cases by using home built low emission spreading equipment (e.g. trailing hose).

The choice of low emission spreading method should be targeted to local context. For example, injection is not feasible in stony soils. Some techniques can be used in growing crops or no till fields (e.g. trailing hoses) while others can be used only on arable land before planting (e.g. direct incorporation).

The principles of low emission application are the following:

(a) Decreasing the surface area where emissions can take place, i.e., through band application, injection or incorporation;
(b) Decreasing the time that emissions can take place, i.e., through rapid incorporation of manure into the soil, immediate irrigation or rapid infiltration;
(c) Decreasing the source strength of the emitting surface, i.e., through lowering the pH (though adding sulphuric acid) and NH_4 concentration of the manure (through dilution).

The above principles are applicable to both solid manure and slurry application. However, the effectiveness of the techniques is higher for slurry than for manures. In the case of solid manure, the most feasible technique is rapid incorporation into the soil and immediate irrigation.

In terms of calculating the costs, the following is taken into account (see Chap. 6 for more details and case studies):

(a) Depreciation of the applicator;
(b) Return on investments;
(c) Added tractor costs and labour;
(d) Operation and maintenance.

In order to ensure that costs are at the lower end of the ranges indicated in Table 10.4, and therefore to ensure net cost saving, farmers should: (a) ensure high use of capital-intensive equipment (e.g. equipment sharing), and (b) ensure the slurries retain a high ammonium content (i.e. have previously been kept using low emission storage).

The co-benefits (not ranked) which can be seen from low NH_3 emission application techniques are:

(a) Uniformity of application;
(b) Consistency of crop response to manure;
(c) Decreased odour emissions;
(d) Decreased biodiversity loss;
(e) Increased palatability of herbage;
(f) Less risk of drift into sensitive areas (e.g. reduced risk of N and P pollution to water courses);
(g) Less visible to public.

Not all of these co-benefits have been included in the cost estimates, given the difficulty of valuing these factors. Further details on these aspects are discussed in Chap. 6.

10.2.6 Fertilizer Application to Land

The costs of NH_3 emission reduction techniques associated with fertilizer application can be seen in Table 10.5, for which the reference method is surface broadcast

Table 10.5 Estimated costs of ammonia emission reduction techniques for application of urea-based fertilisers and ammonium carbonate, ammonium sulphate and ammonium phosphate fertilizers and their associated emission reduction levels

Fertilizer type	Application techniques	Emission reduction (%)	Cost (€ per kg NH_3-N saved)
Urea and urea ammonium nitrate	Injection	>80	−0.5–1
	Urease inhibitors	>30	−0.5–2
	Injection or incorporation following surface application	>50	−0.5–2
	Surface spreading with irrigation	>40	−0.5–1
Ammonium sulphate and phosphate	Injection when applied to carbonate containing soils with high pH	>80	0–4
	Incorporation following surface application when applied to carbonate containing soils with high pH	>50	0–4
	Surface spreading with irrigation	>40	0–4
Ammonium carbonate	Prohibition of use as a mineral fertilizer (if replaced with injected urea or ammonium nitrate)	>90	−1–2

application of urea-based fertilizer. Similar measures apply to the reduction of emissions from other fertilizers which lose a significant amount of NH_3 following application, ammonium bicarbonate and if applied to high pH soils (pH > 6.5), including ammonium sulphate and –ammonium phosphate fertilizers.

Where ranges in the costs are shown, these relate to the farm size (i.e. economies of scale), soil conditions and climate (where relatively dry conditions lead to high emission reduction). The mean costs are estimated to be in the lower half of the range suggested when application is undertaken by contractors or when high emission fertilizers such as urea are substituted with low NH_3 emission fertilizers such as ammonium nitrate or calcium ammonium nitrate.

Abatement of ammonia emissions from fertiliser application methods reflects the following principles:

(a) Decreasing the surface area where the emissions can take place, i.e., through band application, injection and incorporation;
(b) Decreasing the time over which emissions can take place, i.e., through rapid incorporation of fertilizers into the soil or via irrigation;
(c) Decreasing the source strength of the emitting surface, for example through the use of urease inhibitors, or blending and the use of acidifying substances;
(d) A full prohibition on specific types of high emission fertilisers (such as ammonium (bi)carbonate).

Therefore the costs of techniques used to lower the overall emissions (further details of which can be found in Chap. 6) relate to the following:

(a) Depreciation costs, associated with additional investments related to the low-emission applicator;
(b) Return on investments;
(c) Use of heavier tractors and more labour time;
(d) Additional operation and maintenance costs.

The potential benefits of these techniques are the following:

(a) Decreased fertilizer costs;
(b) Decreased application costs where urea is applied in a combined seeding and fertilizing system;
(c) Decreased biodiversity and human risk associated with reduced emissions.

In order to ensure that costs are in the lower end of the ranges illustrated in Table 10.5, the method should be applied with best practice in both timing and placement. Net cost savings can be achieved with both large-scale approaches (e.g. urea injection in no-till row crops, as increasingly practiced in Canada) and small-holder farmers (e.g. urea deep placement by hand in wetland rice, as widely practiced in Bangladesh, see IFDC 2012; Sutton et al. 2013).

10.3 Key Messages from the Expert Workshop: Costs of Ammonia Abatement and the Climate Co-benefits

This section details other key messages emerging from the Expert Workshop in Paris. These messages have also been reported to the Working Group on Strategies and Review (WGSR) of the CLRTAP (see Box 10.1 and UNECE 2011a).

The Expert Workshop concluded that in many cases the costs for the abatement techniques were cheaper than previously estimated and communicated to the UNECE LRTAP Convention, for example from the GAINS model (Klimont and Winiwarter 2011). In some cases costs estimated decreased by more than a factor of five. It was concluded that the main reason for this was that increased experience in implementing measures now existed and a wider range of measures were now available. In many cases the costs were in the range of -1 to $+5$ € per kg of NH_3-N abated, which is a similar range to the cost of several options for nitrogen oxides abatement. In addition, by counting the value of nitrogen saved, good local practice with many measures offered negative costs (i.e. cost savings) of up to -1 € per kg NH_3-N abated.

As noted in the previous section it was estimated that costs of improving nitrogen use efficiency through "integrated farm nitrogen management" was in the range of -1.0 to 1.0 € per kg NH_3-N saved. Greater costs were associated with a higher ambition level.

It was reported that costs using "improved feeding" strategies (again as reported in the previous section) were approximately -0.5 to 0.5 € per kg NH_3-N saved. Higher ambition levels again related to higher costs.

In the case of abatement measures for animal housing, the highest cost measures were for retrofitting low emission technologies into existing buildings. However, if the focus of emissions abatement strategies were on the new (or largely rebuilt) buildings, then these costs could largely be avoided. Higher costs were estimated for techniques with greater abatement potential, for example air scrubbing technology were estimated to cost 2–10 € per kg NH_3-N saved, while lower costs were estimated for methods that involved modest design changes such as partially slatted floors (0–6 € per kg NH_3-N saved) for pigs.

In poultry houses, the cheapest abatement methods involved keeping the manure dry by ventilation and avoiding spillage of water, with estimated costs of 0–3 € per kg NH_3-N saved. Operating costs for scrubbers are greater for poultry than pig houses due to higher dust emissions for poultry, but conversely provide significant co-benefits in simultaneously reducing emissions (and agronomic losses) of phosphorus and other nutrients. It was noted that current technology does not allow acid scrubbing of exhaust air from naturally ventilated cattle houses.

The abatement techniques appeared more cost effective for manure storage than for livestock housing. For example costs of covering new slurry stores on large farms ranged from 0.5 to 4 € per kg NH_3-N saved (depending on technique and abatement effectiveness). For existing outside stores for slurry and solid manures (on large farms), the costs were lower, at 0.5–2 € per kg NH_3-N saved.

Costs for measures to reduce emissions from land application of slurry and solid manure were in the range of 0.1–5 € per kg NH_3-N saved, where the smallest costs were seen for immediate incorporation (where feasible, for example on bare arable land or on tilled fields). Farm size had the greatest effect on the cost estimates, as improved economies of scale are possible on larger farms. The same outcome could be achieved by equipment sharing between farms or the use of specialist contractors with access to low-emission spreading equipment. Using a cost optimised approach (with the GAINS model), it was estimated that for smaller farms that share equipment or use contractors low emission slurry application costs would be typically less than 1 € per kg NH_3-N saved.

In the case of low emission application of urea-based fertilizers the costs were estimated to be relatively small, in the order of 0–1.5 € per kg NH_3-N saved. There were also a range of techniques available including urease inhibitors, drilling into the soil, coated fertilisers and the choice to use ammonium nitrate based fertilisers. There is also potential through good practices and developing economies of scale (e.g. future price reductions for inhibitors) to deliver significant cost savings.

It was highlighted throughout that a reduction in nitrogen emissions as ammonia represents a saving of nitrogen within the farming system. As such these savings should be recognized at an equivalent economic value to that of the mineral nitrogen fertilizer (which equates to around 1 € per kg N, depending on current fertiliser prices).[1] Bearing this in mind, and the fact that many of the costs estimated (especially in the case of larger farms or when contractors are used for manure spreading), are between 0.5 and 2 € per kg NH_3-N saved, highlights that many measures can be of net financial benefit to farmers. This conclusion becomes even more clear once the other co-benefits of the suggested low emission techniques are recognized. These include reduced odour, more consistent fertilizer usage, reduced losses of other nutrients (notably phosphorus) to air and watercourses and the value of improved agronomic flexibility. Since the value of these co-benefits was not quantified here, the benefits of low emission measures should be considered as being conservative.

The co-benefits for climate management which can be achieved from ammonia abatement should not be overlooked. In the farms where nitrogen is managed more efficiently, less overall nitrogen inputs will be required – which will lead to less carbon dioxide loss related to the manufacture of synthetic fertilizer. However, more importantly in this setting is the potential to decrease overall emissions of nitrous oxide (N_2O) from agriculture. Whilst the use of low emission manure spreading techniques, can increase nitrous oxide emissions at the field-scale, that potential trade-off was considered not to be significant in the broader picture (i.e. where overall N losses are minimised, and the conserved N is retained in the

---

[1] Generally the value of manure nitrogen is expressed with fertilizer equivalence value which is significantly less than for mineral fertilizer (e.g. 75 %). However, in the case of emission reduction 100 % fertilizer equivalent value is appropriate since this represents a reduction in losses from the system, so long as manures are applied at agronomically suitable periods.

farming system). The reason for this is twofold: firstly, fertilizer inputs can be optimized with smaller inputs to take account of more efficient nitrogen use associated with reduced nitrogen losses; Secondly, the decrease in N losses from ammonia and nitrate leaching would also lead to a decrease in secondary emissions of nitrous oxide at the landscape scale (outside the farming systems). This way of thinking represents a notable transition from emphasizing pollution swapping at the field scale, to emphasis of overall improvement in nitrogen use efficiency at the landscape scale. This allows reduced inputs of new nitrogen fertilizers and therefore gives the potential to significantly reduce N_2O emissions.

Following the Expert Workshop, in co-operation with the Centre for Integrated Assessment and Modelling (CIAM, which also sits under the LRTAP Convention), the information from the workshop was used to update the GAINS model for use in cost-optimization analysis (see Chap. 9 for further information on the GAINS model). This work was then also presented to the LRTAP Executive Body in December 2011 (UNECE 2011c) and further updated in 2012 (Wagner et al. 2012), outlining projected costs for ammonia abatement through to 2020, for three ambition levels and a cost optimized scenario.

The ambition levels were defined in draft Annex IX texts developed by the Task Force, with a final draft submission to the 49th Session of WGSR of the LRTAP Convention (UNECE 2011b). The calculations were made using the GAINS model (after updates initiated from the discussions at the Expert Workshop on costs, Wagner et al. 2012). As with the findings of the Expert Workshop, the costs calculated were in many cases much lower than previous estimates. Overall, a medium level ambition option (Option B) was estimated to cost 467 million € per year across the UNECE region to reduce ammonia emissions by 2020 to 4,263 ktonnes per year, which is 635 ktonnes per annum lower than the baseline emissions. This equates to a mean unit cost of €0.74 per kg NH_3-N abated across the UNECE region.

It is relevant to compare the costs of ammonia abatement that result from the present scientific update with the costs of nitrogen oxides abatement, as estimated in the GAINS model. One way to do this is to compare the ratio of mitigation costs in further abatement to the estimated financial benefits of taking action. These benefits, to human health, ecosystems and climate, were first estimated at a European scale in the European Nitrogen Assessment, with updated estimates reported by van Grinsven et al. (2013). A key outcome of this analysis is shown in Fig. 10.1, which compares the benefit-cost ratio of further NH_3 and NO_x abatement across the European Union beyond that currently agreed for 2020. Unit costs increase for more ambitious measures so the benefit-cost ratio decreases at higher ambition levels. Fig. 10.2 shows that around 1,100 kt of further NH_3-N abatement could be achieved with a benefit-cost ratio >1, while only about 300 kt of further NO_x-N could be achieved while ensuring the benefit-cost ratio > 1. This comparison, which distinguishes the relative environmental impacts of NH_3 and NO_x emissions, reflects the fact that most of the low-cost measures for NO_x emission abatement have already been taken, so that the remaining available measures become more expensive. By comparison, at a European scale, many of the low-cost measures for NH_3 mitigation have yet to be taken, meaning that the "low hanging fruit" of easy actions are still available.

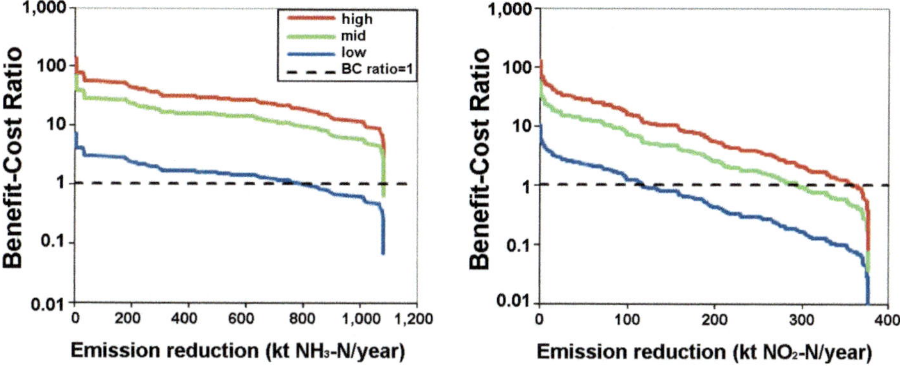

Fig. 10.2 Estimated benefit-cost-ratio of further NH_3 (*left*) and NO_x (*right*) abatement beyond current commitments for 2020. A larger emission reduction is estimated for NH_3 than for NO_x, while ensuring a benefit-cost ratio >1, reflecting that many of the "low hanging fruit" for NH_3 abatement have not so far been implemented and are therefore still available (Figure from Grinsven et al., reprinted with permission, copyright (2013) American Chemical Society)

10.4 Priority Measures for Ammonia Abatement Including Cost Effectiveness

To inform the development of Annex IX of the Gothenburg Protocol (UNECE 2013a, b), TFRN was requested to provide a ranked list of priority measures for ammonia emission reduction. The costs workshop played a vital part in this activity as one of the four criteria on which the measures were ranked was '[measures which are] cost neutral or have a low cost to farmers, especially when taking account of their co-benefits' (UNECE 2011b). The further criteria were as follows:

- Availability and applicability of the measures across the UNECE region;
- Focusing on sectors where the application of measures provided a significant contribution to ammonia emissions reduction;
- Need for long-term capacity-building.

The list of priority actions represents the experts recommended *"Top 5 Measures"* if ammonia emissions are to be reduced significantly across Europe. While these measures were particularly identified in the context of informing a possible revision of Annex IX, they also have a general relevance. The highest priority is given first:

1. **Low-emission application of manures and fertilizers to land**, including:

    (a) Low emission application of slurry and solid manure from cattle, pigs and poultry. Available measures include immediate or fast incorporation into the soil, trailing hose, trailing shoe and other band spreading and injection methods, and slurry dilution via irrigation;

    (b) Low-emission application of urea fertilizers. Available measures include immediate or fast incorporation into the soil, coated pellets, urease inhibitors and fertilizer substitution;

2. **Animal feeding strategies to reduce nitrogen excretion**. Available measures include: (a) low-protein phase feeding on pig and poultry farms; and (b) low-protein supplement feeding of cattle during housing, and improved nitrogen and grazing management of grazed grassland targeted to improve nitrogen use efficiency;

3. **Low emission techniques for all new stores** for cattle and pig slurries and poultry manure. Available measures include covers on all new slurry tanks, use of floating covers or slurry bags, prohibition of the building of new open slurry lagoons and keeping stored poultry manure dry;

4. **Strategies to improve nitrogen use efficiencies and reduce nitrogen surpluses**. The priority target is to establish nitrogen balances on demonstration farms or through on-farm demonstration, as a basis to monitor improvements in nitrogen use efficiency. The focus on initial demonstration would develop technical capacity across the UNECE region to allow wider use of nitrogen budgeting approaches on all major farms at a later date (e.g. after 2020);

5. **Low emission techniques in new and largely rebuilt pig and poultry housing**. Available measures include improved building designs, reducing the area of manure exposed to the air, keeping poultry litter dry and chemical scrubbing of exhaust air. The priority here is for new and largely rebuilt farms, as the costs of ammonia mitigation can be incorporated into overall rebuilding costs, when this is already planned. By comparison, retro-fitting of existing farm buildings is significantly more expensive.

When presenting this list it was also noted that it may be more cost-effective if packages of measures were implemented, rather than selecting one measure to implement. For example it makes little sense to cover manure in storage if it was then applied to the land without the use of low emission techniques. Conversely, as has been noted above, low emissions manure spreading provides the maximum financial benefit to farmers when the manure retains its maximum ammonium content as a result of using low-emission manure storage practices.

10.5 Progress with Review of Annex IX of the Gothenburg Protocol

As noted in previous sections, the outcomes based on the discussions of the Expert Workshop contributed to the preparation of options for revision of Annex IX of the Gothenburg Protocol by the Task Force on Reactive Nitrogen (e.g., UNECE 2011b). The Expert Workshop analysis also provided a greater understanding of potential flexibilities which could be suggested within Annex IX, such as for allowing exemption of mandatory measures for small farms, focussing on new buildings rather than existing buildings, and providing a range of techniques to meet potential abatement targets.

It was found that sometimes the lowest cost and most simple technique may be most attractive, especially where a system can be locally built. For example, the trailing hose is cheaper, more versatile and may be built locally. In addition it does not need stronger tractors and hence may be more acceptable to farmers than the trailing shoe, even though the latter is estimated to be more cost effective in reducing NH_3 emissions. Moreover, experience with the trailing hose and appreciation of co-benefits like uniform application and less conflict about odour, may hasten the adoption of more advanced methods like open slot injection.

Successive versions of new options for Annex IX were provided by the Task Force to the Working Group on Strategies and Review. In the end, the options for revision of Annex IX were presented to the thirtieth session of the Executive Body in May 2012 as part of the process to revise the Gothenburg Protocol. At that time, the Executive Body decided not to make substantive changes to Annex IX and the references to it in the main text of the Gothenburg Protocol (UNECE 2013a, b). This reflected a lack of agreement by the Parties on the level of ambition for a revised Annex IX. However, they did agree that 'The Parties shall, no later than at the second session of the Executive Body after entry into force of the amendment contained in decision 2012/21, evaluate ammonia control measures and consider the need to revise Annex IX' (UNECE 2013a). In effect, revision of Annex IX was agreed to be 'unfinished business'. Through this process, the importance of a revised Annex IX in reducing ammonia emissions with low costs has been clearly highlighted. It will again be a subject for the Convention to discuss in the future. This process will benefit substantially from the present revision of ammonia abatement costs, which highlights the availability of many low-cost options with the opportunity for efficiency and cost savings by farmers, with significant wider co-benefits.[2]

Acknowledgments We gratefully acknowledge financial support from the UK Department for Environment, Food and Rural Affairs (Defra), the European Commission through the ÉCLAIRE FP7 project, the UK Natural Environment Research Council, together with Agriculture and Agri-food Canada.

---

[2] At the time of publication, amendments to Annex IX have not entered into force (https://treaties. un.org/Pages/ViewDetails.aspx?src=TREATY&mtdsg_no=XXVII-1-h&chapter=27&lang=en). However, the adoption of the amendment has been recorded. 'In accordance with article 13, paragraph 3, of the Gothenburg Protocol, [the] Amendment [contained in decision 2012/2] shall enter into force on the ninetieth day after the date on which two thirds of the Parties to the Gothenburg Protocol have deposited with the Depositary their instruments of acceptance thereof.' (UNECE 2013b).

References

Bittman S, Dedina M, Howard CM, Oenema O, Sutton MA (eds) (2014) Options for ammonia mitigation: Guidance from the UNECE Task Force on Reactive Nitrogen. Centre for Ecology & Hydrology, Edinburgh

Erisman JW, Bleeker A (2011) Workshop report on Nitrogen and climate: interactions of reactive nitrogen with climate change and opportunities for integrated management strategies. Available via TFRN. http://www.clrtap-tfrn.org/webfm_send/439. Accessed 15 Aug 2014

IFDC (2012) Fertilizer Deep Placement (FDP) Brochure, International Fertilizer Development Center. http://issuu.com/ifdcinfo/docs/fdp_8pg_final_web?e=1773260/1756718. Accessed 15 Aug 2014

Klimont Z, Winiwarter W (2011) Integrated ammonia abatement – modelling of emission control potentials and costs in GAINS. IIASA Interim Report IR-11-027. IIASA, Laxenburg, Austria

Sutton MA, Howard CM, Erisman JW, Billen G, Bleeker A, Grennfelt P, van Grinsven H, Grizzetti B (2011) Assessing our nitrogen inheritance. In: Sutton MA, Howard C, Erisman JW, Billen G, Bleeker A, Grenfelt P, van Grinsven H, Grizzetti B (eds) The European Nitrogen Assessment. Cambridge University Press, Cambridge

Sutton MA, Bleeker A, Howard CM, Bekunda M, Grizzetti B, de Vries W, van Grinsven HJM, Abrol YP, Adhya TK, Billen G, Davidson EA, Datta A, Diaz R, Erisman JW, Liu XJ, Oenema O, Palm C, Raghuram N, Reis S, Scholz RW, Sims T, Westhoek H, Zhang FS (2013) Our nutrient world: the challenge to produce more food and energy with less pollution. Global Overview of Nutrient Management, Centre for Ecology and Hydrology, Edinburgh

Task Force Reactive Nitrogen Website (2014) Mission Page. http://www.clrtap-tfrn.org/mission. Accessed 15 Aug 2014

UNECE (2001) UNECE Framework Code for good agricultural practice for reducing ammonia EB.AIR/WG.5/2001/7. http://www.unece.org/fileadmin/DAM/env/documents/2014/AIR/WGSR/eb.air.wg.5.2001.7.e.pdf. Accessed 15 Aug 2014

UNECE (2011a) Report by the co-Chairs of the Task Force on Reactive Nitrogen ECE/EB.AIR/WG.5/2011/6. http://www.unece.org/fileadmin/DAM/env/documents/2011/eb/wg5/WGSR48/ECE_EB.AIR_WG.5_2011_6_E.pdf. Accessed 15 Aug 2014

UNECE (2011b) Report by the co-Chairs of the Task Force on Reactive Nitrogen ECE/EB.AIR/WG.5/2011/16. http://www.unece.org/fileadmin/DAM/env/documents/2011/eb/wg5/WGSR49/ece.eb.air.wg.5.2011.16.e.pdf. Accessed 15 Aug 2014

UNECE (2011c) Ammonia reductions and costs implied by the three ambition levels proposed in the Draft Annex IX to the Gothenburg protocol Informal Document 5. http://www.unece.org/fileadmin/DAM/env/documents/2011/eb/ebbureau/Informal_document_N_5.pdf. Accessed 15 Aug 2014

UNECE (2013a) 1999 protocol to abate acidification, eutrophication and ground-level ozone to the convention on long-range transboundary air pollution, as amended on 4 May 2012 ECE/EB.AIR/114. http://www.unece.org/fileadmin/DAM/env/documents/2013/air/eb/ECE.EB.AIR.114_ENG.pdf. Accessed 15 Aug 2014

UNECE (2013b) Decision 2012/2 Amendment of the text of and annexes II to IX to the 1999 protocol to abate acidification, eutrophication and ground-level ozone and the addition of new annexes X and XI ECE/EB.AIR/111/Add.1. http://www.unece.org/fileadmin/DAM/env/lrtap/full%20text/ECE_EB.AIR_111_Add1_2_E.pdf. Accessed 15 Aug 2014

UNECE (2013c) Guidance document on national nitrogen budgets, ECE/EB.AIR/119. http://www.unece.org/fileadmin/DAM/env/documents/2013/air/eb/ECE_EB.AIR_119_ENG.pdf. Accessed 15 Aug 2014

UNECE (2014) Guidance document on preventing and abating ammonia emissions from agricultural sources, ECE/EB.AIR/120. http://www.unece.org/fileadmin/DAM/env/documents/2012/EB/ECE_EB.AIR_120_ENG.pdf. Accessed 15 Aug 2014

UNECE Website (2014a) General information page. http://www.unece.org/env/lrtap/multi_h1.html. Accessed 15 Aug 2014

UNECE Website (2014b) Executive body information page. http://www.unece.org/env/lrtap/executivebody/welcome.html. Accessed 15 Aug 2014

UNECE Website (2014c) WGSR information page. http://www.unece.org/env/lrtap/workinggroups/wgs/welcome.html. Accessed 15 Aug 2014

Van Grinsven HJM, Holland M, Jacobsen BH, Klimont Z, Sutton MA, Jaap Willems W (2013) Costs and benefits of nitrogen for Europe and implications for mitigation. Environ Sci Technol 47:3571–3579. doi:10.1021/es303804g

Wagner F, Winiwarter W, Klimont Z, Amann M, Sutton MA (2012) Ammonia reductions and costs implied by the three ambition levels proposed in the Draft Annex IX to the Gothen-burg protocol. CIAM report 5/2011 – Version 1.2

Westhoek H, Lesschen JP, Rood T, Wagner S, De Marco A, Murphy-Bokern D, Leip A, van Grinsven H, Sutton MA, Oenema O (2014) Food choices, health and environment: effects of cutting Europe's meat and dairy intake. Glob Environ Chang. doi:10.1016/j.gloenvcha.2014.02.004

Index

A

Abatement, 2, 4, 5, 8, 17, 19, 26, 53–72,
   75–107, 113–135, 138, 153–155, 158,
   159, 186, 187, 189–198, 200, 203, 205,
   206, 209, 212–217, 219, 221, 234–237,
   239, 240, 243, 245–250, 253, 254, 256,
   263–279

Acidification, 2, 93–94, 101, 103, 104, 137,
   139, 149, 150, 160, 221, 233, 255,
   264, 265

Aerosol, 2, 76

Air chemistry, 2, 37, 77, 142, 148, 234

Air quality, 37

Ammonia (NH_3)
   abatement, 5, 19, 53–72, 75–107, 128, 153,
      155, 190–198, 206, 209, 212, 221, 234,
      250, 253, 254, 263–279
   emissions, 2–4, 38, 41, 47, 54, 68, 69, 72,
      96, 107, 113–134, 137–161, 177–186,
      188, 191, 193, 196, 199–216, 233–259,
      264–267, 269–271, 273, 276, 277, 279

Ammonium
   concentrations, 93, 221, 226

Atmospheric
   deposition, 27, 141, 172, 175

B

Biodiversity, 9, 157, 160, 161, 264, 272, 273

Biological nitrogen fixation (BNF), 27, 175

C

Carbon sequestration, 138, 143

CEA. *See* Cost effectiveness analysis (CEA)

CLE. *See* Critical level (CLE)

Concentrations, 2, 37–40, 42–44, 47, 48, 77,
   92, 93, 140–142, 147, 160, 179, 184,
   205, 219, 221, 225, 226, 270, 271

Cost benefit analysis (CBA), 105, 107,
   221–224

Cost effectiveness analysis (CEA), 153–157

Critical level (CLE), 76

Critical load (CL), 76

D

Deposition, 2, 27, 76, 141, 172, 175, 183

Dynamic model, 177

E

Ecosystem, 2, 76, 139, 221, 225, 234, 264, 276

EMEP, 190, 191

Emission inventory, 191

Emissions, 2–4, 8–10, 17, 19, 26, 36–38, 40,
   41, 43, 47, 48, 54, 68, 69, 76, 77, 85, 88,
   90–98, 101, 103–107, 114, 137–161,
   177, 178, 181, 187, 188, 190–194,
   199–201, 203, 205, 207, 219, 221,
   225–227, 233–259, 264–279

Eutrophication, 2, 139, 221, 255, 264, 265

F

Farm scale(s), 174, 265, 267–268

G

GHG. *See* Greenhouse gas (GHG)

Gothenburg protocol (GP), 3, 5, 8, 19, 26, 177,
    190, 191, 193, 233, 234, 237, 242, 255,
    256, 265, 277–279
Grassland, 45, 47, 129, 143, 145, 170,
    199–227, 245, 246, 278
Greenhouse gas (GHG), 2–4, 48, 105–107,
    137–161, 225, 227, 234, 264–266
Ground level, 187, 188, 265

I
IPCC, 106, 142, 144, 225

L
Leaching, 9, 10, 17, 90, 149, 153, 154,
    177, 276
Legislation, 68, 191, 194, 196, 199, 200,
    222, 259
Livestock, 12, 15, 17, 36, 48, 49, 54, 69, 70,
    72, 103, 134, 138, 140–143, 146–148,
    152, 157, 160, 171, 173, 174, 177, 178,
    180, 185, 197, 198, 217, 235, 236, 238,
    240, 241, 243, 245–248, 250, 252,
    267–270, 274
Local scale, 264
Long range Transboundary Air Pollution
    Convection (LRTAP), 2–5, 8, 234, 235,
    265, 274, 276

M
Measurements, 97, 98, 146
Mitigation, 7, 76, 77, 92, 93, 96, 138, 139,
    146, 153–155, 157, 160, 161, 187,
    191, 194, 196, 221–224, 227, 256,
    269, 276, 278
Model(s), 3, 14, 45, 68, 133, 141, 191,
    192, 196, 217, 233–259, 266,
    274–276

N
National Emissions Ceilings (NEC)
    Directive, 235
Nitrate (NO_3^-) 2, 9, 36, 141, 148, 151,
    153–155, 176–177, 199, 221,
    223, 226, 235, 237, 264, 272, 273,
    275, 276
Nitrate leaching, 9, 177, 276
Nitrite (NO_2^-), 141
Nitrogen cascade, 138, 264
Nitrogen cycle, 8, 266
Nitrogen oxides (NO_x), 1, 2, 255, 264, 274,
    276, 277

Nitrous oxide (N_2O), 9, 10, 36, 80, 94, 106,
    114, 138–153, 160, 170, 225–227, 234,
    235, 264, 275, 276
NO_3^-. *See* Nitrate (NO_3^-)
NO_x. *See* Nitrogen oxides (NO_x)

O
Organic nitrogen, 43
Ozone
    ground level, 265

P
Particulate matter (PM)
    $PM_{2.5}$, 2
Policy (policies)
    agricultural, 176
    environmental, 138

R
Reactive nitrogen (N_r), 2, 4, 138, 152, 251,
    264–266, 278
Regional scale(s), 264

S
Secondary aerosol, 2

T
Task Force on Reactive Nitrogen (TFRN), 2,
    114, 122, 251, 253–254, 256, 265, 266,
    277, 278

U
United Nations (UN), 5, 171, 235, 264, 265
United Nations Economic Commission for
    Europe (UNECE), 2, 3, 5, 205, 221, 234,
    237, 251, 256, 264–266, 268, 269, 274,
    276–279

V
Volatilisation, 4, 26, 36, 43, 140, 144, 147, 152

W
WGSR. *See* Working Group on Strategies &
    Review (WGSR)
Woodland, 146
Working Group on Strategies & Review
    (WGSR), 265, 266, 274, 276